PHYSICS
MATTERS
NICK ENGLAND

Charles Haines, Henry Ror

HODDER AND STOUGHTON
LONDON SYDNEY AUCKLAND TORONTO

Contents

Preface

This book is about Physics and the part that it plays in our lives every day. *Physics Matters* has been written specifically for GCSE courses. I hope that you'll find it lively, interesting and helpful for your studies, and that it'll help you to understand how important Physics is in the world around us.

There are twelve sections in the book and each one is divided into units — mostly two page spreads. There are questions at the end of each unit to help you to understand the topic that you've just read about. At the end of each section there are more questions, some from sample GCSE examination papers. These will help you to revise all the units in that section. If you want to learn about a particular topic you can look it up either in the contents list at the beginning of this book, or in the index at the end.

I am particularly grateful to the following people who have helped me with the writing and production of this book: David Mackin and Jim Bennetts, who read the manuscript and offered invaluable advice about the content and style; Caroline Evans, Lesley Robertson, Marion Lowe and my wife, who between them turned my untidy writing into the book as you see it; Paul Barratt and David Harrison, whose advice over the last ten years has influenced the book more than they realise. Finally, I would like to thank Dr. David Newsome, Master of Wellington, who was kind enough to give me some time off to write *Physics Matters*.

Nick England
Crowthorne

SECTION A

Introduction

Physicists try to explore and explain the universe in which we live

1 Introducing Physics

A double rainbow over the world's largest radiotelescope array at Socorro in New Mexico. There are 25 radio dishes which can be moved to various positions along a 21 kilometre long railway line

Cloud chamber tracks like these tell us about the structure of atoms

What is physics?

People have always been curious. We have always wanted to know why something happens or how something works.

Why do things fall to the ground? Why does the Moon seem to change shape during the month? Why do we have winter and summer? Why do rainbows appear? How do magnets work? What goes on inside the sun to make it hot? Physics is all about trying to answer questions like these. When you study physics you are trying to understand the world about you.

Our ancestors started to ask questions like those thousands of years ago. Although we have found some answers, there are still a lot of unanswered questions for the modern physicist. Nuclear physicists do not understand everything about the insides of atoms. Astronomers are still making many new discoveries about stars and galaxies.

Physics and technology

Having read the last few paragraphs, you may feel that physics is not much use. What good it is knowing about rainbows? However, physics has lead to many important advances in technology. Here are a few examples of how physics has changed your life.

- In 1831 Michael Faraday discovered that he could make an electrical current in a wire by moving it close to a magnet. Thanks to Faraday's work, you can now turn on a light in your home, by flicking a switch.
- In 1898 Marie Curie discovered the radioactive material radium. At first nobody realised that radioactive materials could be useful, but now radioactive substances are widely used in hospitals. We also use radioactive materials in nuclear power stations to produce electricity, and rightly or wrongly, in nuclear weapons.

- In 1947 a team of researchers led by William Shockley developed the transistor. This is the key to all modern electronic technology. The transistor turned the computer into a small and fast calculating machine. Transistors can be miniaturised and thousands of them can be put onto a small silicon chip. Computers are now widely used in industry, banking and the home. Without the invention of the transistor, computers would still be too expensive to put into schools.

Why study physics?

Physics is not something that happens just at school. It happens all around you. Physics and its effect on technology is part of our lives.

It is quite likely that you will not study physics (or science) after your GCSE. But studying physics will prepare you for life in a world which is full of science and technology. If you do study physics after GCSE level, then you may become one of the country's doctors or engineers. Studying physics together with chemistry and biology leads to many exciting scientific careers.

Computers give us a quick and convenient twenty four hour banking service

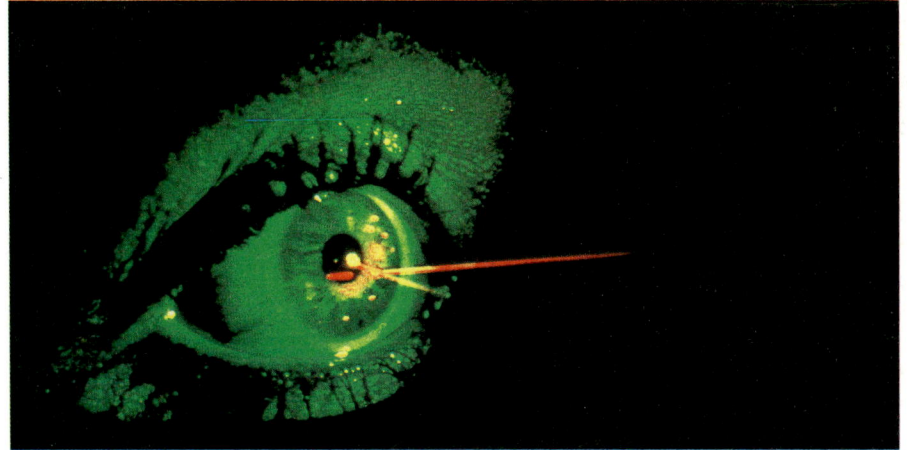

Laser surgery. A single laser beam is aimed at the cornea of the human eye. This technique has become commonplace in modern eye surgery. For example, lasers are used frequently to cure the blindness caused by diabetes

A planeload of happy holiday makers taking advantage of modern technology. However, if you live near a noisy airport, you will know about some of the problems that this technology brings

Questions

1 Thirty years ago computers were very large. They used to fill a whole room or more. By 1965 computers were small enough to fit into spacecraft. What was it that allowed us to make smaller computers?

2 Explain three ways in which physics has improved medical care.

3 (a) Some scientific inventions have been helpful and some have been a nuisance. Name two useful inventions and two that have been dangerous to us.

(b) Do you think that science has more advantages than disadvantages? Are we better off with science than without it?

2 Large and Small

Most people, when fully grown are about 1.7 m tall. So 1 km seems large to us and 1 cm seems small. It is difficult for us to imagine anything much smaller than the thickness of a hair—about 1/1000 cm across. Our eyes cannot see things much smaller than this. Also it is hard to imagine anything bigger than a mountain. The Earth is bigger than a mountain, but we are too close to it to understand how large it is.

However, physicists talk about tiny atoms, less than a millionth of a millimetre across. They talk about stars billions and billions of kilometres away. It is difficult to understand what these numbers mean.

In Figure 1 you can see a series of diagrams, where each distance gets bigger. The distances between planets and stars are extremely large. The population of the world is 5000 million. Everybody in the world, lying head to toe, would only cover 1/10 of the distance to the Sun. The distances to stars and galaxies are even more enormous.

Figure 2 will help you to realise just how small atoms are. An atom is about a ten thousand millionth of a metre across (1/10 000 000 000 m). Each diagram shows something that is about 100 times smaller than the one before.

Figure 1
Getting larger. Starting with a football pitch, the scale of each diagram gets bigger by one hundred times, until we reach the Solar System. Then, because space is so large, we have to increase the scale by 10 000 times to fit it all in

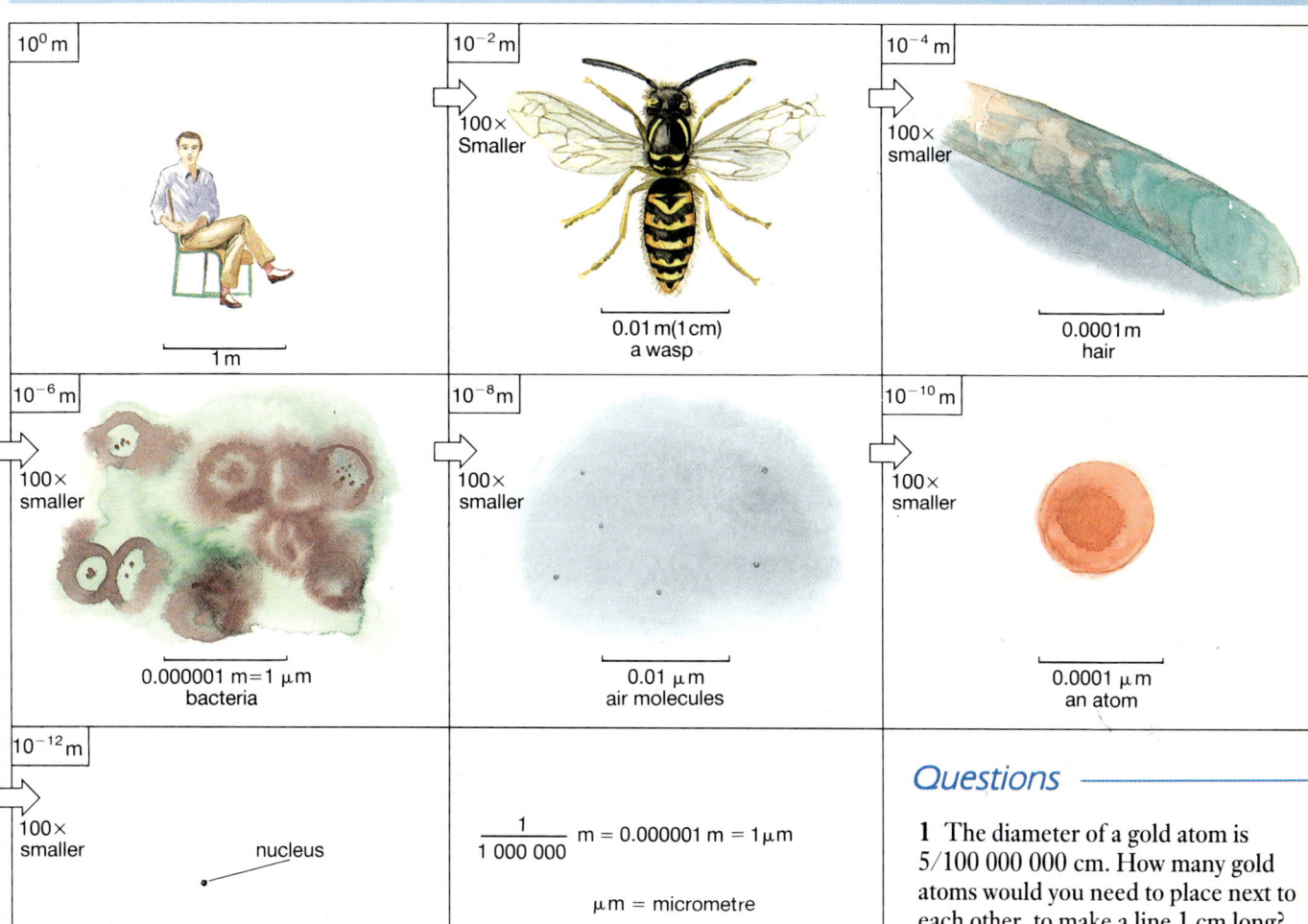

Figure 2
Getting smaller. Starting with the size of ourselves, the scale of each of these diagrams gets one hundred times smaller

Standard form

As you can see, some of the numbers we meet are either very large or very small. The nucleus of an atom has a diameter about 0.000 000 000 000 001 m. Galaxies are 100 000 000 000 000 000 000 km apart. To make these numbers easier to write we use **standard form**.

In standard form $1\,000\,000 = 10^6$ $1\,000 = 10^3$

$$\frac{1}{1\,000\,000} = 10^{-6} \quad \frac{1}{1\,000} = 10^{-3}$$

So we can write the diameter of a nucleus as 10^{-15} m, and the separation of galaxies as 10^{20} km.

Questions

1 The diameter of a gold atom is 5/100 000 000 cm. How many gold atoms would you need to place next to each other, to make a line 1 cm long?

2 Put the following in order of size, starting with the smallest and finishing with the largest: a particle of dust, a grain of salt, a mercury atom, the thickness of a thread of cotton.

3 Explain what each of the following is: (i) a planet (ii) a star (iii) a galaxy.

4 Melanie decides to make a model to show an atom and its nucleus. She decides to use a ping-pong ball (diameter 3 cm) to show the nucleus.
(a) Look at Figure 2. Is an atom 10, 100, 1000 or 10 000 times bigger than its nucleus?
(b) Work out the size Melanie's model atom needs to be.

5 Look at Figure 1. Roughly how many football pitches, end to end, could be fitted betweeen the North and South Poles?

6 The mass of the Earth is about 10^{25} kg. The mass of an atom is about 10^{-26} kg. Approximately how many atoms are there in the Earth?

3 Looking for Patterns

The galaxy is found in this constellation	Distance of galaxy (millions of light years*)	Speed of galaxy (km/s)
Virgo	72	1 200
Perseus	400	
Ursa Major	900	15 000
Corona Borealis	1 200	20 000
Bootes	2 400	40 000
Hydra		60 000

Table 1
* 1 light year = 10 million million (10^{13}) km. This is the distance light travels in one year.

A small group of galaxies. A computer has colour coded the picture to highlight their speeds. The red galaxies are moving away from us faster than the blue galaxies

Figure 1
The speed of galaxies is proportional to their distance away from us, according to Hubble's results

Scientists are always doing experiments. The reason for this is that experiments provide us with information. When we have some information we can start to look for patterns. This is an important skill that you are expected to learn during your GCSE course. Below you can read about a scientist who used his information, or **experimental data**, to look for patterns.

Edwin Hubble (1889–1953)

By about 1930 astronomers realised that our sun was part of an enormous galaxy of stars. They also realised that there were millions of other galaxies in the universe. By looking carefully at the light sent out by galaxies, Edwin Hubble realised the other galaxies were moving away from us. Table 1 shows the distance of some galaxies in various constellations, and their calculated speeds.

Hubble's data suggest that the speed of a galaxy is proportional to its distance away from us. This means that if the distance is doubled, so is the speed. Bootes is twice as far away as Corona Borealis; the speed of the galaxy in Bootes is twice as fast as the speed of the galaxy in Corona Borealis.

Hubble's results are best displayed in the form of a graph (Figure 1). You can see that the data produce a straight line that passes through the **origin**, (the point 0,0).

Hubble stated this law: the speed, v, of a galaxy is directly proportional to its distance, d, away from us.

This is written mathematically as:

$$v \propto d$$

$$\text{or } v = Hd$$

H is the constant of proportionality, or the **Hubble Constant**. The constant is the gradient (or slope) of the graph.

$$\text{The gradient} = \frac{20\,000 \text{ km/s}}{1200 \text{ million light years}}$$
$$= 17 \text{ km/s per million light years}$$

Once Hubble had worked out his law he could make predictions. From the graph he could predict the speed of the galaxy in Perseus is 6 500 km/s. He also extended his graph. (This is called **extrapolation**). This allowed him to calculate how far away the most distant galaxies are. For example Hydra, going away from us with a speed of 60 000 km/s, must be about 3 600 million light years away.

Tips for graph work

Below are some useful hints on how to deal with any data you collect in experiments:
- Present your data in a table, with a heading on each column. The heading shows what quantity you are measuring and its units.
- When you draw graphs label the axes clearly. Give the units of the quantity you are plotting.
- Choose the scale so that it is easy to plot the points, and so your graph is large and easy to read.
- When you have plotted the points, draw the best line through them. This line is not always a straight line—it is often a smooth curve. Not all the points will be exactly on the line. You should, though, ignore any obvious experimental errors (Figure 2).

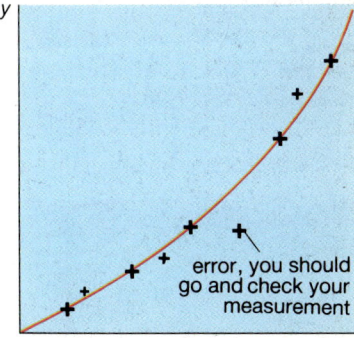

Figure 2
Not all graphs produce a straight line

Questions

1 In which of the graphs is *y* proportional to *x*?

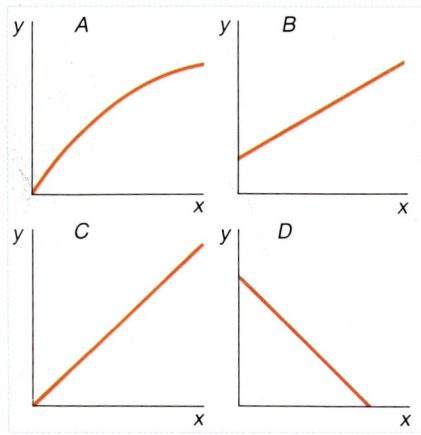

2 Quasars are very bright star-like objects. Astronomers now think they are galaxies forming. They are some of the most distant objects in the universe. Use Figure 1 to work out the distance to a quasar moving at (i) 50 000 km/s (ii) 200 000 km/s.

3 Skylab was an American satellite that was in orbit around the Earth in the 1970s. The astronauts on board could not weigh themselves using ordinary scales because they felt weightless. So they measured their mass by sitting in a chair that swung from side to side. How long one completed swing took depended on the astronaut's mass.
(a) Plot a graph of the time for 10 swings (*y*-axis) against the astronaut's mass (*x*-axis). (Note that the time depends on the mass. We plot the *dependent variable* on the *y*-axis, and the *independent variable* on the *x*-axis.)
(b) Draw a smooth curve through the points you have plotted.
(c) Is the time proportional to the mass?
(d) Use your graph to calculate Ed's mass.
(e) Why did the astronauts measure 10 swings rather than one?

astronaut	mass (kg)	time for 10 swings
Al	80	17.9
Buzz	70	16.7
Charlie	100	20.0
Doc	60	15.5
Ed		16.0

4 Units of Measurement

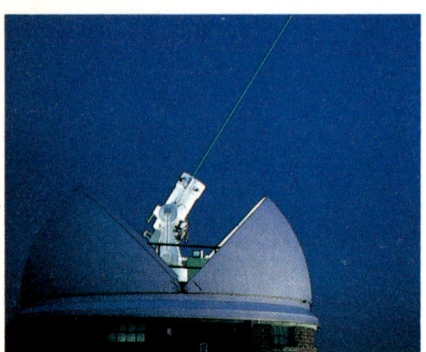

The height of an orbiting satellite is found by measuring the time for a pulse of laser light to return to this laser-ranging telescope after it has been reflected from the satellite. Why do we need to know the speed of light to do this?

SI units

Until the 1960s in Britain, the most commonly used system of units was based on feet, pounds and seconds. We used to measure temperature in degrees fahrenheit rather than degrees celsius. That system of units was difficult to use, and, more importantly, most other countries used a different system. It is much more sensible to use the same system of measurements everywhere. Then we know what everyone is talking about. We now use the **International System of units** (SI units for short). The three most basic SI units are:

- the **metre** (m) to measure length.
- the **kilogram** (kg) to measure mass.
- the **second** (s) to measure time.

Later in this book you will meet other units, such as the newton (N) and the joule (J). These units are based on the SI units, kilogram, metre and second. When you use units like the newton or joule, you must make sure your measurements are in kilogram, metre and second.

Length

The mass of this mountain is about a million million kilograms

As you already know (pages 4 and 5) scientists have to measure a wide range of lengths or distances. Here are some useful units of length based on the metre.

- 1 kilometre (km)   = 1000 m
- 1 centimetre (cm)   = 1/100 m
- 1 millimetre (mm)   = 1/1000 m
- 1 micrometre (μm) = 1/1 000 000 m
- 1 nanometre (nm)   = 1/1 000 000 000 m

Mass

The SI unit of mass is the kilogram, but it is useful to have other units to measure small or large masses.

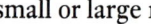

1 megatonne = 1 000 000 tonne	The mass of a large building
1 tonne = 1000 kg	About the mass of a car
1 kilogram (kg)	The mass of a bag of sugar
1 gram = 1/1000 kg	The mass of a raspberry
1 milligram = 1/1 000 000 kg	The mass of a fly

Time

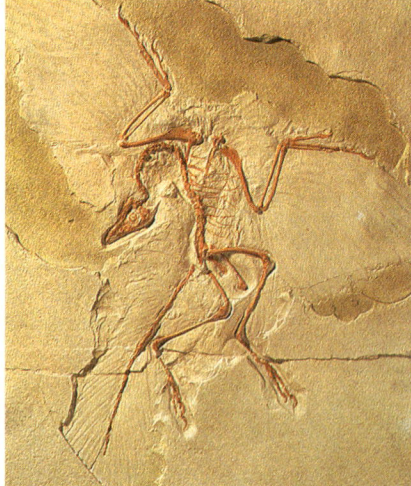

This archaeopteryx fossil is about 170 million years old

Some useful units of time based on the second are:

- 1 hour = 3600 s
- 1 minute = 60 s
- 1 millisecond (ms) = 1/1000 s
- 1 microsecond (μs) = 1/1 000 000 s

Area

To calculate area you use this formula:
area = length × width
   For this rectangle:

$A = 5 \text{ cm} \times 3 \text{ cm}$

$\quad = 15 \text{ cm}^2$

The area in m^2 is:

$A = 0.05 \text{ m} \times 0.03 \text{ m}$

$\quad = \dfrac{15}{10\,000} \text{ m}^2$

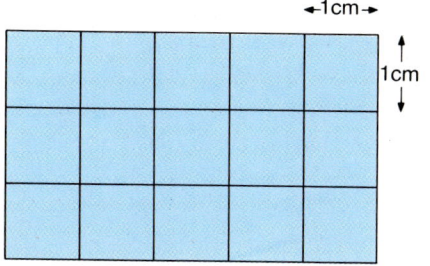

Note: $1 \text{ cm}^2 = \dfrac{1}{10\,000} \text{ m}^2$

Volume

To calculate volume you use this formula:
Volume = height × length × width
For this block of wood:

$V = 3 \text{ cm} \times 5 \text{ cm} \times 4 \text{ cm}$

$\quad = 60 \text{ cm}^3$

The volume in m^3 is:

$V = \dfrac{3}{100} \text{ m} \times \dfrac{5}{100} \text{ m} \times \dfrac{4}{100} \text{ m}$

$\quad = \dfrac{60}{1\,000\,000} \text{ m}^3$

Note: $1 \text{ cm}^3 = \dfrac{1}{1\,000\,000} \text{ m}^3$

   You should be careful when you calculate volumes. You could have measured (with a ruler) these lengths for the wood block above:
length = 5.1 cm; width = 4.1 cm; height = 3.1 cm
So volume = 3.1 cm × 5.1 cm × 4.1 cm
$\quad\quad\quad = 64.821 \text{ cm}^3$
You should round this off to 64.8 cm³. To write 64.821 cm³ means that you have measured the block's volume really accurately, but you haven't.
   Volumes of liquids are often measured in **litres** (l) or **millilitres** (ml).

- 1 litre = 1000 cm³ = $\dfrac{1}{1000}$ m³

- 1 ml = 1 cm³ = $\dfrac{1}{1\,000\,000}$ m³

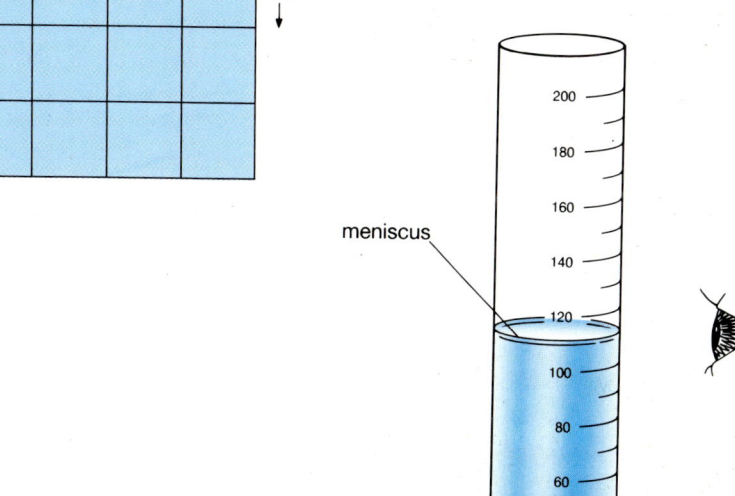

Figure 1
When you measure the volume of a liquid, make sure your eye is level with the bottom of the meniscus

Questions

1 (a) What is your height in m?
(b) What is your mass in kg?
(c) The speed limit on British roads is 70 miles per hour. What is this speed in km per hour?
2 (a) Estimate today's outside temperature in °C.
(b) Estimate the mass of a tin of baked beans in kg.
(c) Estimate the volume of your bean tin in: (i) cm³, (ii) m³.
3 A student measures these lengths for the sides of a metal block: 3.7 cm; 4.3 cm; 7.7 cm. What is the block's volume?

5 Density

Concorde is made from aluminium to give it a low density and high strength

A tree obviously weighs more than a nail. Sometimes, you hear people say 'steel is heavier than wood'. What they mean is this. A piece of steel is heavier than a piece of wood with the same volume.

To compare the heaviness of materials we use the idea of **density**. Density can be calculated using this equation:

$$\text{density} = \frac{\text{mass}}{\text{volume}} \quad or \quad D = \frac{M}{V}$$

Density is usually measured in units of kg/m^3.

Using density in engineering

Knowing the density of materials is very important to an engineer. This allows her to calculate the mass of building materials.

Example. What is the mass of a steel girder which is 10 m long, 0.1 m high and 0.1 m wide?

$$\text{The volume of the girder} = 10 \text{ m} \times 0.1 \text{ m} \times 0.1 \text{ m}$$

$$= 0.1 \text{ m}^3$$

$$D = \frac{M}{V}$$

$$\text{So } M = D \times V$$

$$= 8000 \text{ kg/m}^3 \times 0.1 \text{ m}^3$$

$$= 800 \text{ kg}$$

Steel is a very common building material because it is so strong. Despite this, in aeroplanes aluminium and titanium are used because they have low densities. It is important to make an aeroplane as light as possible.

Glass fibre is one of the most important modern building materials. It is made by strengthening plastic with glass fibres. Table 2 allows us to compare steel and glass fibre. Glass fibre is actually a little stronger than mild steel. This means a larger force is needed to break it. Glass fibre has a much lower density than steel. This makes it ideal for building small boats. Unfortunately glass fibre cannot be used for very large boats because it bends too much.

material	density (kg/m^3)
gold	19 300
mercury	13 600
lead	11 400
steel	8 000
titanium	4 500
aluminium	2 700
glass	2 500
water	1 000
cork	200
air	1.3
hydrogen	0.09

Table 1

	Steel	Glass fibre
relative strength	40 000	50 000
density (kg/m^3)	8000	2000
$\frac{\text{strength}}{\text{density}}$	5	25

Table 2

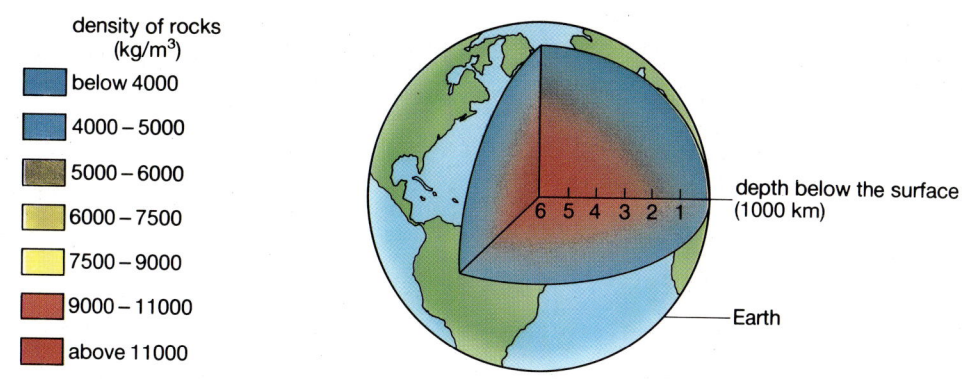

Figure 1

Density of rocks

Geologists are interested in the density of rocks. Figure 1 shows the density of different rocks in the Earth. Blue shows rocks of low density and red shows rocks of higher density. The Earth gets denser towards the centre.

Rocks near the surface of the Earth, such as granite, have densities of about 2700 kg/m^3. Volcanic rocks have higher densities. This is because the lava that is thrown out of volcanoes comes from deep below the Earth's surface, where the density is higher.

Will it float or sink?

The density of an object determines whether it sinks or floats. Cork floats on water, steel sinks. Cork is less dense than water, steel is more dense. An object only floats on a liquid if it is less dense than the liquid.

Geologists use this idea to separate out minerals. Pitchblende is usually found in granite rocks. Pitchblende is valuable because it contains uranium. Granite and pitchblende can be separated because of their different densities. The mixture of the two materials is crushed up. Then it is all put into bromoform, which is a dense (and poisonous) liquid. The pitchblende sinks but the lighter granite floats.

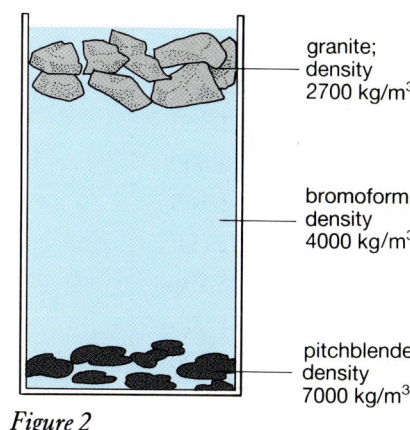

Figure 2

Questions

1 (a) Explain why aluminium and titanium are used to build aeroplanes.
(b) In Table 2, the last row is headed 'strength/density'. Explain why this is an important ratio.

2 Look carefully at Figure 1.
(a) Sketch a graph to show the density of rocks in the Earth (*y*-axis) against depth below the surface (*x*-axis).
(b) The outer part of the Earth is solid; this is called the mantle. The inner part of the Earth is liquid; this is called the core. Use your graph to guess where the mantle meets the core.

3 Carole is a geologist. She wants to work out the density of a rock. First she weighs the rock, then she puts it into a beaker of water to work out its volume.

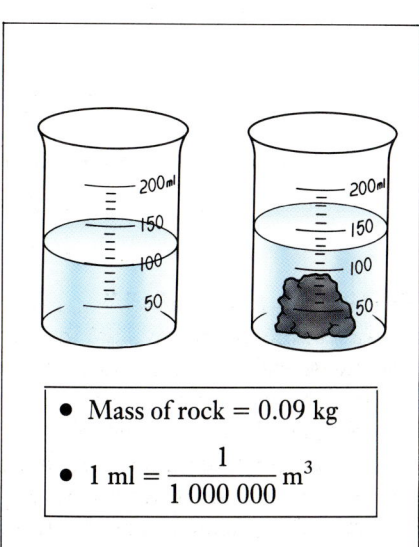

- Mass of rock = 0.09 kg

- 1 ml = $\dfrac{1}{1\,000\,000}$ m^3

(a) Use the diagrams to calculate the rock's volume. Give your answer in m^3. (The markings on the beaker are all in ml.)
(b) Now calculate the rock's density in kg/m^3.

4 What is the volume of:
(a) 1000 kg of aluminium?
(b) 100 kg of cork?

5 White dwarf stars are extremely dense. They have a density of about 100 million kg/m^3. If you had a matchbox full of material from a white dwarf, what would its mass be? (Hint: A matchbox has a volume of about 0.000 05 m^3.)

SECTION A: *STUDY QUESTIONS*

1

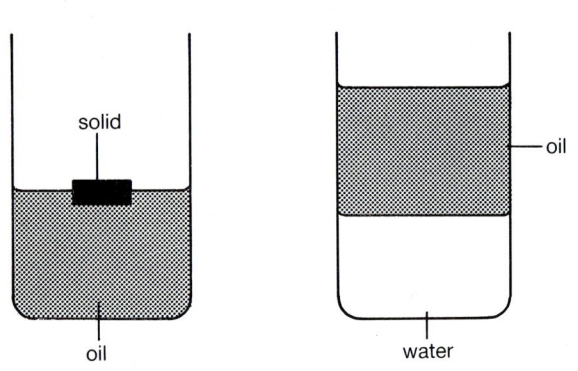

A solid of density 900 kg/m³ floats, as shown, in oil. The oil floats on water (density 1000 kg/m³). The density of the oil, in kg/m³, could be

A 850   B 900   C 950   D 1000   E 1050

NEA

2 The density of water is 1000 kg/m³ and the density of copper is 8900 kg/m³. Which one of the following statements is false?

A Copper is 8.9 times as dense as water

B $\dfrac{\text{The density of a certain volume of copper}}{\text{The density of the same volume of water}} = 8.9$

C $\dfrac{\text{The volume of a certain mass of copper}}{\text{The volume of the same volume of water}} = 8.9$

D $\dfrac{\text{The weight of a certain volume of copper}}{\text{The weight of the same volume of water}} = 8.9$

E $\dfrac{\text{The mass of a certain volume of copper}}{\text{The mass of the same volume of water}} = 8.9$

NI

3 The diagrams represent four measuring cylinders containing liquid. The mass and the volume of the liquid in each are stated.

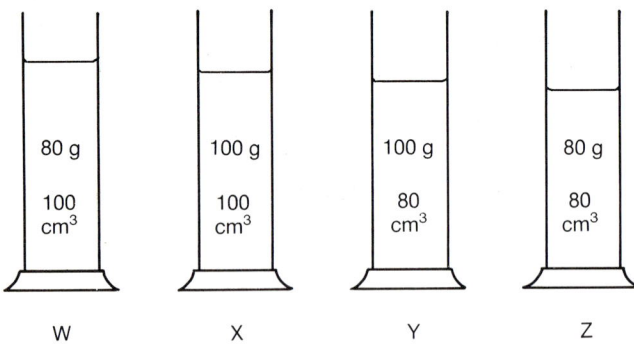

Which two measuring cylinders could contain an identical liquid?

A W and X   B W and Y   C X and Y
D X and Z   E Y and Z   **LEAG**

4 An ordinary teacup contains water to a level 1 cm from its rim.
Which one of the following is the best approximation to the mass of the water in the cup?
A 0.002 kg
B 0.02 kg
C 0.2 kg
D 2 kg
E 20 kg

LEAG

5 Mr Edwards has just bought a gold ornament from Mr Shark who is a shady second-hand dealer. Mr Edwards decides to test whether the ornament is made of gold. The results of his measurements are shown below.

> - Mass of ornament   = 910 g
> - Volume of ornament = 70 cm³

(a) Suggest how Mr Edwards did his measurements.
(b) Calculate the density of his ornament.
(c) The ornament looks gold on the surface. Suggest what Mr Edwards might find if he cuts it in half.

> - Density of gold = 19.3 g/cm³
> - Density of lead = 11.6 g/cm³

6 Gold can be hammered out into very thin sheets. A sheet of gold foil has a mass of 1.93 g. The foil is 20 cm long, and 5 cm wide. How thick is the foil?

7 Angela wants to work out the volume of a wooden block. She measures the length of the sides with a ruler. Use her measurements to calculate the volume of the block.

- Length = 6.7 cm
- Height = 3.4 cm
- Breadth = 5.3 cm

8 A student is asked to find the density of three cubes A, B and C. Each cube is made of a different material. The materials are aluminium, iron and wood.

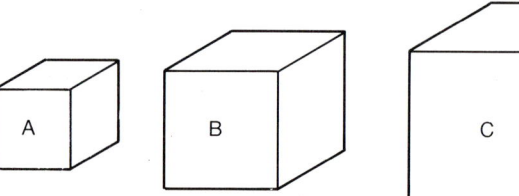

(a) The student is required to find the volume of each cube by a different method as shown below. Show how each volume is calculated.

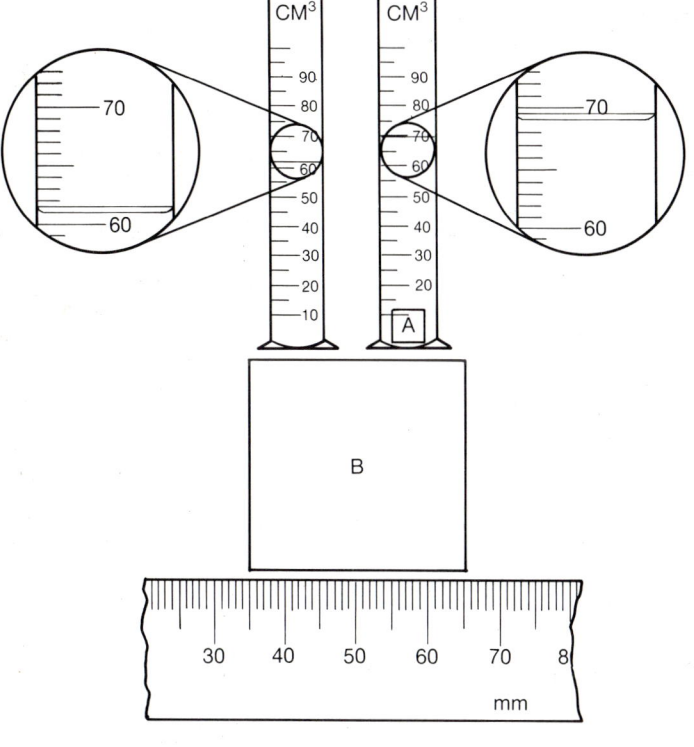

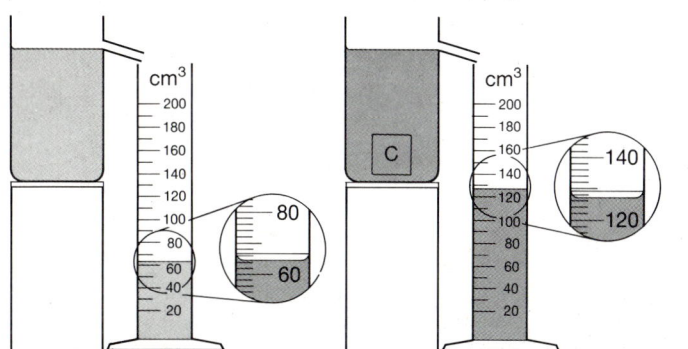

(b) On the basis of these results, complete the table below and identify the materials of the cubes.

Block	Mass (g)	Volume (cm³)	Density (g/cm³)	Material
A	20			
B	13.5			
C	512			

NEA

9 What are the readings on the instruments below?

(a)

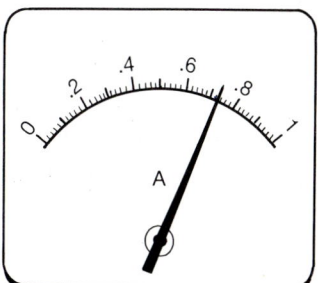

(b)

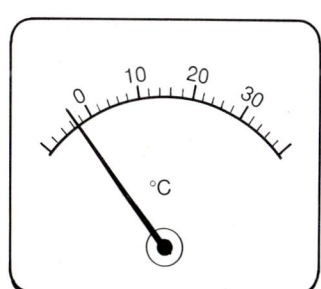

(c)

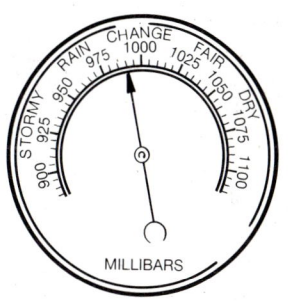

(d)

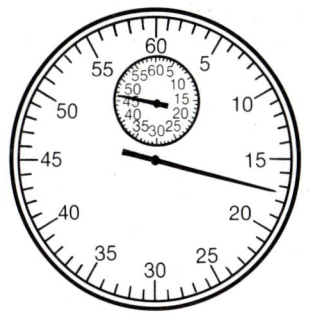

(e)

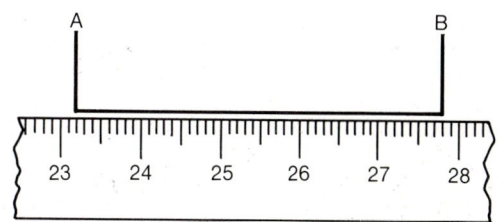

NEA

10 An experiment investigates the displacement of five marbles of equal size by placing them in a tube floating in a liquid in a measuring cylinder. The results are shown in the diagrams below.

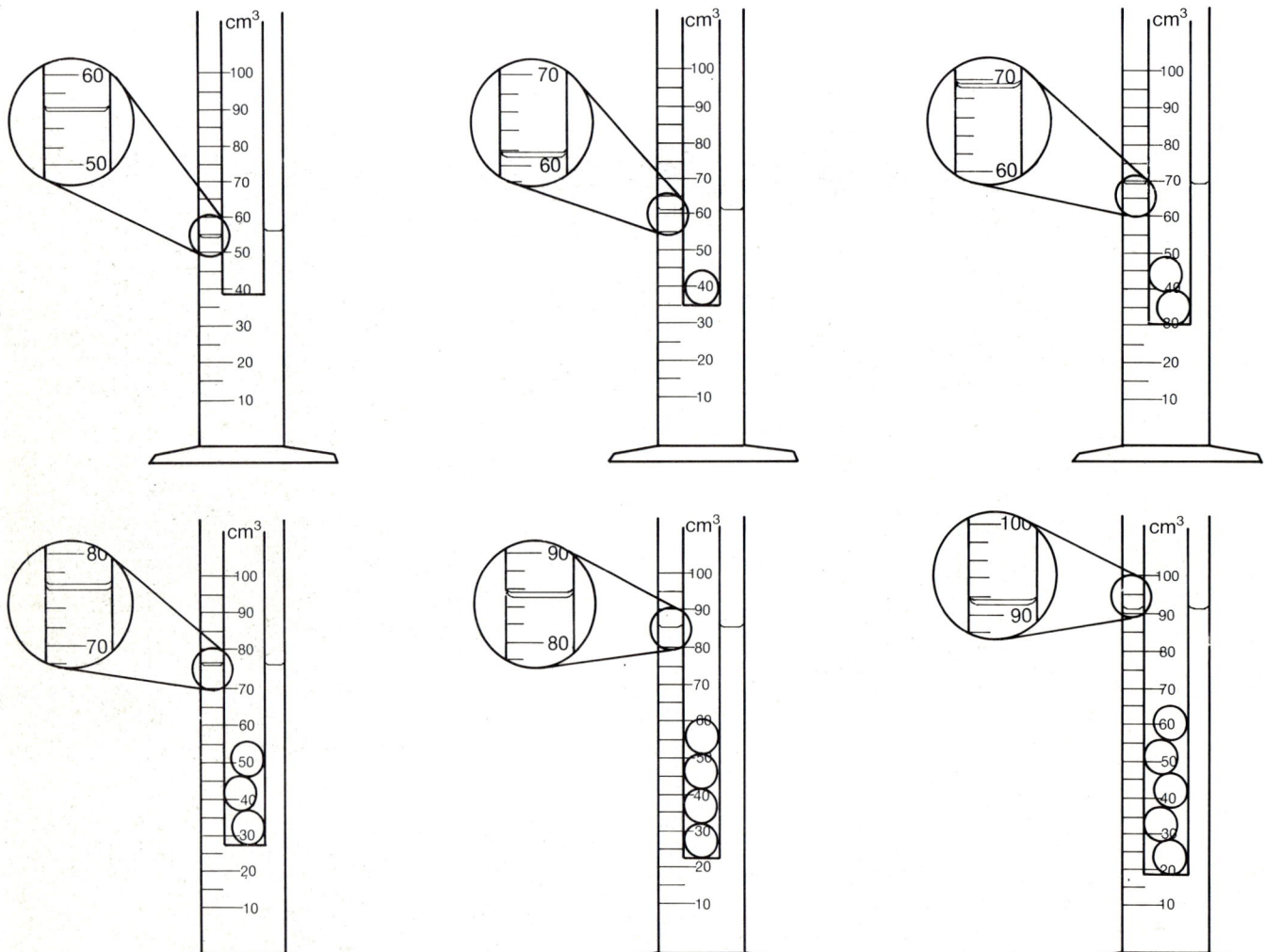

(a) Determine the reading on the measuring cylinder at each stage and record your results in a table like the one below.
(b) (i)   Which marble (1-5) gave the greatest displacement?
(ii) Which marble (1-5) gave the least displacement?

(iii) What was the average displacement of the five marbles?
(iv) How do you account for the fact that marbles of the same volume give different displacements?

NEA

Number of marbles	Reading on measuring cylinder (cm^3)	Displacement (cm^3)
0		0
1		
2		
3		
4		
5		

SECTION B

Forces

If this spider could make her web out of steel of a similar thickness it would probably break when a fly hit it

1 Forces Near and Far

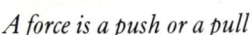

A force is a push or a pull

What is a force?

A force is a push or a pull. Whenever you push or pull something you are exerting a force on it. The forces that you exert can cause three things:

- You can change the shape of an object. You can stretch or squash a spring. You can bend or break a ruler.
- You can change the speed of an object. You increase the speed of a ball when you throw it. You decrease its speed when you catch it.
- A force can also change the direction in which something is travelling. We use a steering wheel to turn a car.

The forces described so far are what we call **contact forces**. Your hand touches something to exert a force. There are also **non-contact forces**. Gravitational, magnetic and electric forces are non-contact forces. These forces can act over large distances without two objects touching. The Earth pulls you down whether or not your feet are on the ground. Although the Earth is 150 million km away from the Sun, the Sun's gravitational pull keeps us in orbit around it. Magnets also exert forces on each other without coming into contact.

The size of forces

The unit we use to measure force is the **newton**. A force of 1 newton (1 N) is defined in terms of how quickly that force can change the speed of a 1 kg mass (see page 53). The box below will help you to get the feel of the size of several forces.

- The pull of gravity on a fly = 0.001 N
- The pull of gravity on an apple = 1 N
- The frictional force slowing a rolling football = 2 N
- The force required to squash an egg = 50 N
- The pull of gravity on you = 500 N
- The frictional force exerted by the brakes of a car = 5000 N
- The push from the engines of a rocket = 1 000 000 N

Two important forces

The pull of gravity and friction are two forces that we notice every day of our lives.

Weight is the name that we give to the pull of gravity on an object. Near the Earth's surface the pull of gravity is 10 N on each kilogram. We say that the Earth's gravitational field strength is 10 N/kg.

Example. What is your weight if your body has a mass of 50 kg?

$$\text{weight} = \text{pull of gravity}$$
$$= 50 \text{ kg} \times 10 \text{ N/kg}$$
$$= 500 \text{ N}$$

Friction is the contact force that slows down moving things. Friction can also prevent stationary things from starting to move when other forces act on them. Figure 1 helps you to understand why frictional forces occur. Any surface is not perfectly smooth. If you look at a surface through a powerful microscope you will be able to see that it has many rough spikes and edges. When two surfaces move past each other these rough spikes catch onto each other and slow down the motion.

Friction is often a nuisance because the rubbing between two surfaces turns kinetic (motion) energy into heat. Some ways of reducing friction are shown in Figure 2. Sometimes, though, friction is useful. Brakes work by friction to slow down cars. Also, when you walk, the frictional forces between your foot and the floor push you forward.

The frictional drag from the sea on this hovercraft is reduced by a cushion of compressed air.

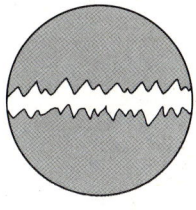

Figure 1
How two surfaces appear when seen through a powerful microscope.

Figure 2
Reducing friction

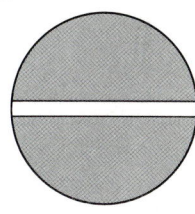

(a) If the surfaces are highly polished, friction is less.

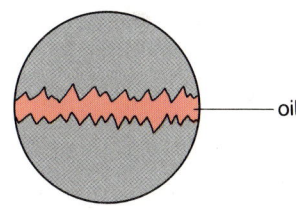

(b) A layer of oil between two surfaces acts as a cushion to stop the edges catching.

Questions

1 In the diagrams (right) some forces are acting on various objects. An arrow such as this: → 5 N means that a force of 5 N acts to the right. For each diagram state the effect that the forces produce.

2 What is the weight of a 2 kg bag of sugar?

3 One of the world's most important ways of reducing friction was invented by the ancient Egyptians in 3000 BC. What is it called?

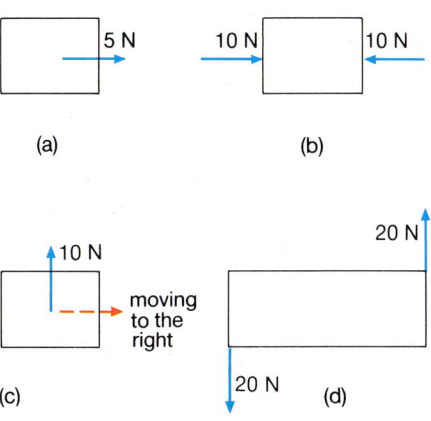

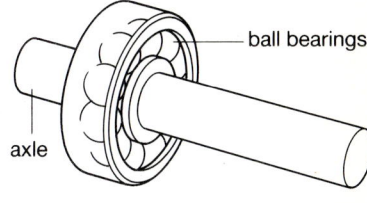

(c) Steel balls reduce friction by allowing surfaces to roll over each other.

2 More about Forces

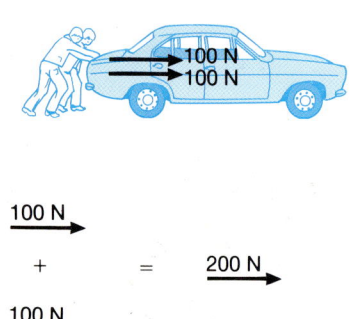

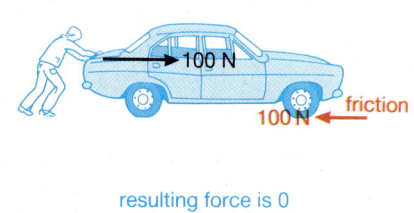

Figure 1

Figure 2

Vectors and scalars

If you want to move a chair you have to give it a push. The direction in which the chair moves depends on the direction of your push. A hard push will move the chair quickly. So both the *direction* and the *size* of a force are important.

Force is a **vector**. Vector quantities have both size and direction. Other examples of vector quantities are: velocity (wind 50 km/h north), displacement (2 paces backwards), acceleration and momentum (see section C).

A quantity that has only a size is called a **scalar**. Some examples of scalar quantities are: mass (2 kg of oranges), temperature (20°C), and energy (100 J).

Adding forces

Adding scalars is always easy: 3 kg of apples + 3 kg of apples = 6 kg of apples. Adding vectors can be a little harder. Here are three examples to show you how to do it:

- When two forces act in the same direction they add up to give a large *resultant* or *net force*. In this case 100 N + 100 N = 200 N. (Figure 1.)
- In the second example the driver has left the hand brake on (Figure 2). Your push of 100 N to the right is cancelled out by a frictional force acting to the left. The resultant force is zero.
- In Figure 3 you can see two tugs pulling a passenger liner into port. To work out the resultant force in this case we need to make a scale drawing. The first step is to choose a scale. We will make a length of 1 cm represent a force of 100 000 N. The lines representing the forces are then drawn to scale. (These are the blue lines in Figure 3(b).) Then two more lines are drawn to complete a parallelogram (the black lines). The resultant force can now be worked out by drawing a line across the diagonal of the parallelogram (the thick blue line). In this example, the line is 6 cm long, so the resultant force on the liner is 600 000 N. You can see that the direction of this resultant force pulls the liner straight ahead.

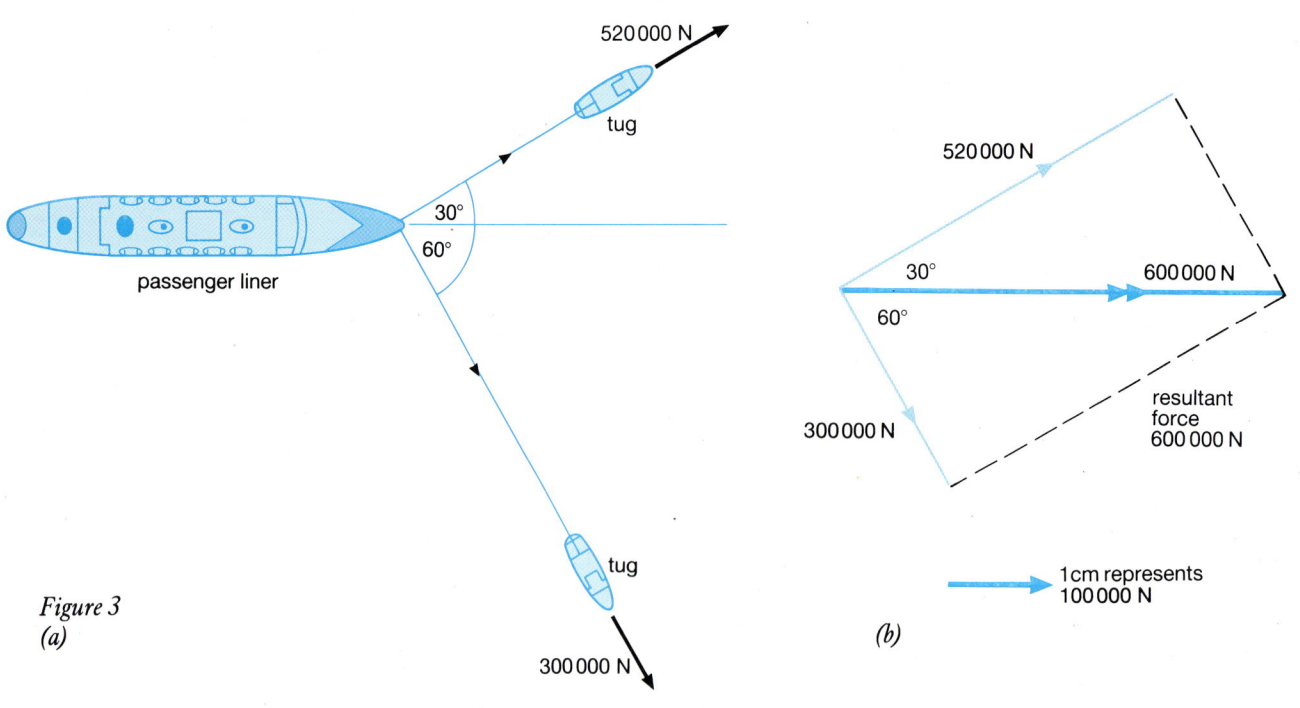

Figure 3
(a)

(b)

Newton's third law

Newton's third law states that 'to every force there is an equal and opposite force'. So forces always occur in pairs. This sounds an easy law to apply, but there are a few tricks. Here are some examples:

Canberra coming into harbour, assisted by tugs. Unlike the large liner, the tugs can change direction and speed very easily

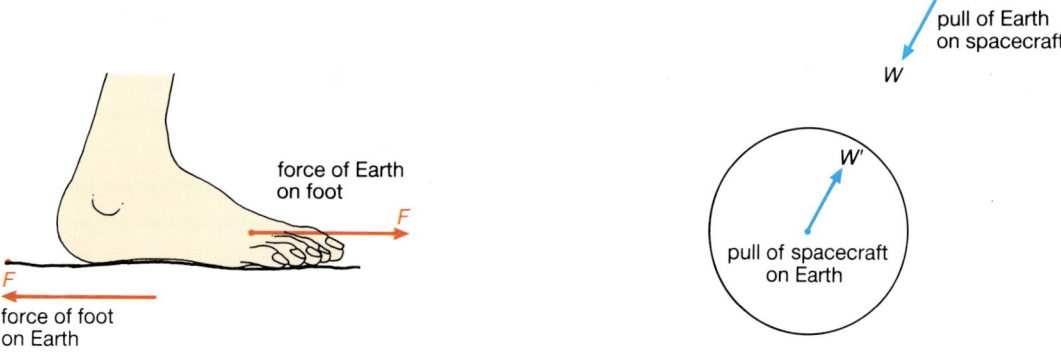

force of Earth
on foot

F

F

force of foot
on Earth

pull of Earth
on spacecraft

W

W'

pull of spacecraft
on Earth

Figure 4

Figure 5

- When you walk, the Earth pushes your foot forwards; your foot pushes back on the Earth (Figure 4).
- A spacecraft returning to Earth is pulled downwards (Figure 5). The spacecraft pulls the Earth back. This means that as the spacecraft moves, the Earth moves too. But the Earth is so big that it moves only a tiny amount—far too little for us to notice.

• Figure 6 shows a man sitting on a chair. There are two pairs of forces to think about. The Earth's gravitational pull on the man is W. The man's gravitational pull on the Earth is W^1. The chair exerts a reaction force R on the man. He exerts a force R^1 on the chair.

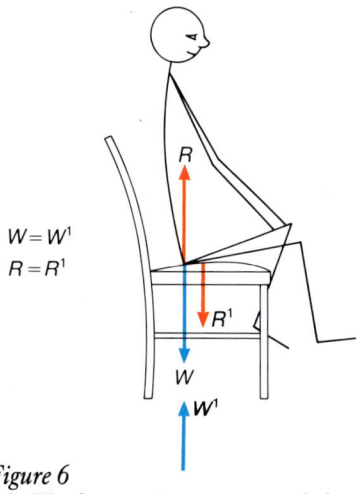
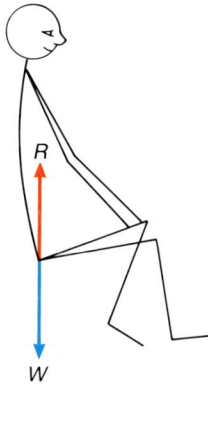

Figure 6
(a) The forces acting on a man sitting on a chair

(b) The man stays where he is because R = W.

Questions

1 What is the resultant force on this rocket?

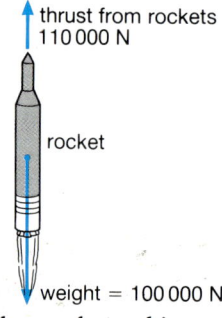

thrust from rockets
110 000 N

rocket

weight = 100 000 N

2 The diagram shows a large ship being pulled by two tugs. As the ship moves through the water there is a frictional force that acts on it. Work out the resultant force on the ship.

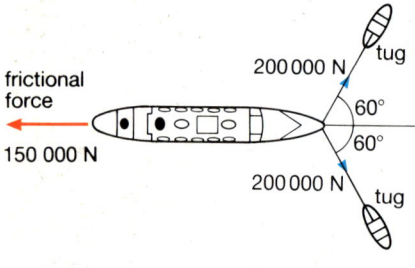

frictional force
150 000 N

200 000 N tug

60°
60°

200 000 N tug

3 Tom, Wesley and Alan go on a pub crawl. They start at the Eastgate and finish at the Randolph. At each pub each of them drinks 0.5 litre of beer.
(a) What is the total amount of beer consumed during the crawl?
(b) What is their displacement from the Eastgate by the end of the crawl?

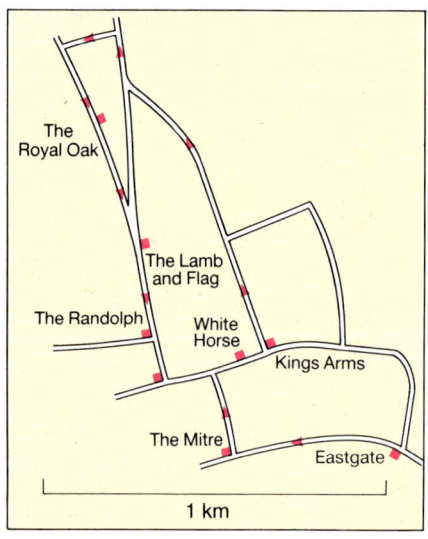

4 (a) Below you can see two sumo wrestlers, Taiho and Toshimitsu, locked in combat. Make a sketch of the diagram and mark in as many pairs of forces as you can.

Taiho   Toshimitsu

(b) Taiho wins the bout by pushing Toshimitsu down. Koki, who was watching the fight says: 'Taiho is the stronger so Taiho's push on Toshimitsu was bigger than Toshimitsu's push on Taiho'. Explain what is wrong with this comment.

3 Stretching

Figure 1 shows a simple experiment to investigate the behaviour of a spring. The spring stretches when a load is hung on the end of it. The increase in length of the spring is called its *extension*. You can plot a graph of load against extension. You will find that it produces a straight line such as *AB* in Figure 2. This shows that the extension is proportional to the load. A material that behaves in this way is said to obey Hooke's Law:

> Extension is proportional to the load

At point *B* the spring has reached its *elastic limit*. Hooke's Law is no longer obeyed. Over the region *AB* of the graph the spring shows elastic behaviour. This means that when the load is removed from the spring, it returns to its original length and shape.

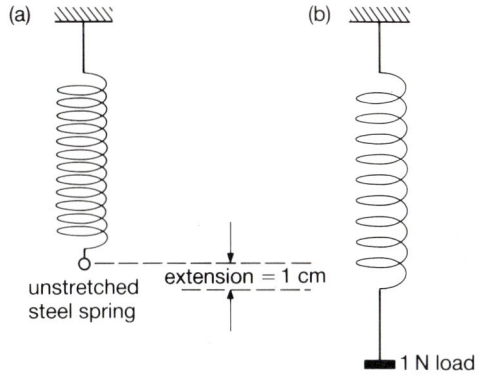

Figure 1

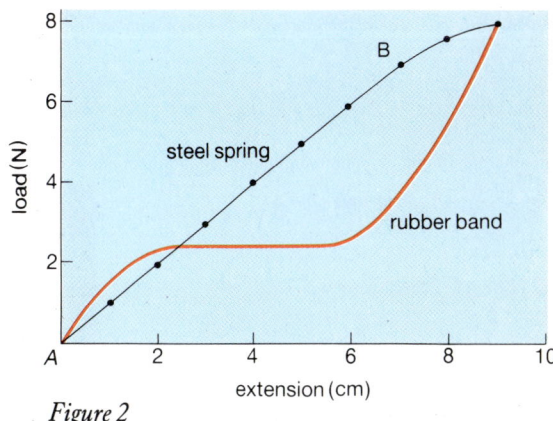

Figure 2

However, if a load of more than 7 N (beyond point *B*) is applied to this spring, it changes its shape permanently. When the load is removed it does not return to its original shape. This is called *plastic deformation*. When you bounce a hard rubber ball it deforms *elastically*. How do you think Plasticene deforms?

Only some materials obey Hooke's Law. You can see from Figure 2 that a rubber band certainly does not.

Car journeys would be very bumpy without suspension. Most cars, such as this Volvo, use springs, but some have fluid suspensions. What are the red and white triangles showing?

Figure 3

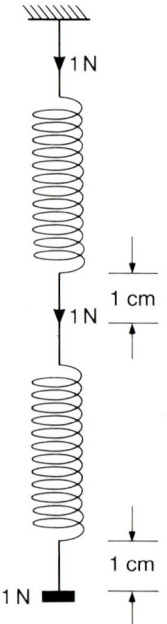

(a) The total extension is 2 cm

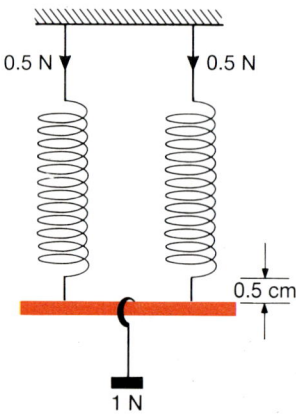

(b) Each spring extends 0.5 cm

Steel girders form the framework of large modern buildings. This photograph shows the Renault centre, in Swindon

Figure 3 shows what happens when a load is put onto two springs. When you put two springs in series each one is pulled by the force of 1 N. Each spring is extended by 1 cm, and the total extension is now 2 cm (Figure 3(a)). When springs are put in parallel (side by side) each one supports half of the load. Each spring only extends by 0.5 cm (Figure 3(b)).

Building with steel

Steel is important in many industries. Steel is needed for the manufacture of tools, machines and vehicles. Steel girders are used in the construction of large buildings and bridges. An engineer must work out exactly how much load a girder can support before building starts.

This is the sort of problem that a construction engineer might have to solve. Tests in the laboratory on a sample of steel have produced the load/extension graph shown in Figure 4. Gillian, the engineer, needs to know if the girder can support a load of 18 000 N.

Gillian knows from the graph that the steel sample starts to deform plastically when it supports a load of 2000 N. The steel girder in the building must not deform in this way.

When working out what the girder can support, the length does not matter. The same force acts through every part of the girder. However, the area of the girder is important. The girder has an area ten times as big as the sample (Figure 4(a)). It can support ten times the load. So the maximum load that the girder can support is 20 000 N.

However, Gillian would not want to put 18 000 N on to this girder, because it does not leave much room for error. Engineers like to have a built-in safety factor.

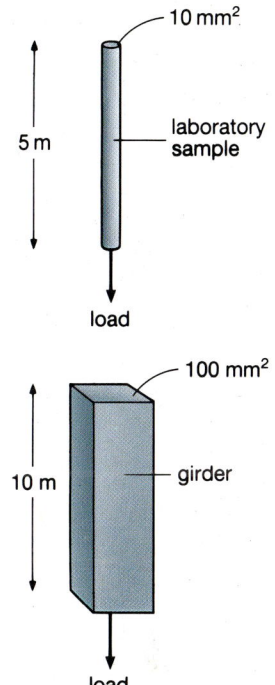

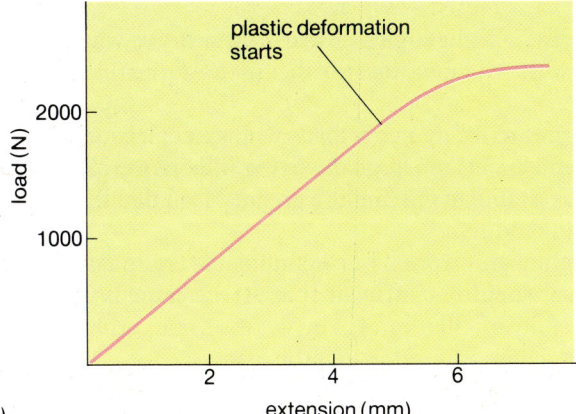

Figure 4(a)
The load/extension graph for a sample of steel in the laboratory can be used to predict how a steel girder will behave.

Figure 4(b)

Questions

1 In diagram (i) below: the 2 N weight extends the spring by 4 cm. Diagram (ii) shows an arrangement with three more of the same type of spring. How far will the point X move upwards when the 4 N weight is removed?

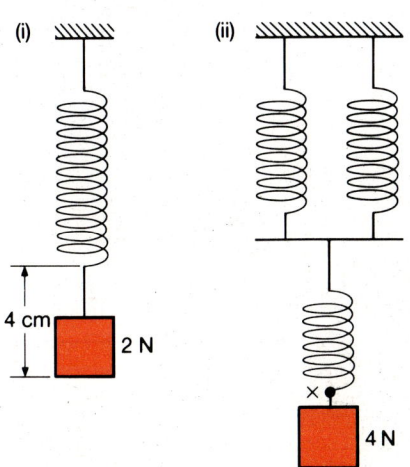

2 Figure 2, in the text, shows the force extension graphs for a rubber band and a spring. Use this graph to work out:

(i) the extension caused by a force when the band and spring are in series

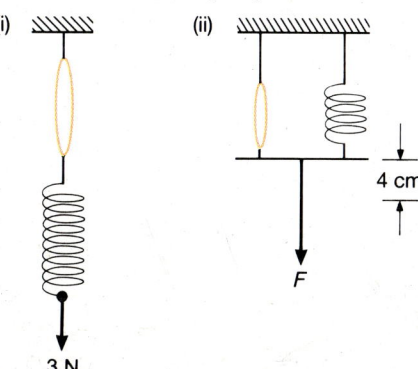

(ii) the force needed to produce a 4 cm extension when they are in parallel.

3 The diagram to the right shows a steel girder under test. Gillian needs to know how far the beam bends when it is loaded. The table shows the sag, h, in the middle, for different loads.

(a) Plot a graph of load (y-axis) against h (x-axis).
(b) Gillian made a mistake in one of her measurements of h. Which measurement is wrong and what should it have been?
(c) Using the same axes sketch a graph to show how a longer beam will sag under the same loads.
(d) Which side of the beam is being stretched and which side is being squashed?

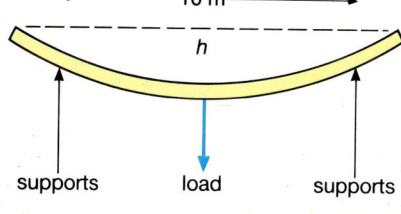

sag, h (cm)	1.3	3.2	5.2	7.0	9.1
load (N)	1000	2500	4000	5500	7000

4 Turning Forces

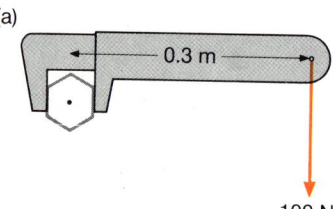

Figure 1

(a)

(b)

(a) Turning moment
   = 100 N × 0.3 m
   = 30 Nm

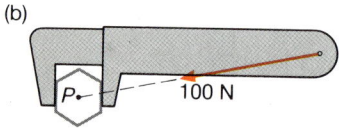

(b) A force acting through the nut has no turning moment.

If you have ever tried changing the wheel on a car, you will know that you are not strong enough to undo the nuts with your fingers. You need a spanner to get a larger turning effect. A tight nut will need a long spanner and a large force.

The size of the turning effect of a force about a point is called a **turning moment**.

> Turning moment = force × perpendicular distance of the force from the point

Perpendicular means 'at right angles'. Figure 1 shows you why this distance is important. If you push the spanner towards the nut (Figure 1(b)) you get no turning force at all.

The same idea applies to lifting heavy loads with a mobile crane (Figure 2). If the turning effect of the load is too large the crane will tip over. So, inside his cab, the crane operator has a table to tell him the greatest load that the crane can lift for a particular working radius.

Table 1 shows you how this works. For example, the crane can lift a load of 60 tonnes safely with a **working radius** of 16 m. If the crane is working at a radius of 32 m, it can only lift 30 tonnes. You get the same turning effect by doubling the working radius and lifting half the load.

Working radius (m)	Maximum safe load (tonnes)	load × radius (tonne × m)
12	80	960
16	60	960
20	48	960
24	40	960
28	34	960
32	30	960
36	27	960

Table 1
A load table for a crane operator

0   10 m

Figure 2
A mobile crane

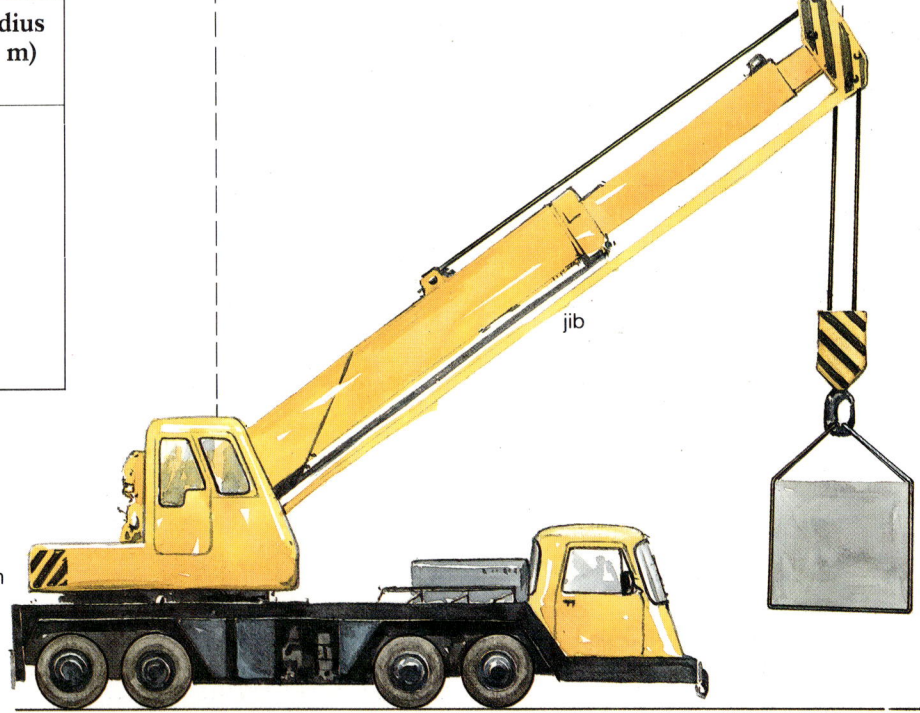

Balancing

In Figure 3 you can see three children playing on a see-saw. Jaipal and Dominic are sitting at the ends, 2 m from the pivot. Where would Mandy sit to balance the see-saw?

When the anticlockwise turning effect of Jaipal is equal to the clockwise turning effect of Mandy and Dominic, then the see-saw balances.

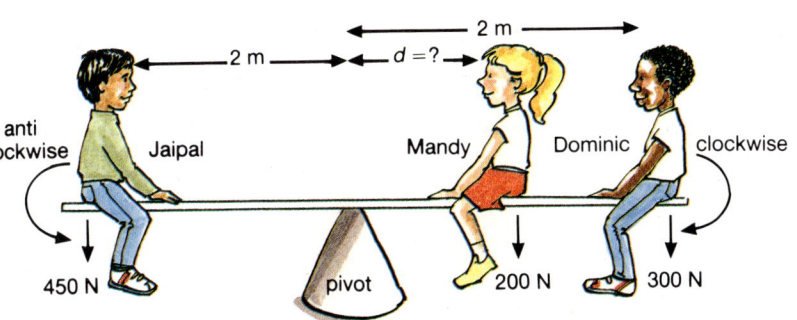

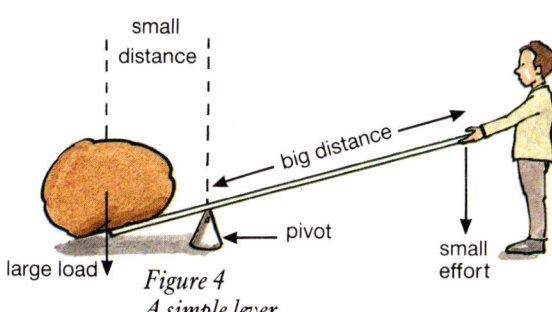

small distance

big distance

large load

pivot

small effort

Figure 4
A simple lever

Figure 3
How is the see-saw balanced?

Jaipal's anticlockwise turning moment = 450 N × 2 m
$$= 900 \, \text{Nm}$$

Dominic's and Mandy's turning moment = 300 N × 2 m + 200 N × d

So if the see-saw is to balance:

$$900 \, \text{Nm} = 600 \, \text{Nm} + 200 \, d$$

$$\text{So } 200 \, d = 300 \, \text{Nm}$$

$$d = 1.5 \, \text{m}$$

Mandy must sit 1.5 m away from the pivot.
When the see-saw is balanced we say it is in **equilibrium**.
   In equilibrium:

The sum of the anticlockwise turning moments	=	The sum of the clockwise turning moments

We use this idea of balancing moments to our advantage to increase leverage.
   In Figure 4 the rock can be lifted when its turning effect is balanced by the turning effect of the man's push. With a long lever a small effort can lift a large load.

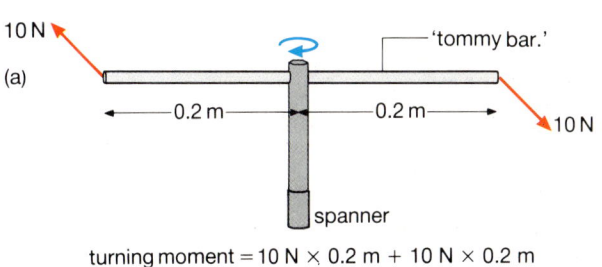

(a)

10 N

'tommy bar.'

0.2 m   0.2 m

10 N

spanner

turning moment = 10 N × 0.2 m + 10 N × 0.2 m
                        = 4 Nm

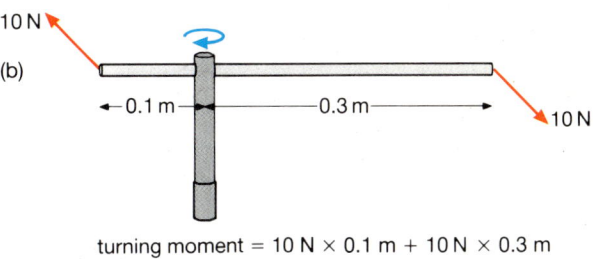

(b)

10 N

0.1 m   0.3 m

10 N

turning moment = 10 N × 0.1 m + 10 N × 0.3 m
                        = 4 Nm

Figure 5
When two equal forces act to produce a turning effect, they are called a couple. A couple causes a rotational effect, but the resultant force on the bar is zero. Notice also that whatever the position of the bar relative to the spanner, the same turning effect is produced.

Questions

1 (a) When you cannot undo a tight screw, you use a screwdriver with a large handle. Explain why.
(b) Explain why door handles are not put near hinges.
(c) Where do you have to put the spare penny to balance the ruler?

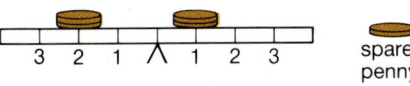

3 2 1   1 2 3   spare penny

2 Look at the crane in Figure 2.
(a) What is meant by working radius?

(b) Use the scale in Figure 2 to calculate the working radius of the crane.
(c) Use the data in Table 1 to calculate the greatest load that the crane can lift safely in this position.
(d) Laura, a student engineer, makes this comment: 'You can see from the driver's table that the crane can lift 960 tonnes, when the working radius is 1m.' Do you agree with her?
3 Alice decides she would like to join the other children on the see-saw in

Figure 3. She has a weight of 170 N. How far will Mandy have to move to balance the see-saw if Alice sits (i) above the pivot (ii) on Dominic's lap?
4 What turning moment, about the nut, does this 200 N force have?

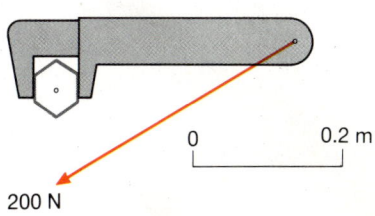

0   0.2 m

200 N

5 Stability

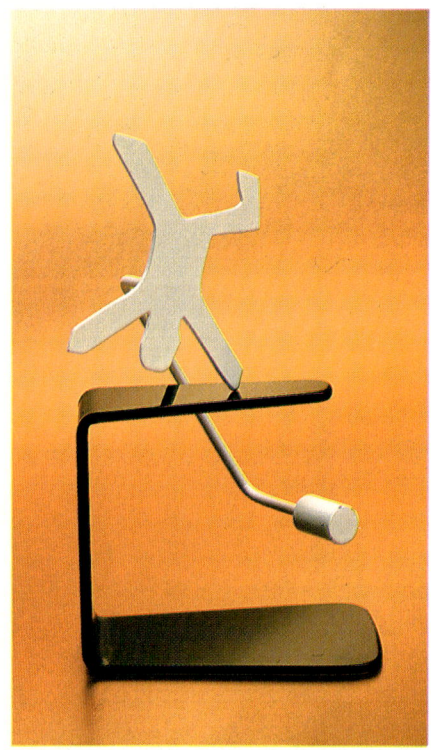

This rocking toy is stable because its centre of gravity is below the pivot

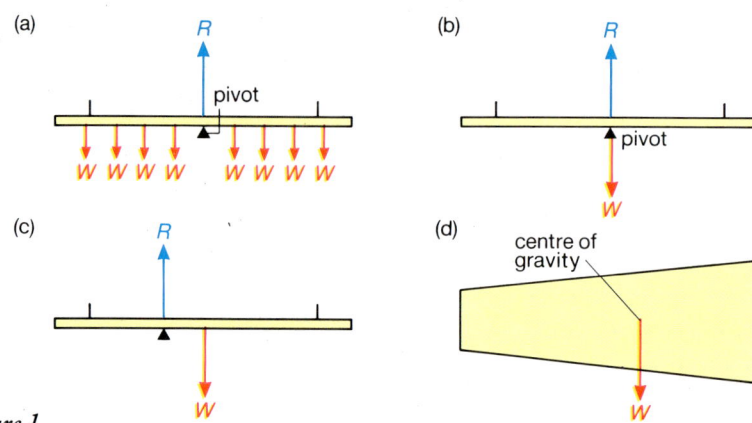

Figure 1

Centre of gravity

Figure 1(a) shows a see-saw that is balanced about its midpoint. Gravity has the same turning effect on the right-hand side of the see-saw, as is has on the left-hand side. The resultant turning effect is zero.

The action of the weight of the see-saw is the same as a single force, W, that acts downwards through the pivot (Figure 1(b)). This force has no turning moment about the pivot. As the see-saw is stationary, the pivot must exert an upwards force R on it, which is equal to W.

When the see-saw is not pivoted about its midpoint, the weight will act to turn it (Figure 1(c)).

The point that the weight acts through is called the **centre of gravity**. The centre of gravity of the see-saw lies at its midpoint because it has a regular shape. In Figure 1(d) the centre of gravity lies nearer the thick end of the shape.

Jumping higher

When you stand up straight your centre of gravity lies inside your body. But in some positions you can get your centre of gravity outside your body.

This idea is most important for pole vaulters and high jumpers (Figure 2(a)). Suppose a high jumper's legs can provide enough energy to lift her centre of gravity to a height 2 m above the ground. If she can make her centre of gravity pass under the bar, she can clear a bar higher than 2 m (Figure 2(b)).

Why do Fosbury-floppers usually win?

Figure 2
(a) Centre of gravity inside the body

(b) Centre of gravity outside the body

Equilibrium, stability and toppling

Something is in equilibrium when both the resultant force and resultant turning moment on it are zero. We talk about three different kinds of equilibrium, depending on what happens to the object when it is given a small push.

- A football on a flat piece of ground is in **neutral equilibrium**. When given a gentle kick, the ball rolls, keeping its centre of gravity at the same height.
- A tall thin radio mast is in **unstable equilibrium**. It is balanced with its centre of gravity above its base. However, a small push (from the wind) will move its centre of gravity downwards. To prevent toppling the mast is stabilised with support wires, as shown in the photograph.
- A car is in **stable equilibrium** (Figure 3(a)). When the car is tilted a little the centre of gravity is lifted, (b). In this position the action of the weight keeps the car on the road. In (c) the centre of gravity lies above the wheels; the car is in a position of unstable equilibrium. If the car tips further (d), the weight now provides a turning moment to topple the car. Cars are more stable if they have a low centre of gravity and a wide wheel base.

The support wires on this radio mast stop it from falling

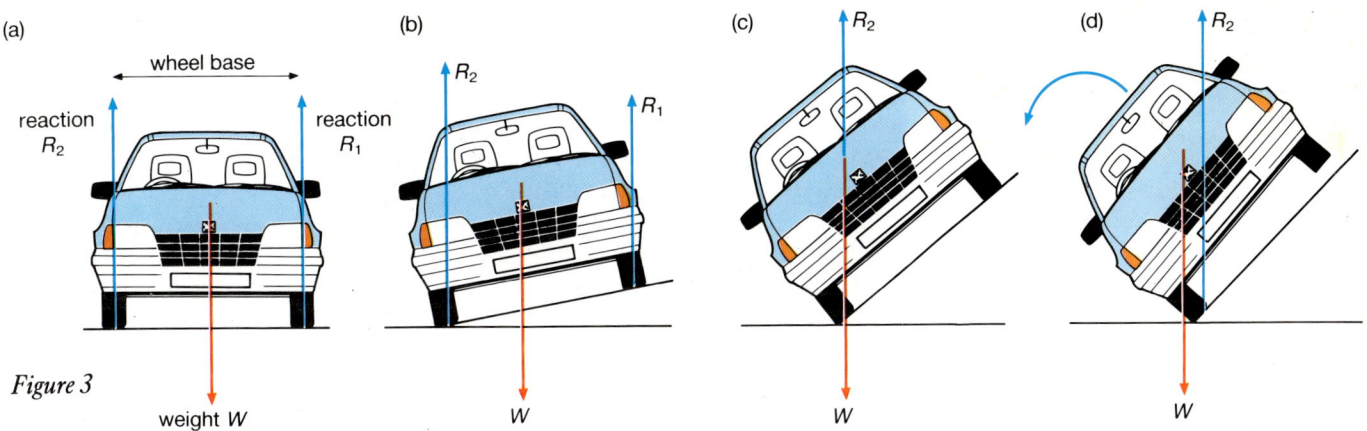

Figure 3

(a) wheel base / reaction R_2 / reaction R_1 / weight W

(b) R_2 / R_1 / W

(c) R_2 / W

(d) R_2 / W

Questions

1 You are in the business of manufacturing lager glasses. You want to make your profits as big as possible. So you decide to produce a very unstable glass, which will get knocked over very easily. Then people will have to buy more glasses. Discuss which shape of glass will be most suitable for your purpose.

  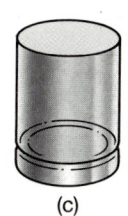

(a)   (b)   (c)

2 The diagram to the right shows a cantilever bridge which is constructed from three spans of concrete beams. The centre span has a weight of 6 MN, and each of the other arms has a weight of 10 MN. The arrows show the forces that act on the right-hand arm. (It is 20 m long.)

(a) Where is the centre of gravity of the right-hand arm? Why is it not in the middle of the section.

(b) Work out the total anticlockwise turning moment about A, produced by the 3 MN force acting on D and the 10 MN force acting at B.

(c) The bridge is in equilibrium, so what clockwise turning moment about A must the force R_1 produce?

(d) Now work out the size of R_1.

(e) Since the bridge is stationary, $R_1 + R_2$ must equal the sum of the downwards forces. So how big is R_2?

(f) Explain why the pillars at C and E need to be built more strongly than the other two.

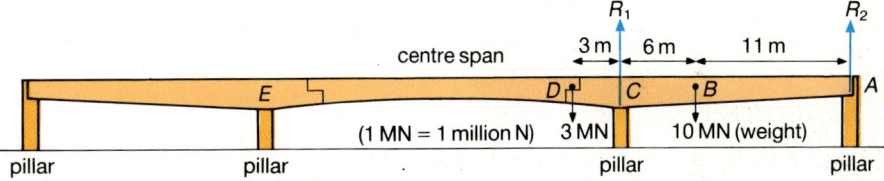

centre span / R_1 / R_2 / 3 m / 6 m / 11 m / E / D / C / B / A / (1 MN = 1 million N) / 3 MN / 10 MN (weight) / pillar / pillar / pillar / pillar

6 Pressure

You always choose a sharp knife when you want to chop up meat or vegetables ready for cooking. A sharp knife has a very thin edge to the blade. This means that the force which you apply is concentrated into a very small area. We say that the pressure under the blade is large.

$$\text{pressure} = \frac{\text{force}}{\text{area}} \qquad P = \frac{F}{A}$$

The unit of pressure is **N/m²** or **pascal (Pa)**; $1\,\text{N/m}^2 = 1\,\text{Pa}$.

Pressure points

The photograph shows a physics teacher lying on a bed of nails. How can he lie there without hurting himself? You all know that nails are sharp and will make a hole in you if you tread on one. The teacher has spread his body out though, so that it is supported by a lot of nails. The area of nails supporting him is large enough for it not to hurt (too much!).

Stiletto heels are pressure points. They concentrate the wearer's weight into a very small area. This can make it difficult to walk across soft surfaces (such as grass), and stiletto heels often damage polished floors

What point is he trying to make here?

When Gillian, an engineer, designs the foundations of a bridge she must think about pressure. In the example shown in Figure 1, the bridge will sink into the soil if it causes a pressure greater than 80 kN/m². What area must the foundations have to stop this happening?

The bridge has a weight of 1.2 MN, so each pillar will support 0.6 MN.

$$P = \frac{F}{A}$$

$$\text{So} \quad A = \frac{F}{P}$$

$$= \frac{0.6\,\text{MN}}{80\,\text{kN/m}^2} = \frac{600\,000\,\text{N}}{80\,000\,\text{N/m}^2} = 7.5\,\text{m}^2$$

Figure 1

What would happen if these foundations were poor?

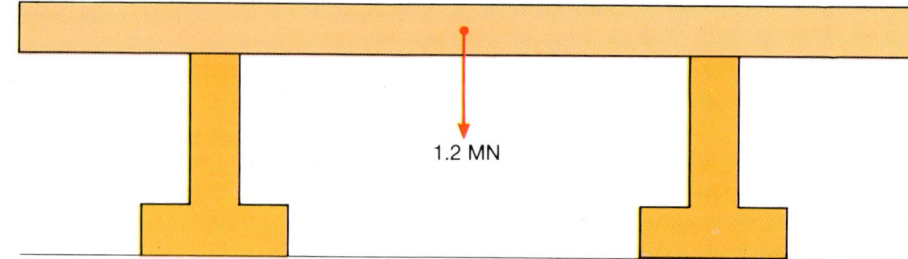

1.2 MN

Crushing pressure and breaking stress

Engineers also have to think about the pressure that a material can withstand without breaking; this is called the **crushing pressure**. A lot of modern buildings are made out of concrete, which is cheap and easy to produce. Concrete is very strong when it is compressed (squashed) but weak when it is stretched. When materials are stretched we say that they are **under stress**. We define stress like this:

$$\text{stress} = \frac{\text{stretching force}}{\text{area}}$$

material	crushing pressure MN/m^2	breaking stress MN/m^2
concete	70	1
cast iron	70	10
steel	400	200

Table 1

So stress is rather like pressure, but it stretches something rather than squashing it.

If a concrete beam is going to be stretched (under tension) in a building it is strengthened with steel bars. As you can see from Table 1, steel is far stronger than concrete under tension. A breaking stress of 200 MN/m^2 means that a steel bar of area 1 m^2 could support 200 MN before breaking.

Questions

1 (a) When a pressure of 4 N/mm^2 is applied to your skin it hurts. In the nail bed shown in the photograph each nail has an area of 1 mm^2. The teacher admits to having a weight of 650 N. What is the smallest number of nails that he must lie on, if he is not to be hurt?
(b) Explain why he has to be very careful getting on and off the nail bed.
2 Look carefully at the bridge in Figure 1.
(a) Although Gillian got her calculation right, the chief engineer, Mr Kruszewki, thinks that each foundation ought to have a base area of 30 m^2. Why might he think this?
(b) Use the information in Table 1 and Figure 1 to work out the smallest area that the concrete pillars can have if they are to hold up the bridge safely. What do you think Mr Kruszewki would say about your answer?
3 The diagram in the next column shows a concrete beam.
(a) Make a copy of the diagram and show which parts of the beam are stretched and which are compressed.
(b) Explain the positioning of the steel bars.

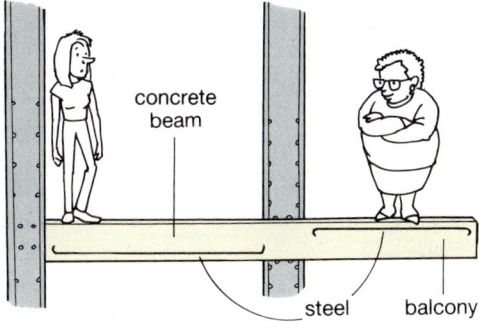

concrete beam

steel   balcony

4 Opposite you can see Boris, Vladimir and Leonid. Boris is twice as tall, twice as long and twice as wide as Vladimir, but they have the same density.
(a) How may times more massive than Vladimir is Boris?
(b) The area of Boris's paw is greater than the area of Vladimir's paw. By how many times is it bigger?
(c) Is the pressure bigger under Boris's paws or under Vladimir's paws? By what factor does the pressure differ under the two cats' paws?
(d) Leonid the lion is a distant cousin to Boris. Lions, as you know, are much

bigger than domestic cats. Can you use the result of part (c) to explain why a lion's legs are proportionately thicker than a cat's?

Boris (cat)

Vladimir (kitten)

Leonid (lion)

7 Pressure in Liquids

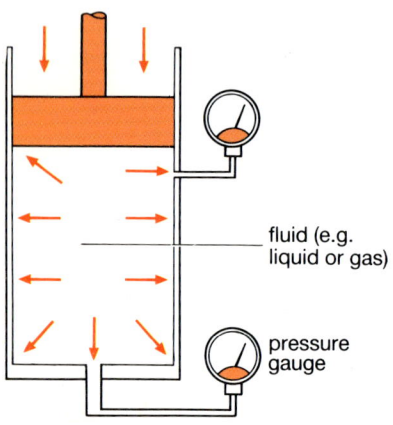

Figure 1
The pressure in a fluid acts equally in all directions

Transmitting pressures

When you hit a nail with a hammer the pressure is transmitted downwards to the point. This happens only because the nail is rigid.

When a ruler is used to push a lot of marbles lying on a table, they do not all move along the direction of the push. Some of the marbles give others a sideways push. The marbles are behaving like a fluid.

In Figure 1 you can see a cylinder of fluid which has been squashed by pushing a piston down. The pressure increases everywhere in the fluid, not just by the piston. The fluid is made up of lots of tiny particles, which act like marbles, to transmit the pressure to all points.

Hydraulic machines

We often use liquids to transmit pressures. Liquids can change shape, but they hardly change their volume when compressed. Figure 2 shows how a hydraulic jack works.

A force of 50 N presses down on the surface above A. The extra pressure that this force produces in the oil is:

$$P = \frac{F}{A}$$

$$= \frac{50 \text{ N}}{10 \text{ cm}^2} = 5 \text{ N/cm}^2$$

The same pressure is passed through the liquid to B. So the upwards force that the surface above B can provide is:

$$F = P \times A$$

$$= 5 \text{ N/cm}^2 \times 100 \text{ cm}^2 = 500 \text{ N}$$

With this machine you can lift a load of 500 N, by applying a force of only 50 N. Figure 3 shows another use of hydraulics.

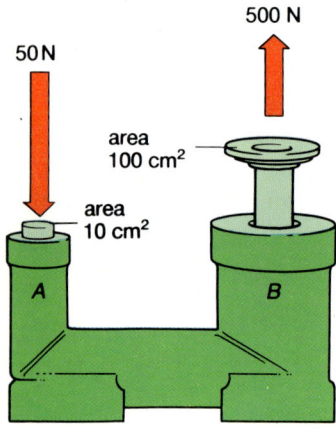

Figure 2
The principle of a hydraulic jack

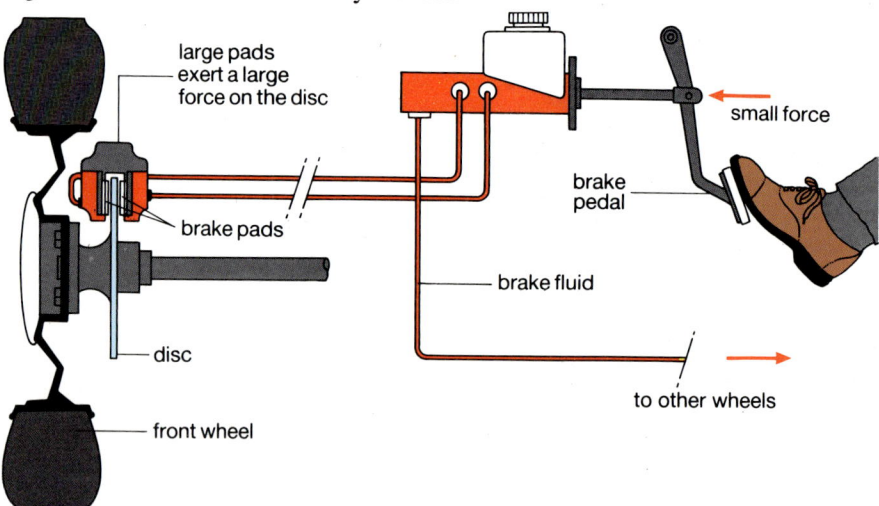

Figure 3
Cars use a hydraulic braking system. The foot exerts a small force on the brake pedal. The pressure created by this force is transmitted by the brake fluid to the brake pads. The brake pads have a large area and exert a large force on the wheel disc. The same pressure can be transmitted to all four wheels.

9 Floating and Flying

You can probably lie on your back and float in a swimming pool. There are two forces that are acting on you. Your weight acts downwards and the water provides an upwards push, called an upthrust. Without this upthrust you would sink. Water provides this upthrust because the pressure of water underneath you is greater than the pressure on top.

Figure 1 shows how you can measure the size of an upthrust. The metal block has a weight of 3 N (Figure 1(a)). When it is lowered into the displacement can, the balance reading falls from 3 N to 2 N (Figure 1(b)). This means that the water provides an upthrust of 1 N on the block. When the block is submerged some water is displaced. In this case, the weight of water displaced is 1 N.

The upthrust on an object when all or part of it is submerged in a fluid, is equal to the weight of water displaced. This is **Archimedes' Principle**.

The salt water in the Dead Sea is denser than the water in other seas. This makes you float higher so that it is easier to read a newspaper than to swim!

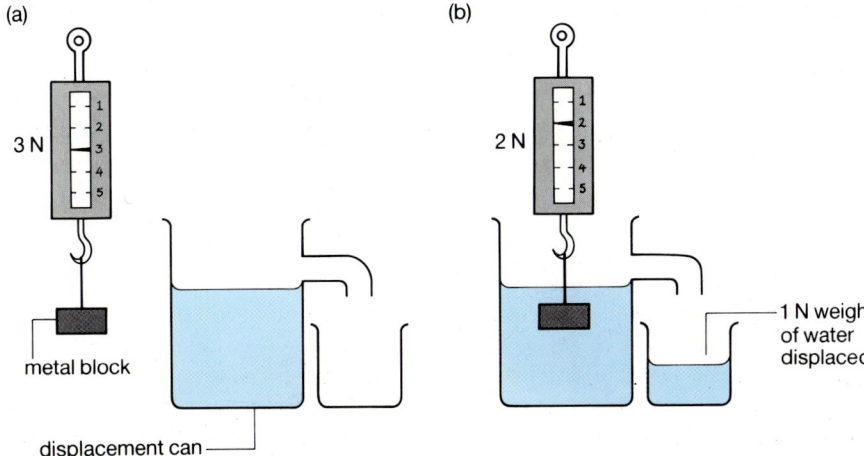

(a)

3 N

metal block

displacement can

(b)

2 N

— 1 N weight of water displaced

Figure 1

Floating

Figure 2 shows two blocks of wood of different densities, which are floating. Since these blocks are in equilibrium there can be no resultant force acting on them. This means that the upthrust from the fluid is equal to the weight of the wood.

So when something floats, the weight of the object equals the weight of fluid displaced.

In Figure 2(a) the block of wood floats with only half of its volume submerged. What is its density?

$$\text{weight of wood} = \text{weight of water displaced}$$

$$\text{So mass of wood} = \text{mass of water displaced}$$

$$\text{The volume of wood} = 2 \times \text{volume of water displaced.}$$

This means that, in comparison with water, the wood has the same mass, but is twice the volume. Its density is therefore half that of water, so it is 500 kg/m^3.

In Figure 2(b) the density of the wood is the same as that of water. The wood will float if its density is the same or less than that of water. It will sink if its density is greater (Figure 2(c)).

When you are describing the size of a ship it is usual to talk about its **displacement**. A ship that has a displacement of 100 000 tonnes, displaces that amount of water. This means that the ship's mass is also 100 000 tonnes. If the ship takes on board 10 000 tonnes of cargo it must displace a total of 110 000 tonnes of water to float.

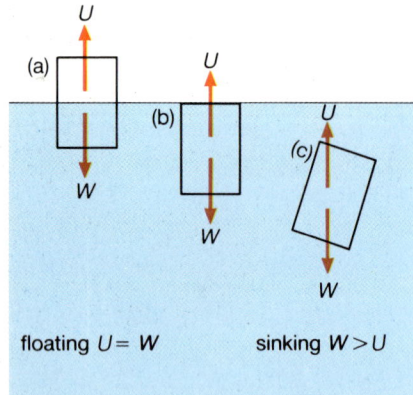

floating $U = W$ sinking $W > U$

Figure 2

The mass of the supertanker is 88 000 tonnes. How is this measured?

Flying

A hot air balloon floats upwards. This is because hot air is less dense than cold air. Aeroplanes, however, rely on a different principle to help them fly. This is **Bernouilli's Principle**.

You can demonstrate Bernouilli's Principle for yourself. When you hold a piece of paper by the edge it hangs down limply. When you blow hard over the top of the paper it lifts upwards. The reason for this is that the pressure in a fast-moving stream of air is lower than it is in still air.

Aeroplanes' wings are designed to make air flow further and faster over the top of the wings (Figure 3). The pressure of air on the bottom of the wing is now greater than the pressure on the top. This provides the wing with the lift to take off.

A sail on a yacht provides another application of Bernouilli's Principle (Figure 4). Fast moving air flowing past the sail gives it some lift. Part of this lift, Y, pushes the boat forwards. The other part of the lift, X, tends to push the boat sideways. To resist this sideways push you lower the keel of the yacht.

This hot air balloon took Richard Branson and Per Lindstrom across the Atlantic. The hot air in the balloon is less dense than the atmosphere surrounding it, so the balloon is able to lift its passengers off the ground

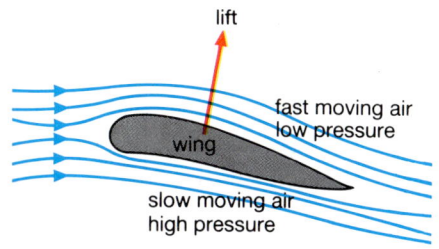

Figure 3
The lift on an aeroplane wing

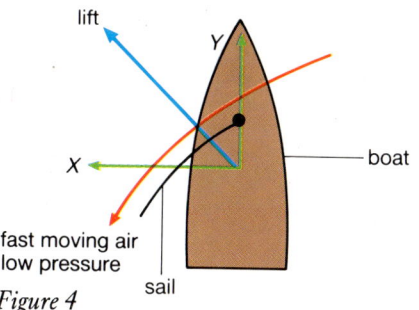

Figure 4
View of a yacht from above.

Questions

1 The diagram below shows part of a car's carburettor. The idea is to make an air-petrol mixture by feeding petrol into the fast-moving air stream.
(a) At which point is the air pressure lowest *A*, *B* or *C*?
(b) Why is it necessary to have a constriction at *B*?

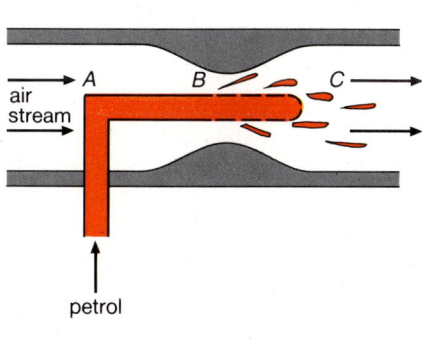

2 A long cylinder hangs from a spring balance. As the cylinder is lowered into some water, measurements are taken from the spring balance, for different values of *d*, (see diagram and table opposite).
(a) Plot a graph of the balance reading (*y*-axis) against the distance *d* (*x*-axis).

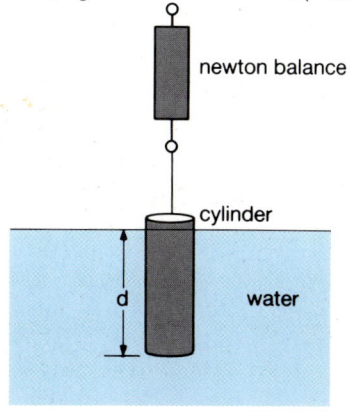

(b) Use your graph to work out the length of the cylinder.
(c) What is the weight of the cylinder?
(d) When the cylinder is fully submerged what is the upthrust on it?
(e) Work out the density of the cylinder.

Distance, *d* (cm)	Force on balance (N)
0	51
10	46
20	41
30	36
40	32
50	32
60	32

SECTION B: *STUDY QUESTIONS*

1 Using a pulley as shown, a force of 20 N is required to keep the 2 kg mass stationary.

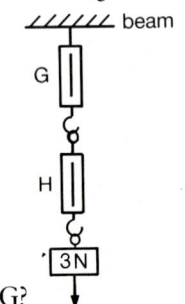

If the weight of the pulley is small enough to ignore, what is the force exerted on the ceiling?

A 40 N upwards   **B** 20 N upwards
C 40 N downwards   **D** 22 N downwards
E 20 N downwards

MEG

2 The diagram shows a forcemeter G attached to a beam. G holds a second forcemeter H which holds a 3 N weight. The forcemeters each have a weight of 2 N.

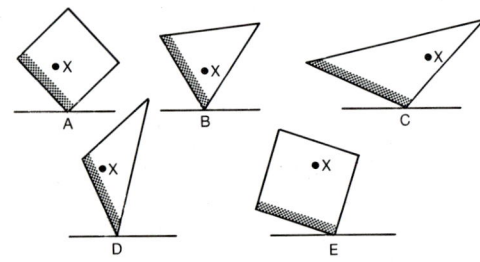

What is the reading on G?
A 1 N   **B** 2 N   **C** 3 N   **D** 5 N   **E** 7 N

MEG

3 The diagrams show five objects, held in the positions shown. In each case, X is the centre of gravity of the body and the shaded part is the base. Which object will *not* topple on to its base when it is released?

4 Popeye is blowing into the sail in his boat to push it forwards. Is this possible?

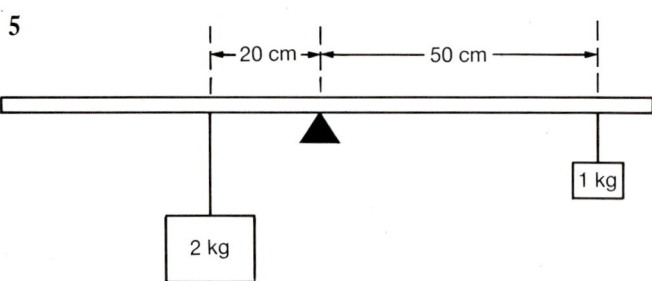

5

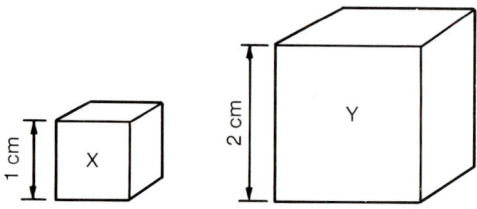

The diagram shows a uniform beam pivoted at its mid-point, from which two masses are suspended. If a third mass is added to balance the system, its size and distance from the pivot may be

	Mass (kg)	Distance (cm)
A	1	10
B	1	30
C	2	10
D	2	30

SEG

6 Two cubes X and Y are made from the same material. The sides of cube X are 1 cm and the sides of cube Y are 2 cm.

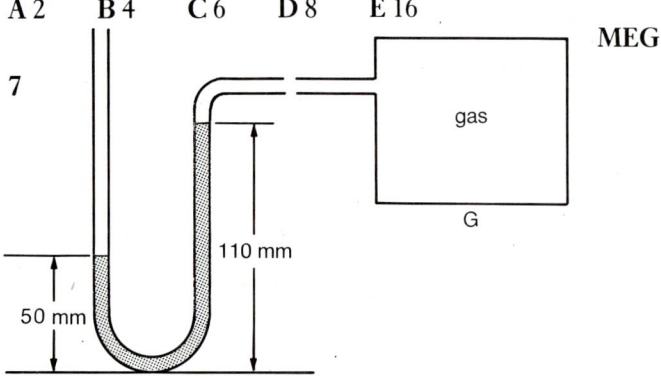

The cubes stand on a flat surface. How many times larger is the pressure exerted by Y than that exerted by X?
A 2   **B** 4   **C** 6   **D** 8   **E** 16

MEG

7

The diagram illustrates an experiment to measure the pressure of the gas inside the container G using a mercury manometer. The heights of the columns, measured from the bottom of the manometer, are given in the diagram. If the height of a mercury barometer is 750 mm, the pressure of the gas in G, in mm of mercury, is
A 60   **B** 690   **C** 800   **D** 810   **E** 860

NI

8 Below, you can see two wells. Oil, salt-water and natural gas are trapped by oilproof layers of rock.

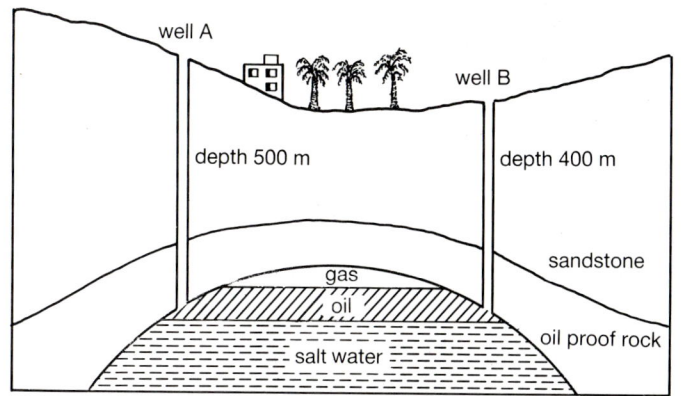

(a) (i) Why is the salt water lying below the oil?
(ii) Why is the natural gas lying above the oil?
(b) When well A is drilled, oil shoots 10 m above the ground. When well B is drilled, oil shoots much higher above the ground. Explain why oil shoots much higher from well B.
(c) As the oil escapes from the wells, what happens to (i) the volume of the natural gas; (ii) the pressure of the natural gas?
(d) After a few months, oil can still be obtained from well A but it has to be pumped out. Explain why the oil does not shoot out by itself.

LEAG

9 (a) State clearly *two* conditions for a body to be in equilibrium.
(b) The diagram below shows a mousetrap. The idea is that as the mouse walks towards the cheese he tips the plank up and then falls conveniently near to Boris the cat. Go through the following calculations to see how far along the plank the mouse can go before it tips up.

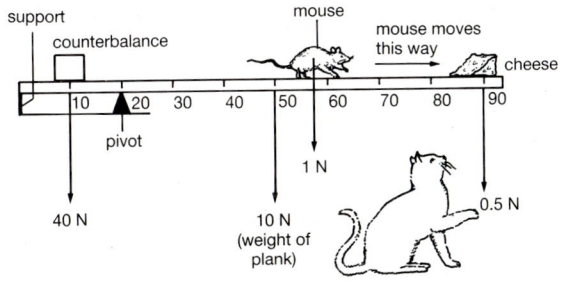

(i) Calculate the turning moment of the counter balance about the pivot.
(ii) The weight of the plank may be taken to act through the 50 cm mark. Calculate the turning moment of the cheese and the plank together.
(iii) Use your answers to (i) and (ii) to tell Boris where the mouse will over balance the trap.
(c) Use the information in the diagram to calculate the upwards force that the pivot exerts on the plank, just as the 'trap' overbalances.

10 The diagram represents a solid metal ring.

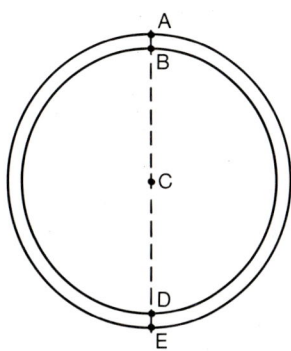

Which point **A, B, C, D** or **E** is closest to the centre of mass of the ring?

LEAG

11 (a) (i) State what is meant by *pressure*.
(ii) A suitable unit for pressure is the pascal (N/m^2). State another unit used to measure pressure.
(b) Explain briefly, in terms of pressure, why
(i) the sharp edge of a knife, and not the blunt edge, is used for cutting.
(ii) the thickness of the wall of a dam, used to store water, is greater at the base than at the top.
(iii) a tin can collapses when the air is removed from it.
(iv) an aneroid barometer may be used as an altimeter in an aircraft.
(c) A simple type of hydraulic braking system is shown in the diagram below.

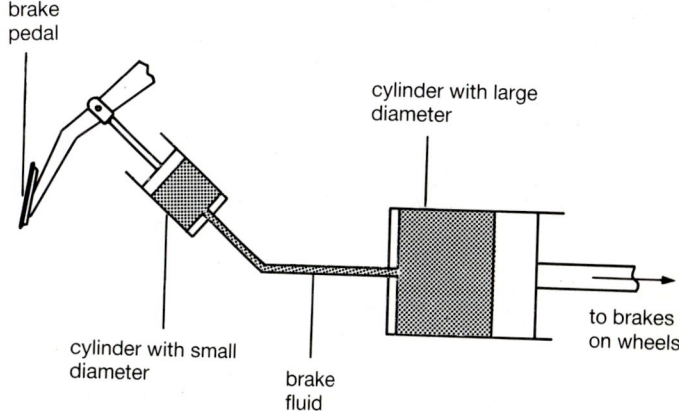

The areas of cross-section of the small cylinder and large cylinder are 0.0004 m^2 and 0.0024 m^2 respectively. The brake pedal is pushed against the piston in the small cylinder with a force of 90 N.
(i) Determine the pressure exerted on the brake fluid.
(ii) Determine the force exerted by the brake fluid on the piston in the large cylinder.

NI

12 Bristol airport lies about 150 km due west of Heathrow airport.

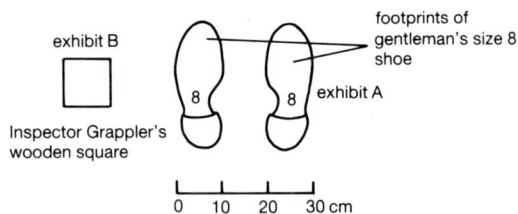

(a) On a windless day a light aircraft sets off from Heathrow to Bristol. The plane flies at 250 km/h. How long does it take to get there?

(b) On the return journey a strong wind blows due north. The wind speed is 50 km/h. If the pilot sets off to fly east to Heathrow, in which direction will he end up going? Draw a diagram to illustrate your answer.

(c) He corrects his course so that he actually flies eastwards. In which direction must he point the plane? How long does his journey take? (You might find it useful to make a scale drawing.)

13 Mike is worried about his weight. But he is so heavy that his scales cannot measure him any more. However, he has had an idea. He knows that atmospheric pressure ($100\,000$ N/m^2) supports a column of mercury 75 cm high. He has attached a hot water bottle to a mercury manometer. When he stands on it the mercury moves as shown.

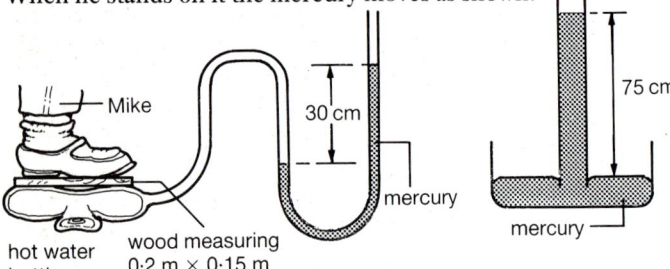

(a) What extra pressure does Mike cause?

(b) What is the area of the wood that Mike stands on?

(c) Now work out Mike's weight. What is his mass in kg? Unfortunately the bottle burst just as Mike was reading the height of the mercury. So, he has devised another method of weighing himself. He has borrowed a Newton balance that can measure weights up to 500 N. The diagram shows his next weighing experiment.

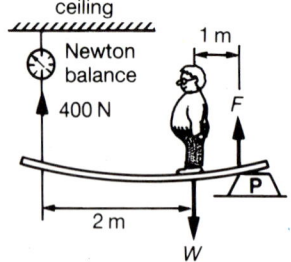

(d) Use the diagram to work out Mike's weight.

(e) How big is the force F?

14 Colonel Carruthers has been murdered at Campbell Castle. Inspector Grappler of the Yard has been sent to investigate. He finds old Carruthers dead in the library. Outside the library window there are some footprints in the flower bed (exhibit A). This is an important clue; the inspector realises that he can estimate the suspect's height.

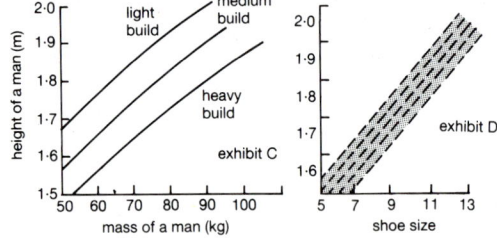

Inspector Grappler carries out an experiment in the flower bed. He uses a wooden square (exhibit B) and some weights. He piles the weights onto the square and measures how far the square sinks into the flower bed. His results are below.

Mass of the inspector's weights (kg)	Depth of the hole made by the square (mm)
5	7
10	14
15	20
20	25
25	30

(a) Work out the area of the suspect's shoes.

(b) Inspector Grappler measured that the shoes had sunk 23 mm into the flower bed. Use the data in the table and your answer to part (a) to make an estimate of the suspect's mass.

(c) Exhibit C shows roughly how the height of a man depends on his mass. Exhibit D shows roughly how a man's shoe size depends on his height. Use these graphs to make an informed guess of the suspect's height.

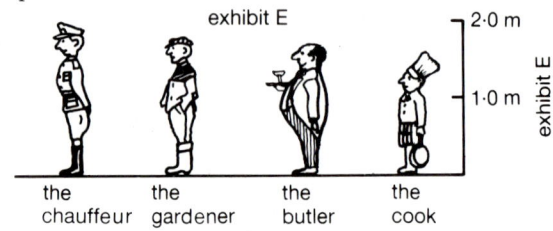

(d) Of Colonel Carruthers' servants, which one do you suspect?

SECTION C
Forces in Motion

Surfers make use of the energy provided by the waves to travel at high speed

1 How Fast Do Things Move?

Ben Johnson breaks the world record to run 100 metres in 9.83 seconds. What was his average speed?

Moving object	Speed (m/s)
glacier (Rhonegletsher)	0.000001
snail	0.0005
human walking	2
human sprinter	10
express train	60
Concorde	600
Earth moving round the Sun	30 000
light and radio waves	300 000 000

This snail would take nearly two and a half days to go 100 metres!

When you travel in a fast car you finish your journey in a short time. When you travel in a slow car your journey takes longer. We might say that the speed of a car is 100 kilometres per hour (100 km/h) which means that the car will travel a distance of 100 kilometres in one hour. We can write a formula connecting distance, speed and time:

$$\text{average speed} = \frac{\text{distance travelled}}{\text{time taken}} = \frac{d}{t}$$

We write average speed because the speed of the car may change a little during the journey. When you travel along a motorway your speed does not remain exactly the same. You slow down when you get stuck behind a lorry and speed up when you pull out to overtake a car.

Example. At top speed a guided-missile destroyer travels 170 km in three hours. What is its average speed in m/s?

$$\text{average speed} = \frac{d}{t}$$

$$= \frac{170 \times 1000 \text{ m}}{3 \times 3600 \text{ s}}$$

$$= 16 \text{ m/s}$$

remember 1 km = 1000 m;
1 hour = (60 × 60 s) = 3600 s

Velocity

It is not just the speed that is important when you go on a journey. The direction matters as well. Figure 1 shows three possible routes taken by a helicopter leaving London. The helicopter travels at 300 km/h. So in one hour the helicopter can reach Liverpool, Paris or Brussels depending on which direction it travels in.

When we want to talk about a direction as well as a speed we use the word **velocity**. For example, when the helicopter flies towards Liverpool, we can say that its velocity is 300 km/h, on a compass bearing of 330°. Velocity is a vector, speed is a scalar (see page 18).

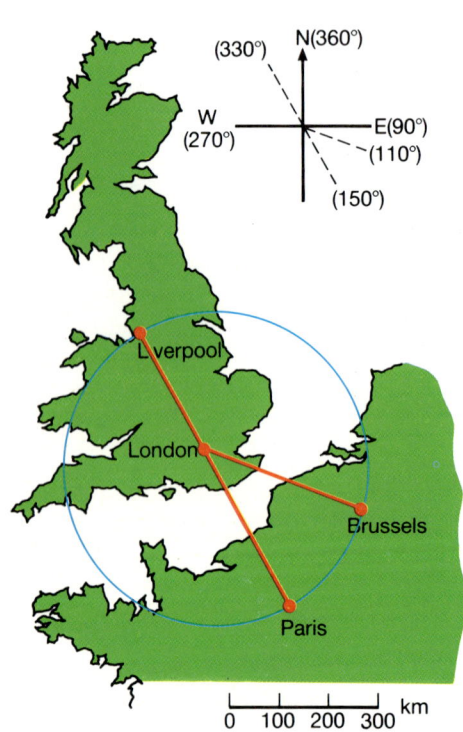

Figure 1

Time-tabling railway trains

You have probably sat on a slow train in a railway station wondering why the train was taking so long to start. As likely as not you would have been waiting for an express train to overtake you. There is often only one line between railway stations. The trains must be time-tabled so that the faster express trains can overtake a slower train that has stopped at a station.

Suppose you are a passenger train planning officer working for British Rail. You have to make sure that a slow train going from Reading to Chippenham does not get in the way of the London to Cardiff express train. The slow train travels at an average speed of 70 km/h and the express train at an average speed of 120 km/h.

Plotting a graph of the distance travelled by these trains against time helps you to solve the problem. You can see from Figure 2 what happens if both trains leave their stations at 0900 h. The slow train travels the 70 km to Swindon, and the faster train travels from London to Swindon, in the same time. You can also see from the graph that there is a slow train to Newport leaving Bristol Parkway at 1015 h that is going to get in the way of the express train.

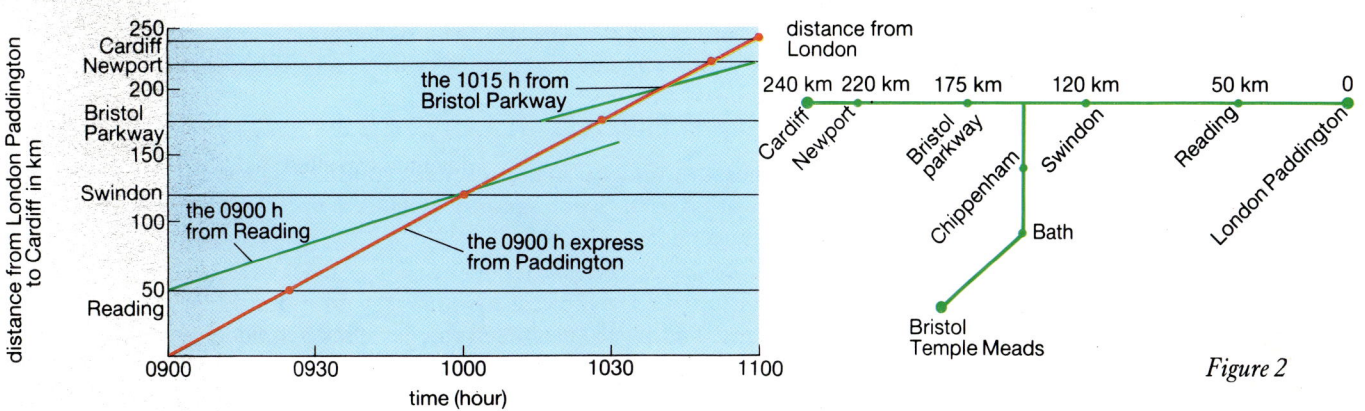

Figure 2

Questions

1 The world record for running 1500 m is held by Saïd Aouita. He ran that distance in 3 minutes and 29 seconds. What was his average speed?

2 When Richard Noble broke the land speed record in *Thrust II* in October 1983 he travelled at an average speed of about 1000 km/h. How long did it take him to travel 1 km?

3 Sketch a graph of distance travelled (*y*-axis) against time (*x*-axis) for a train coming into a station. The train stops for a while at the station and then starts again.

4 This question refers to the train time-tabling shown in Figure 2.
(a) How can you tell from the graph that the express train travels faster than the other two?

(b) At what time does the express train reach Bristol Parkway?
(c) At what time should the 1015 h from Bristol leave so that it meets the express train at Newport?

5 Ravi, Paul and Tina enter a 30 km road race. The graph shows Ravi's and Paul's progress through the race.
(a) Which runner ran at a constant speed? Explain your answer.
(b) What was Paul's average speed for the 30 km run?
Tina was one hour late starting the race. During the race she ran at a constant speed of 15 km/h.
(c) Copy the graph and add to it a line to show how Tina ran.
(d) How far had Tina run when she overtook Paul?

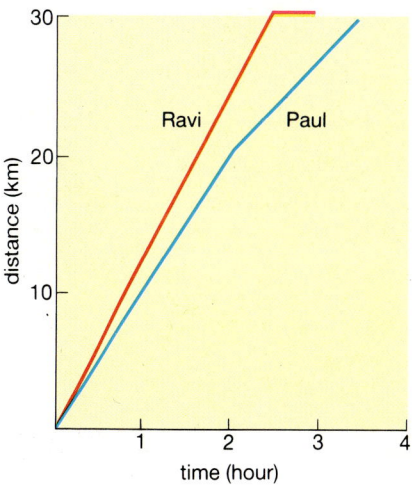

2 Acceleration

Drag cars accelerate to a high speed in a short time

When a car is speeding up, we say it is *accelerating*. When it is slowing down we say it is *decelerating*. A deceleration can be thought of as a negative acceleration.

You may be interested in driving a sports car which can accelerate quickly away from the traffic lights when they turn green. A car that accelerates rapidly reaches a high speed in a short time. For example, a sports car speeds up to 45 km/h in 5 seconds. A truck speeds up to 45 km/h in 10 seconds. We say the acceleration of the car is twice as big as the truck's acceleration.

You can work out the acceleration of the car or truck using the formula:

$$\text{acceleration} = \frac{\text{change of velocity}}{\text{time}}$$

$$\text{acceleration of sports car} = \frac{45 \text{ km/h}}{5 \text{ s}}$$

$$= 9 \text{ km/h per second}$$

$$\text{acceleration of the truck} = \frac{45 \text{ km/h}}{10 \text{ s}}$$

$$= 4.5 \text{ km/h per second}$$

Some more about acceleration

We usually measure velocity in metres per second. This means that acceleration is usually measured in **m/s per second**. What is the acceleration of our sports car in m/s per second?

$$45 \text{ km/h} = \frac{45\,000 \text{ m}}{3600 \text{ s}} = 12.5 \text{ m/s}$$

$$\text{So acceleration} = \frac{\text{change of velocity}}{\text{time}}$$

$$= \frac{12.5 \text{ m/s}}{5 \text{ s}} = 2.5 \text{ m/s per second}$$

or

$$2.5 \text{ m/s}^2 \text{ (metres per second squared)}$$

Velocity/time graphs

You have already seen that distance/time graphs can be helpful when looking at problems involving moving cars or trains. It can also be helpful to plot graphs of velocity against time.

Figure 1 shows the velocity of a cyclist as she cycled through a town. The following things happened in her journey.
(1) For the first 20 seconds, as she came into the town, she travelled at a constant velocity.
(2) Then she started to cycle up a hill, so she slowed down.
(3) At the top of the hill she came to some traffic lights. She had to wait for them to change.

(4) When the traffic lights changed to green she accelerated away.
The slope of a velocity/time graph is useful, because it shows how quickly the cyclist changes velocity. If the slope is steep, she is changing velocity (accelerating or decelerating) quickly.

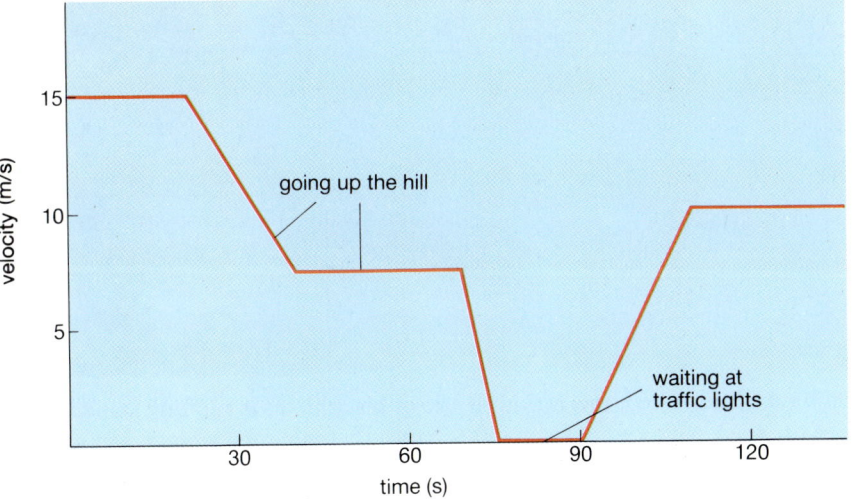

This car has been in a serious crash. When a fast-moving, heavy object like a car decelerates rapidly, it exerts a large force on anything that tries to stop it. It is the reaction to this force that damages the car (and its passengers)

Figure 1

Questions

1 This question refers to the graph in Figure 1.
(a) What was the cyclist's velocity after 60 s?
(b) How long did she have to wait at the traffic lights?
(c) Which was larger, her deceleration as she stopped at the traffic lights, or her acceleration when she started again? Explain your answer.

2 Below is a table showing how the speed, in km/h, of Ayrton Senna's McLaren racing car changes, as he accelerates away from the starting grid at the beginning of the Brazilian grand prix.
(a) Plot a graph of speed (*y*-axis) against time (*x*-axis).
(b) Use your graph to estimate the acceleration of the McLaren in km/h per second at: (i) 16 s, (ii) 1 s.

	Starting speed (m/s)	Final speed (m/s)	Time taken (s)	Acceleration (m/s²)
Cheetah	0	30	5	6
Second stage of a rocket	450	750	100	
Aircraft taking off	0		30	2
Car crash	30	0		−150

3 Copy the table above and fill in the missing values.

4 In America drag cars are designed to cover distances of 400 m in about 6 seconds. During this time the cars accelerate very rapidly from a standing start. At the end of 6 seconds, a drag car reaches a speed of 150 m/s.
(a) What is the drag car's average speed?
(b) What is its average acceleration?

5 Describe the motion in the following two cases:

speed (km/h)	0	35	70	130	175	205	230	250	260	260
time (s)	0	1	2	4	6	8	10	12	14	16

3 Observing Motion

Athletes use photographs like this one to analyse their motion so that they can improve their performance

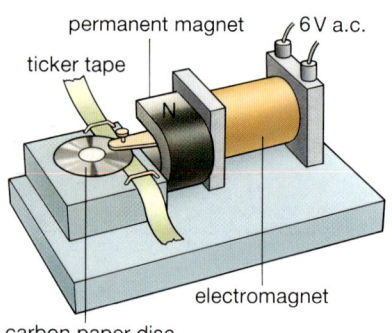

Figure 1

A trainer who wants to know how well one of his athletes is running stands at the side of the track with a stop watch in his hand. By timing the athlete over a set distance, the trainer knows how fast he is running. But this only tells the trainer an average speed. If he wants to learn more, he needs to look at the athlete's movements over very short time intervals. The best way to look at something that is moving is to film it. This is what sports scientists do to check on an athlete. By looking at each frame of film the scientists can see if the athlete is wasting any effort in his running action.

Another way to look at motion is to use multiflash photography. On the next page you can see a photograph of a golfer hitting a ball. In this technique the golfer swings his club in a darkened room, while a lamp flashes on and off at regular intervals. When the images of the club are close together, it is moving slowly. When the images are far apart, the club is moving quickly.

Ticker timer

For studying motion in the lab we use a **ticker timer** (Figure 1). A ticker timer has a small hammer that vibrates up and down 50 times per second. The hammer hits a piece of carbon paper which leaves a mark on a length of tape.

Figure 2 shows you two tapes that have been pulled through the timer. You can see that the dots are close together over the region *PQ*. Then the dots get further apart, so the object moved faster over *QR*. Then the movement slowed down again over the last part of the tape, *RS*. Since the timer produces 50 dots per second, the time between dots is $\frac{1}{50}$ s or 0.02 s. So we can work out the speed:

$$\boxed{\text{speed} = \frac{\text{distance between dots}}{\text{time between dots}}}$$

Between *P* and *Q*, speed $= \dfrac{0.5 \text{ cm}}{0.02 \text{ s}} = 25 \text{ cm/s } or \ 0.25 \text{ m/s}$

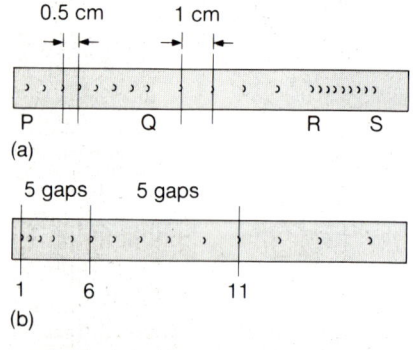

Figure 2

In tape B the dots get further and further apart, so the object attached to this tape was accelerating all the time.

Measuring acceleration

Figure 3 shows how you might measure the acceleration of a trolley moving down a slope. When you have let the trolley go down the slope the tape stuck to it will look a bit like tape B in Figure 2. It is helpful to cut up the tape into 5-tick lengths. You do this by cutting through the first dot, then the sixth and eleventh and so on. Each length of tape is the distance travelled by the trolley in $\frac{1}{10}$ s (5 dots means $5 \times \frac{1}{50}$ seconds). You can then use your pieces of tape to make a graph and see how the trolley moved. If your pieces of tape form a straight line then the acceleration was constant (see Figure 4).

Cutting up your tape like this is like plotting a graph of speed against time; the steeper the slope the greater the acceleration.

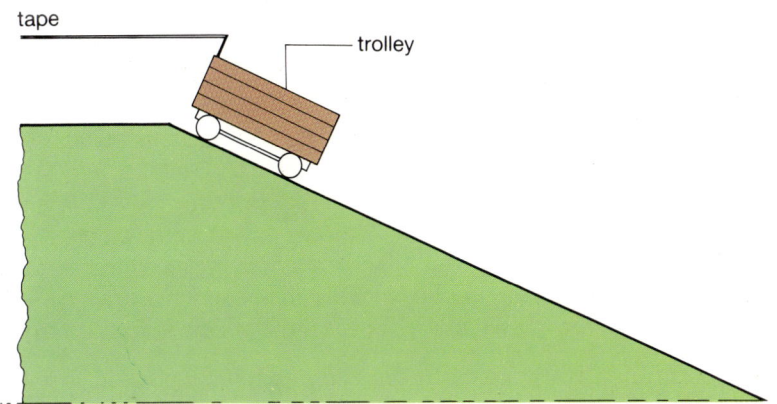

Figure 3

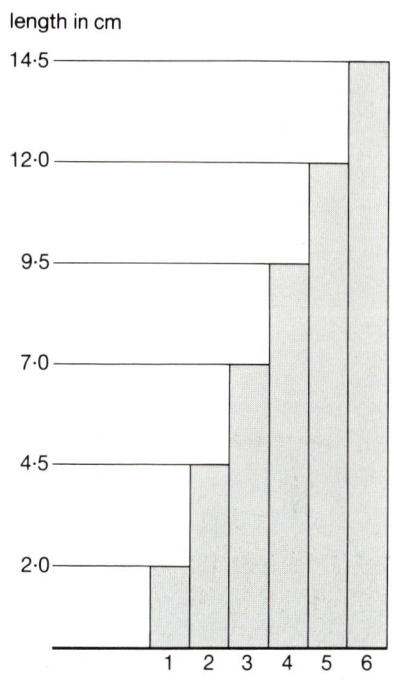

length in cm

Figure 4
Each length of tape shows the movement of the trolley over 0.1 s

Questions

1 Work out the speed of the tape in Figure 2(a) in: (i) the region *QR*, (ii) the region *RS*.

2 (a) Why can you tell from Figure 4 that the trolley accelerated down the slope at a constant rate?
(b) Work out the average speed of the trolley in: (i) interval 4, (ii) interval 6.
(c) What is the time between the middle of interval 4 and the middle of interval 6?
(d) Now use your results from parts (b) and (c) to calculate the acceleration of the trolley.

3 Examine the photograph of the golfer. Where is his club moving fastest? Explain your answer.

4 In aircraft the airspeed, *v*, is usually measured using an airspeed indicator. This indicator makes use of the difference in air pressure between still and fast-moving air. The pressure of fast-moving air is less than the pressure of still air.
(a) Use the data in the table to plot a graph of pressure difference (*y*-axis) against v^2 (*x*-axis). Explain why the graph shows that the pressure difference is proportional to the square of the air speed.

Difference between still and moving air pressures ($N m^2$)	Airspeed, *v* (m/s)
400	40
1600	80
3600	120
6400	160
10000	200

(b) Use the graph to work out the airspeed when the pressure difference is: (i) 2000 N/m^2, (ii) 12 000 N/m^2.

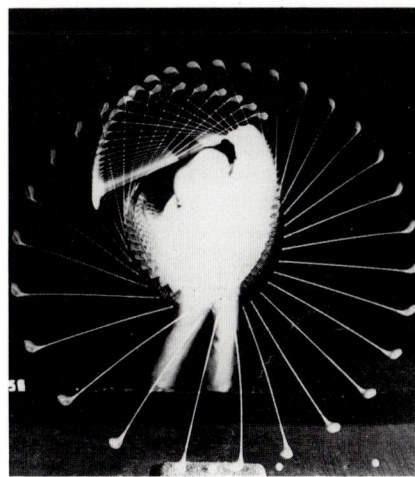

4 Equations of Motion

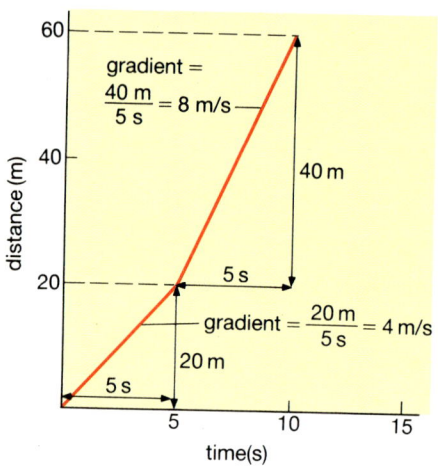

Figure 1
The distance/time graph for Ravi

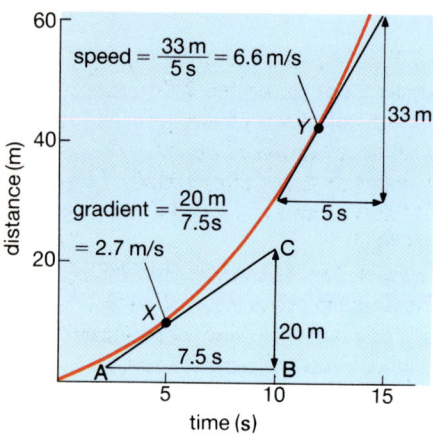

Figure 2
The distance/time graph for Paul

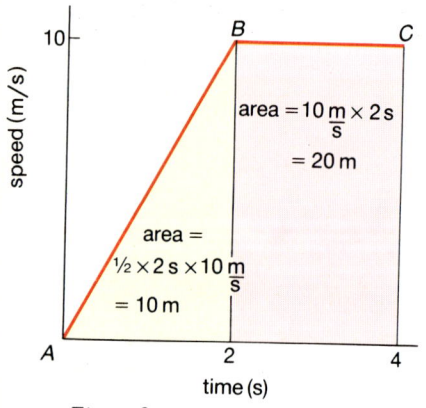

Figure 3
The speed/time graph of a sprinter

More about graphs

Ravi is getting near to the end of a race. He is running at 4 m/s. When he sees the finishing line, he doubles his speed and runs at 8 m/s. Figure 1 is Ravi's distance/time graph. In the first 5 seconds Ravi ran 20 m. In the next 5 seconds he ran 40 m. When Ravi doubled his speed the slope, or **gradient**, of the graph doubled. This leads to an important rule:

> The gradient of a distance/time graph equals the speed

Figure 2 shows how Paul finishes the same race. When he sees the finishing line he starts to accelerate. He accelerates gently, so his speed increases all the time. That is why the gradient of his distance/time graph gets steeper and steeper.

   Example. What is Paul's speed at point X?

(1) Draw a line parallel to the gradient of the graph at X. It does not matter how long this line is.

(2) Draw in the horizontal line AB and the vertical line BC.

(3) Use the scale of the graph to work out that the distance $BC = 20$ m. Work out that the time $AB = 7.5$ s.

(4) Then use the formula $\text{speed} = \dfrac{\text{distance}}{\text{time}} = \dfrac{20 \text{ m}}{7.5 \text{ s}}$

$$= 2.7 \text{ m/s}$$

When a sprinter runs a 100 m race, she does not reach her top speed as soon as the starting pistol is fired. She takes a few seconds to accelerate from rest, up to her top speed. Figure 3 shows this in the form of a speed/time graph. From Figure 3 you can see that she took 2 seconds to accelerate up to 10 m/s. So her acceleration is 5 m/s^2, which is the gradient of the graph.

> The gradient of a speed/time graph equals the acceleration

You can also use the same graph to calculate the distance that she has run. It is easy to work out the distance travelled once she has reached a constant speed of 10 m/s. She runs 10 m each second, so over the region BC of the graph she runs a distance of 10 m/s × 2 s = 20 m, which is the **area** under that part of the graph.

> The area under a speed/time graph equals the distance travelled

We can check this formula for the first 2 seconds of her race. The distance travelled in that time is equal to the area of the triangle under the line AB.

   area of triangle = ½ base × height = ½ × 2 s × 10 m/s

$$= 10 \text{ m}$$

We would have found the same answer if we had used the formula:
distance = average speed × time. The average speed over the first 2 seconds is 5 m/s, halfway between 0 and 10 m/s.

   So d = average speed × time = 5 m/s × 2 s

$$= 10 \text{ m}$$

Equations of motion

We can use the ideas in the last paragraph to produce some formulae to describe the motion of any object moving under a constant acceleration. The formulae that follow do not apply if the acceleration changes during the motion. Figure 4 shows a speed/time graph for a moving object, which accelerates with acceleration, a, in a time, t, from a starting speed, u, to a final speed, v.

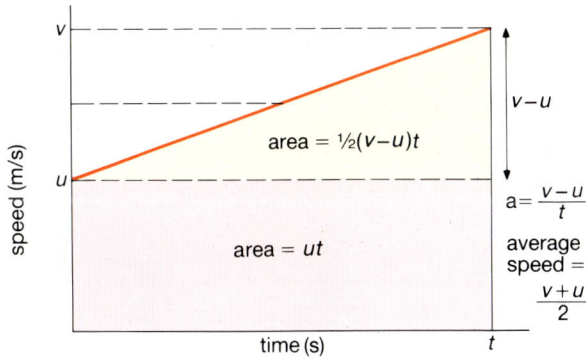

Figure 4
The speed/time graph of an object with constant acceleration

The acceleration is the slope of the graph:

$$\text{So } a = \frac{v - u}{t}$$

$$\text{or } v = u + at$$

The distance travelled is the area under the graph:

$$\text{So } d = ut + \frac{1}{2}(v - u)t$$
$$\text{area of rectangle} + \text{area of triangle}$$

$$\text{but } v - u = at$$

$$\text{Therefore } d = ut + \tfrac{1}{2}at^2$$

The average speed for the journey is half way between u and v.

$$\textbf{average speed} = \frac{\textbf{v} + \textbf{u}}{\textbf{2}}$$

Questions

1 Use Figure 3 to work out how long our athlete took to run her 100 m race.
2 A spacecraft blasts off with an upwards acceleration of 1 m/s^2.
(a) What is its speed after 100 s?
(b) What is its average speed during these 100 s?
(c) How far has it travelled after 100 s?
3 The diagram below shows how the velocity of a balloon (filled with helium) varies once it has left a child's hand. When the velocity is positive the balloon is rising, when the velocity is negative, the balloon is falling.
(a) At which one of the points marked is the balloon's acceleration zero?
(b) At which point did the balloon reach its maximum height?
(c) Make a rough estimate of the balloon's maximum height.

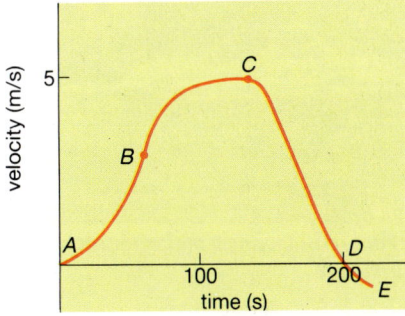

5 Forces and Motion

This unit is about **Newton's first law of motion**. This law says that when a force pushes or pulls an object, its velocity will change. When there is no force, the object remains stationary, or carries on moving in a straight line with a constant speed.

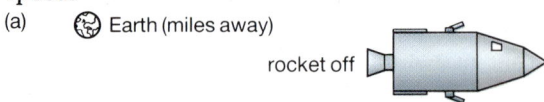

In Figure 1(a) you can see the Russian spacecraft *Vostock*. It is a very great distance from the Earth, so we can forget about any gravitational pull. At the moment *Vostock* is not moving and its rockets are turned off.

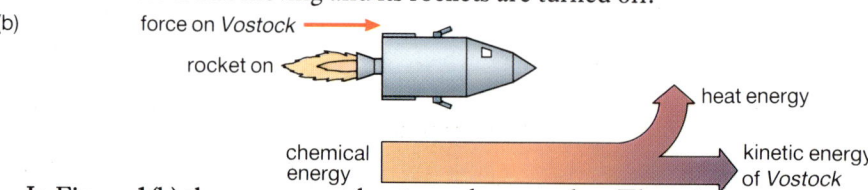

In Figure 1(b) the cosmonaut has turned on a rocket. There is now a force pushing *Vostock* forwards and its speed increases. The stored chemical energy in the fuel is turned into kinetic energy of the moving spacecraft, and heat.

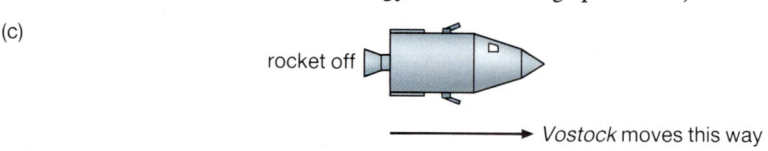

In Figure 1(c) the rocket has been turned off. Again no forces act on *Vostock* and it carries on moving at a constant speed. In space there is no air so nothing gets in the way to slow *Vostock* down. The only way to slow it down is to fire rockets that are pointing forwards. The blast from these produces a backwards force on the spacecraft and its speed decreases (Figure 1(d)).

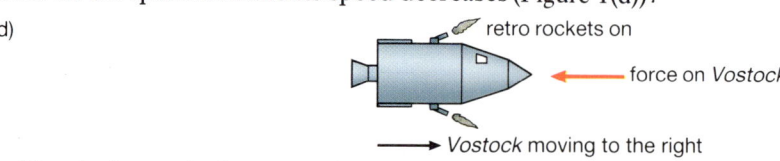

If we look at a similar example of motion on the Earth, things are not quite so simple. The problem is that on Earth there are frictional forces. When a car wheel turns on its axle, the moving surfaces rub together and slow the car down. The car also has to push its way through the air. The air gives rise to another force that slows the car; we call it wind resistance. If you put your hand out of the window of a fast-moving car, you can feel how strong this force is.

When a car is parked on a level road it stays where it is. This is because there is no force to push it along the road. In the top photograph the engine has been started and the car accelerates. But wind resistance begins to act. Wind resistance gets larger as the speed increases.

In the second photograph the car is going along the road at constant speed. The forwards push from the wheels is equal to the backwards push from the wind resistance.

In the third photograph the engine has been switched off. Now only frictional and wind resistance forces push on the car. These slow down the car. In space it is possible to move at a constant speed without using fuel because there are no frictional forces. On the road a car needs energy from fuel to work against wind and frictional resistance.

The resultant force on the car is forwards, so the car accelerates

There is no resultant force, so the car moves at a constant speed

Wind resistance now makes the car decelerate

Streamlining

If you want to go fast you need to make the effect of wind resistance as small as possible. There is a bigger wind resistance on objects which have a large area. Pointed objects can pass easily through air in the same way that a knife will cut through something. Fast cars are carefully designed so that air can flow easily past them; this is called **streamlining**.

You see the same idea in ships. Those that are built for speed are very thin and have pointed bows. Skiers and cyclists crouch so that they go head first into the wind. This makes the area that goes into the wind as small as possible. They also wear very smooth, skin-tight clothing which reduces the effect of the wind.

Here are some examples of streamlining in technology and nature. The smoke streams directed across the car in a wind tunnel help designers improve the car's streamlining. The solid rear wheel of the bicycle is designed to reduce air resistance. The dolphin's streamlined shape allows it to move very quickly through the water

Questions

1 (a) The diagrams below show the forces that act on an aeroplane at different times during a flight. The longer an arrow, the larger the force. For each of the cases below describe how the plane is moving.
(b) Draw another diagram to show the forces acting on the plane just after it has taken off.

2 A student wrote this in an exam: 'When something is moving forwards, there must be a force acting on it. As soon as you turn the engine off in a car or the rocket off in a spaceship, they stop.' The student has made some mistakes; rewrite this paragraph explaining where he went wrong.

3 The data in the table show how the wind resistance on a car varies as the speed increases.
(a) Use the data to work out the fastest the car can travel when the wheels produce a forwards push of 2600 N.
(b) When the car is made more streamlined it travels faster. Explain why.

Speed (m/s)	Wind resistance (N)
0	0
5	600
10	1200
15	1800
20	2400
25	3000

6 Force, Mass and Acceleration

The large force produced by a powerful engine acts on a small mass to give Nigel Mansell's racing car a large acceleration

When a force acts on a car it will accelerate. What affects how big the acceleration is? If you have ever had a car with a flat battery then you will have given the car a push to start. When one person tries to push a car alone then the acceleration of the car is very slow. It takes a long time for the car to increase its speed. When three people give the car a push, it accelerates much more rapidly.

The acceleration is proportional to the force:

$$\text{acceleration} \propto \text{force}$$

You also know from everyday experience that large massive objects are difficult to set in motion. When you throw a ball you can accelerate your arm more quickly if the ball has a small mass. You can throw a cricket ball much faster than you can put a shot. A shot has a mass of about 7 kg and the force that your arm can apply cannot accelerate it as rapidly as a cricket ball.

The acceleration is inversely proportional to the mass:

$$\text{acceleration} \propto \frac{1}{\text{mass}}$$

It is difficult to get large things moving, but it is also very difficult to stop large things when they are moving. The world's largest ship is the *Seawise Giant*; it is 460 m long and has a mass of about 560 000 tonnes. Once a ship like this is moving the captain has to allow about 10 km for the ship to slow down.

The frictional force of water on a large mass like a ship can produce only a small deceleration, so it takes a long time for a ship like this one to slow down

Newton's second law of motion

With some careful experiments in the laboratory, you can see how the size of the acceleration is connected to the size of the force. Figure 1 shows the idea. A trolley placed on a table is accelerated by pulling it with an elastic cord. This is stretched so that it always remains the length of the trolley. The acceleration of the trolley is measured by putting a piece of ticker tape on the back of the trolley.

The first experiment is to increase the force acting on one trolley. You can see how our ticker tape graphs get steeper as the force increases. This means the acceleration is getting bigger. A 1 kg trolley accelerates at a rate of 1 m/s^2 for a force of 1 N; 2 m/s^2 for a force of 2 N; and 3 m/s^2 for a force of 3 N.

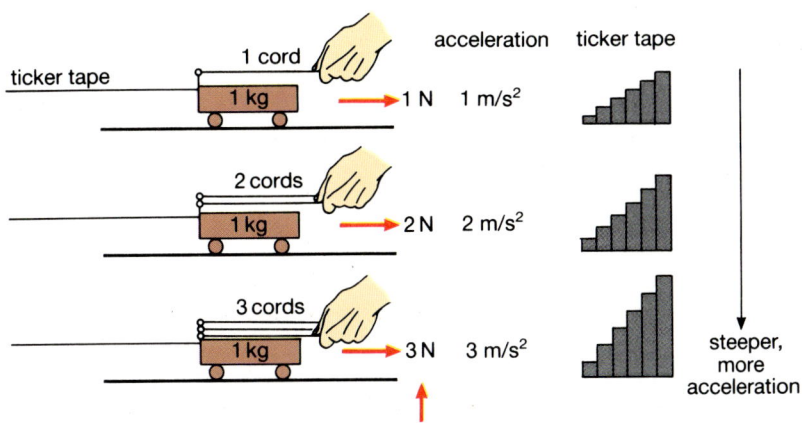

Figure 1
(a) Experiment 1
Keep the mass constant and change the force

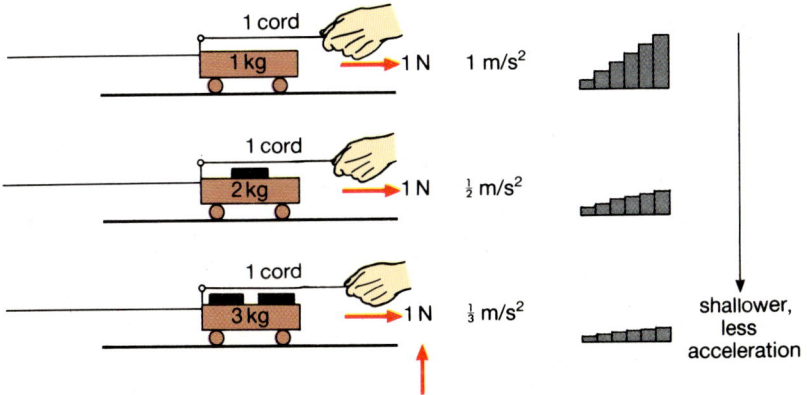

Figure 1
(b) Experiment 2
Keep the force constant and change the mass

In the second experiment (Figure 1(b)), you change the mass of the trolley but you pull with the same force. As the mass increases the ticker tape graphs become less steep. This means the acceleration is getting less. For a force of 1 N acting on a 1 kg trolley the acceleration is $1\,m/s^2$, for a 2 kg trolley the acceleration is $\frac{1}{2}\,m/s^2$, for a 3 kg trolley the acceleration is $\frac{1}{3}\,m/s^2$. For each of the experiments in Figure 1, you can see that the following equation is true:

> Force = mass × acceleration

This is **Newton's second law of motion**. This equation lets us define the Newton in this way: *A force of 1 N will accelerate a mass of 1 kg at a rate of $1\,m/s^2$.*

Example. A sledge has a mass of 10 kg. A boy pulls it with a force of 20 N. What is its acceleration?

$$F = m \times a$$

So

$$a = \frac{F}{m} = \frac{20\,N}{10\,kg} = 2\,m/s^2$$

Questions

1 Trains accelerate very slowly out of stations. Why is their acceleration very much slower than that of a car?

2 The manufacturers of Formula 1 racing cars try to make them as light as possible. Why do you think they do that?

3 This question refers to the experiments described in Figure 1.

(a) How many cords are needed to accelerate a trolley of mass 4 kg at a rate of $0.5\,m/s^2$?

(b) What acceleration is produced by four cords acting on a trolley of mass 3 kg?

(c) The trolley in this experiment has frictional forces acting on it. What effect does friction have on your results?

(d) What can you do to help compensate for friction?

4 A driver travelling on the motorway at 30 m/s takes her foot off the accelerator and slows down to 27 m/s in 6 s.

(a) Calculate the size of her deceleration.

(b) The mass of the car is 1000 kg. Calculate the size of the wind resistance and frictional forces that are acting on the car.

7 Free Fall

The diver will enter the water head first. She is streamlined, so the forces acting on her are small. If she belly flops a large force will slow her down, which will hurt

When you drop something it accelerates downwards, moving faster and faster, until it hits the ground. There is a story that Galileo used the leaning tower of Pisa to demonstrate the effect of gravitational acceleration. He dropped a large iron cannon ball and a small one. Both balls reached the ground at the same time. They accelerated at the same rate of about $10 \, \text{m/s}^2$.

Weight and mass

The size of the Earth's gravitational pull on an object is proportional to its mass. The Earth pulls a 1 kg mass with a force of 10 N and a 2 kg mass with a force of 20 N. We say that the strength of the **Earth's gravitational field**, g, is 10 N/Kg.

The **weight**, W, of an object is the force that gravity exerts on it, this is equal to the object's mass $\times$ the pull of gravity on each kilogram.

$$W = mg$$

The value of g is roughly the same everywhere on the Earth, but away from the Earth it has different values. The Moon is smaller than the Earth and pulls things towards it less strongly. On the Moon's surface the value of g is 1.6 N/kg. In space, far away from all planets, there are no gravitational pulls, so g is zero, and therefore everything is weightless.

The size of g also gives us the **gravitational acceleration**, because:

$$\text{acceleration} = \frac{\text{force}}{\text{mass}} \qquad or \qquad g = \frac{W}{m}$$

Example. What is the weight of a 70 kg man on the Moon?

$$W = mg$$
$$= 70 \, \text{kg} \times 1.6 \, \text{N/kg}$$
$$= 112 \, \text{N}$$

Parachuting

You have read earlier that everything accelerates towards the ground at the same rate. But that is only true if the effects of air resistance are small. If you drop a feather you know that it will flutter slowly towards the ground. That is because the size of the air resistance on the feather is only slightly less than the downwards pull of gravity.

The size of the air resistance on an object depends on the area of the object and its speed:
- the larger the area, the larger the air resistance.
- the larger the speed, the larger the air resistance.

Figure 1 shows the effect of air resistance on two balls, which are the same size and shape, but the red ball has a mass of 0.1 kg and the blue ball a mass of 1 kg. The balls are moving at the same speed, the air resistance is the same, 1 N, on each. The pull of gravity on the red ball is balanced by air resistance, so it now moves at a constant speed. The red ball will not go any faster and we say it has reached **terminal velocity**. For the blue ball, however, the pull of gravity is greater than air resistance so it continues to accelerate.

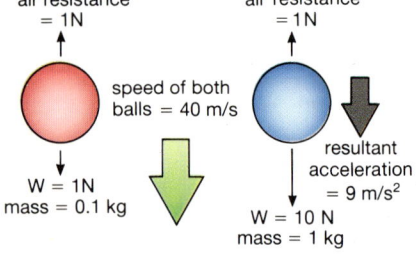

Figure 1
At this instant both balls have a speed of 40m/s. At this speed the weight of the red ball is balanced by air resistance, but the heavier blue ball is still accelerating

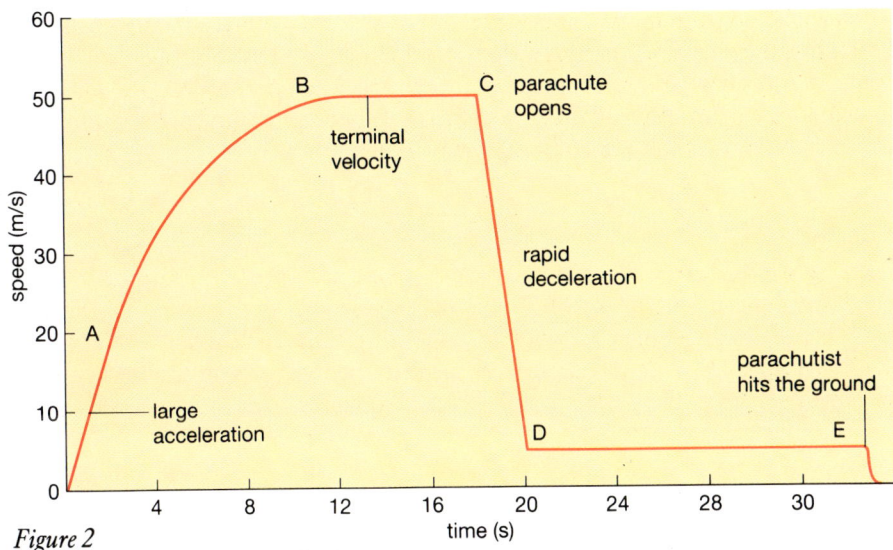

Figure 2
Speed/time graph for a parachutist

The sky diver has reached her terminal velocity. She is not in a streamlined position. How could she go faster?

Figure 2 shows how the speed of a sky diver changes as she falls towards the ground. The graph has five distinct parts:

(1) *OA*. She accelerates at about $10 \, \text{m/s}^2$ just after leaving the aeroplane.

(2) *AB*. The effects of air resistance mean that her acceleration gets less as there is now a force acting in the opposite direction to her weight.

(3) *BC*. The air resistance force is the same as her weight. She now moves at a constant speed because the resultant force acting on her is zero.

(4) *CD*. She opens her parachute at *C*. There is now a very large air resistance force so she decelerates rapidly.

(5) *DE*. The air resistance force on her parachute is the same size as her weight, so she moves with a constant speed until she hits the ground at *E*.

Questions

1 A student wrote 'my weight is 67 kg'. What is wrong with this statement, and what do you think his weight really is?

2 A hammer has a mass of 1 kg. What is its weight (i) on Earth (ii) on the Moon (iii) in outer space?

3 Explain this observation: 'when a sheet of paper is dropped it flutters down to the ground, but when the same sheet of paper is screwed up into a ball it accelerates rapidly downwards when dropped'.

4 This question refers to the speed/time graph in Figure 2.
(a) What was the speed of the sky diver when she hit the ground?
(b) Why is her acceleration over the part *AB* less than it was at the beginning of her fall?

(c) Use the graph to estimate roughly how far she fell during her dive. Was it nearer 100 m, 1000 m, or 10 000 m?

5 The graph (right) shows how the air resistance force on our sky diver's parachute changes with her speed of fall.
(a) What is the resistive force acting on her if she is travelling at a constant speed of 5 m/s?
(b) Explain why your answer to part (a) must be the same size as her weight.
(c) Use the graph to predict how fast these people would fall using the same parachute: (i) a boy of weight 400 N, (ii) a man of weight 1000 N.
(d) Make a copy of the graph and add to it a sketch to show how you think the air resistance force would vary on a

parachute of twice the area of the one used by our sky diver.

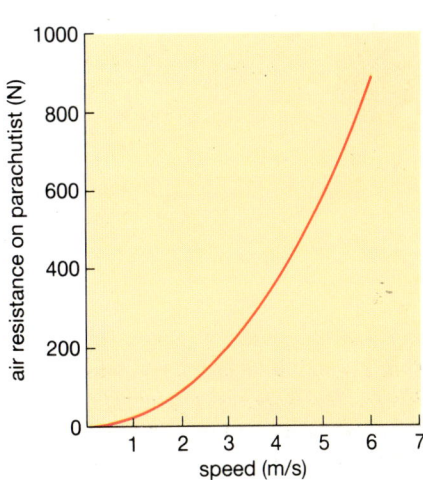

8 Applying Forces

Most of the energy of an impact in a car crash is absorbed by the crumple zones. This photograph shows the crumple zones on a Volvo

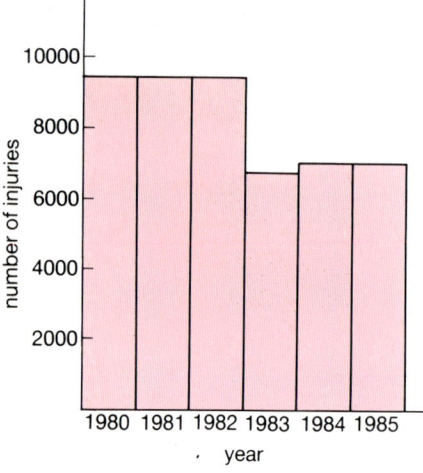

Figure 1
The number of serious injuries to front-seat passengers and drivers in cars and light vans in Great Britain. In January 1983 it became compulsory to wear seat belts in the front seats. The number of passengers and drivers wearing seat belts rose form 45% to 95%. The statistics speak for themselves. Will you drive a car without a seat belt?

Car crashes

In the last unit you read about a parachutist who landed on the ground travelling at 5 m/s. This may not sound like a very high speed but it is quite fast enough for a parachutist to suffer an injury such as a broken ankle. You hit the ground at the same speed when you jump down from a garden wall. To avoid injuring themselves, parachutists are taught to bend their knees on landing. This helps them to decelerate more slowly, because their decrease in speed takes longer. When the deceleration is less, the force acting on the parachutist's legs is less. Now there is a better chance of avoiding injury.

So, when a rapid deceleration occurs a large force acts to cause it. This idea is most important when designing cars with safety in mind. The photograph above shows a typical family saloon car. In the centre of the car is a rigid passenger cell which is designed not to buckle in a crash. However, the front and back of the car are designed to crumple on impact. These structures are called the **crumple zones**. In an accident, the deceleration that the passengers feel will be reduced. This is because of the time taken for the front of the car to crumple. This works in the same way as bending your legs on landing on the ground.

Another factor that has helped road safety is the wearing of seat belts (Figure 1). The photograph below shows what happens to a passenger if he is not wearing a seat belt. When the car stops suddenly he carries on moving because there is no force acting on him. The first thing he hits is the windscreen. He now experiences a large decelerating force which could fracture his skull. Glass from the window will also cut his face very badly.

This is what can happen if you don't wear a seatbelt. The large force produced by rapid deceleration can throw passengers out of a car

Figure 2 shows the deceleration of two passengers in a car crash; Helen was wearing a seat belt and John was not. Helen decelerated over a period of 0.1 s. During this time the front end of the car was crumpling up. John was not wearing a seat belt, so he did not take advantage of the time provided by the crumple zone. Instead he carried on moving until he hit the windscreen; his deceleration was then much faster and the force acting on his head was very large indeed. John was killed; Helen survived.

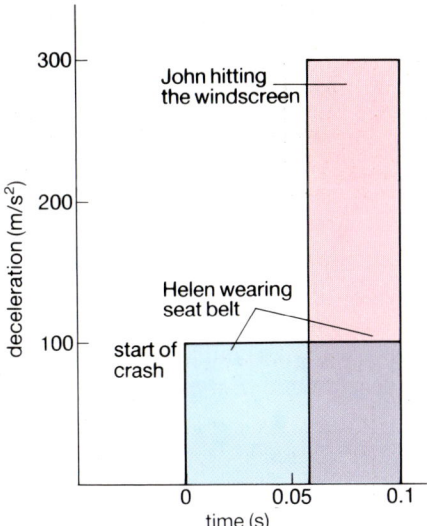

acceleration

$$= \frac{\text{resultant force}}{\text{mass}}$$

$$= \frac{3\,000\,000\,\text{N}}{3\,000\,000\,\text{kg}}$$

$$= 1\,\text{m/s}^2$$

Figure 2
Graph to show approximate decelerations in a car crash

Figure 3
Acceleration of the Saturn V rocket

A Saturn V rocket takes off, accelerated by 33 million newtons of thrust

Overcoming gravity

You have just read that large forces will cause large accelerations or decelerations. Large forces are also used to accelerate large masses. One of our greatest technological achievements has been to land men on the moon in a *Saturn V* rocket. The mass of the rocket was 3 million kg and the thrust from its engines was about 33 million N. What sort of acceleration did this force produce? (see Figure 3).

Questions

1 Explain why it is a good idea to wear your safety belt in a car. At the moment it is not compulsory to have safety belts fitted in the back of your car. Do you think it ought to be compulsory?

2 What is a crumple zone in a car?

3 This question refers to John and Helen's car crash. John's mass was 80 kg and Helen's was 60 kg.
(a) Use Figure 2 to calculate the force that was acting on each of them to slow them down during the crash.

(b) How can you tell from the graphs that before the collision they were travelling at 10 m/s?
(c) Use the information to draw speed/time graphs for each of Helen and John during the crash.
(d) Helen was driving and was responsible for the crash. The police found that she had a large amount of alcohol in her blood. What measures should be taken to stop people from drinking and driving?

4 (a) After take off, a *Saturn V* rocket burns 14 000 kg of fuel per second. How much mass has the rocket lost after 2 minutes?
(b) What is the mass of the rocket after 2 minutes? (The rocket started with a mass of 3 million kg.)
(c) Now calculate the acceleration of the rocket after 2 minutes. The thrust from the engines is 33 million N.

9 Moving in Circles

Two Scottish players combine to force an England winger into touch

In the photograph you can see a rugby international between Scotland and England in progress. England's right wing has the ball and is running to score a try by the corner flag. However, the Scottish winger is running across to stop him, and he succeeds by giving the English winger a push. The sideways push does not slow him down but it changes the direction that he is running in. He is forced out of play, and England fails to score a try.

This account of rugby players demonstrates an important idea. So far, when you have seen forces acting on moving objects, the forces have caused something to speed up or slow down. However, if a force is applied at right angles to the direction of motion then there is no change of speed, but the direction of the motion is changed (Figure 1).

Earlier you met acceleration and you used the equation:

$$\text{acceleration} = \frac{\text{change of velocity}}{\text{time}}$$

When we use the word **velocity** we need to state the direction of motion as well as the speed. The rugby player changes his velocity but does not alter his speed. So it is possible to have an acceleration without changing speed. This is a very difficult idea to understand, but makes sense if you think about this: the player was pushed, and you know that pushes cause acceleration ($F = ma$).

Figure 1

speed 10 m/s

(a) Before the push

speed 10 m/s

(b) After the push the direction in which the man runs is changed. There is no change of speed, but there is a change of velocity

The friction between the tyres and the track provides the force needed to change direction

Circular motion

You can fasten a conker to a string and whirl it around your head in a horizontal circle. The conker moves at a constant speed, but its direction is always changing. The velocity of the conker is changing so it must be accelerating. The pull of the string is the force which accelerates the conker. The direction of this force is always at right angles to the motion of the conker. It is towards the centre of the circular path. It is called **centripetal force** (Figure 2).

The photographs on this page show some examples of centripetal forces making things move in a circular path. When an athlete throws a hammer he turns around two or three times before letting go of the hammer. While he turns round, his arms are providing a centripetal force to keep the hammer moving in a circle. As soon as he lets go, the hammer flies off along the direction it was travelling in.

A satellite in orbit around the Earth provides an interesting example of a centripetal force in action. The Earth's gravitational pull acts on the satellite to keep it in its circular path.

There are three factors that affect the size of the centripetal force acting on an object moving in a circular path. More force is needed if:

- the mass is increased
- the speed is increased
- the radius of the circular path gets less.

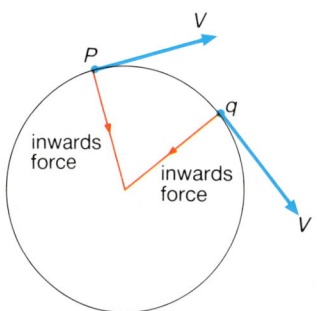

Figure 2
Conker on a string moving around a horizontal circle at constant speed

Igor uses his arms to provide the centripetal force

Questions

1 Explain carefully how it is possible for something to accelerate but not change its speed.

2 The diagram shows a ball attached to a string going around a circle.

(a) Copy the diagram and mark the direction of the force on the ball at *A*.

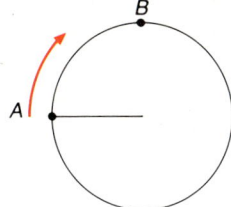

(b) When it gets to *B* the string breaks. In which direction does the ball move?

3 You can read below part of a discussion that Sundip and Jane had about how a spin drier works. Which one of them gives the better explanation? Correct any errors that they have made during their discussion.

Jane: A spin drier turns round very quickly to get the water out of the clothes. It works because when something moves round in a circle there is a force pushing things

outwards, so the water is forced out through the holes in the spin drier.

Sundip: When something moves in a circle there is a force towards the centre. The drier makes the washing go in a circle but the water keeps moving in a straight line; that is why the water gets out.

Jane: What gives the washing a force towards the centre? Why doesn't it give the water the same force?

10 Collisions

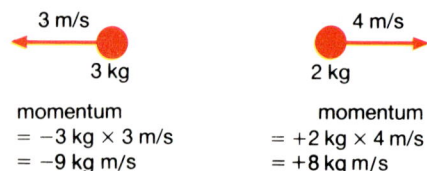

momentum
$= -3\,\text{kg} \times 3\,\text{m/s}$
$= -9\,\text{kg m/s}$

momentum
$= +2\,\text{kg} \times 4\,\text{m/s}$
$= +8\,\text{kg m/s}$

Figure 1
In this diagram we have given anything moving to the right positive momentum, and anything moving to the left negative momentum

Why does a gun recoil when it is fired?

An understanding of forces and momentum can make you into a world champion at the snooker table!

Momentum

Momentum is defined as the product of mass × velocity. We measure it in units of kilogram metres per second (kg m/s).

$$\text{momentum} = m \times v$$

Velocity is a vector quantity, so momentum is too. We must give a direction when we talk about momentum (see Figure 1). Momentum is very useful when we meet problems involving collisions or explosions. There is always as much momentum after a collision as there is before it.

Impulse

When a force, F, pushes a mass, m, we can work out the acceleration using:

$$F = ma$$

But the acceleration, $a = \dfrac{v - u}{t}$

where v is the final velocity, u is the starting velocity and t is the time taken for the velocity to change. The second equation can be substituted into the first to give:

$$F = \frac{m(v - u)}{t}$$

$$\text{or } Ft = mv - mu$$

Ft is called an **impulse** and has units of newton seconds. In words the equation can be expressed as:

$$\text{impulse} = \text{change of momentum}$$

So to cause a change of momentum of 100 kg m/s, for example, we must apply an impulse of 100 Ns. But we could apply a force of 100 N for 1 s, or 1 N for 100 s. Each causes the same change of momentum.

Conservation of momentum

In Section B you met Newton's third law. This law says that when one body pushes against a second body, the second body pushes back with an equal and opposite force. Using this idea we can work out what happens in collisions.

Figure 2 shows two ice hockey players chasing after the puck. The blue player pushes the red player with his stick. The size of the push is 400 N and it lasts for 0.4 s. How fast are the players moving after the push? Since Ft = change of momentum, the red player's momentum increases by 400 N × 0.4 s = 160 Ns. We can calculate his increase in velocity using:

$$\text{increase in momentum} = \text{mass} \times \text{increase in velocity}$$

$$\text{So } 160\,\text{Ns} = 80\,\text{kg} \times \text{increase in velocity}$$

$$\text{increase in velocity} = 2\,\text{m/s}$$

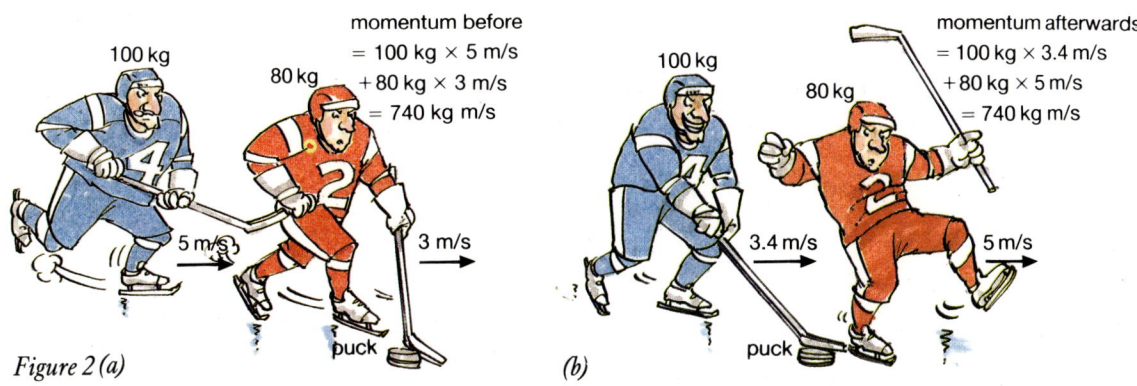

momentum before
= 100 kg × 5 m/s
+ 80 kg × 3 m/s
= 740 kg m/s

momentum afterwards
= 100 kg × 3.4 m/s
+ 80 kg × 5 m/s
= 740 kg m/s

Figure 2 (a)   *(b)*

This means the red player moves with a velocity of 5 m/s after the push. While the stick was touching the red player, there was also a push back on the blue player. The same force of 400 N acted on the blue player for 0.4 s. So his momentum decreased by 160 Ns.

$$\text{His decrease in velocity} = \frac{\text{decrease in momentum}}{\text{mass}}$$

$$= \frac{160\,\text{Ns}}{100\,\text{kg}}$$

$$= 1.6\,\text{m/s}.$$

His final velocity = 5 m/s − 1.6 m/s = 3.4 m/s.

The momentum of one player increases by 160 Ns, the momentum of the second player decreases by 160 Ns. So the total change in momentum was zero. This leads to an important law:

> The momentum before a collision *always* equals the momentum after the collision.

This is known as the **principle of conservation of momentum**. Figure 3 shows another example: you can see that momentum is conserved.

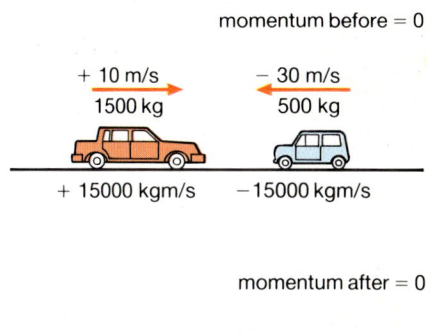

momentum before = 0

+ 10 m/s
1500 kg

− 30 m/s
500 kg

+ 15000 kgm/s   − 15000 kgm/s

momentum after = 0

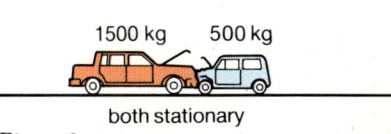

1500 kg   500 kg

both stationary

Figure 3

Questions

1 (a) Explain why momentum can be measured in units of kg m/s or Ns.
(b) What change of momentum is caused by an impulse of 2 Ns?
2 What is the momentum of a runner of mass 70 kg, running at 10 m/s?
3 In each of the following experiments carried out in a laboratory, the two

3 m/s   2 m/s
3 kg   2 kg

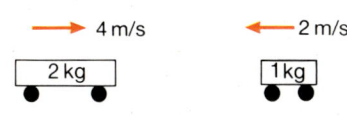

4 m/s   2 m/s
2 kg   1 kg

trollies collide and stick together. Work out the speeds of the trollies after their collisions.

4 In Figure 3 two cars collided head on; each driver had a mass of 70 kg.
(a) Calculate the change of momentum for each driver.
(b) The cars stopped in 0.25 s. Calculate the average force that acted on each driver.

(c) Explain which driver is likely to be more seriously injured.
5 A field gun of mass 1000 kg, which is free to move, fires a shell of mass 10 kg at a speed of 200 m/s.
(a) What is the momentum of the shell after firing?
(b) What is the momentum of the gun just after firing?
(c) Calculate the recoil velocity of the gun.
(d) Why do you think very large guns are mounted on railway trucks?

11 Rockets and Jets

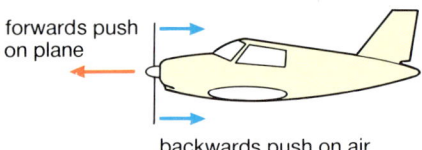

Figure 1
The forces acting on an aeroplane powered by a propellor

Every time you go swimming you demonstrate Newton's third law. As you move your arms and legs through the water, you push the water backwards. But the water pushes you forwards.

The same idea applies to an aeroplane in flight powered by a propellor (Figure 1). The propellor accelerates air backwards, but the air exerts an equal forwards force on the aircraft. As you will see below, rockets and jet engines use the same principle.

Rockets

A propellor takes advantage of air resistance. If air did not exert a force on us as we move through it, a propellor would not work. This is why an ordinary aeroplane cannot fly in space, where there is no air for its propellors to push against. The simplest way to demonstrate the action of a rocket is with an air-filled balloon. If you blow up a balloon and then let it go, it whizzes around the room. It would also whizz around in space where there is no air, because the balloon gets its forwards push from the escaping air. You can explain this by using the principle of conservation of momentum. Before you let go of the balloon, the momentum of the balloon and air is nothing. Once the air is allowed to escape from the balloon the *total* momentum is still zero, but the balloon has forwards momentum and the air has backwards momentum (Figure 2).

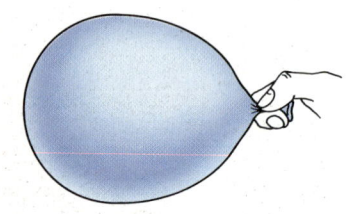

Figure 2
(a) No momentum

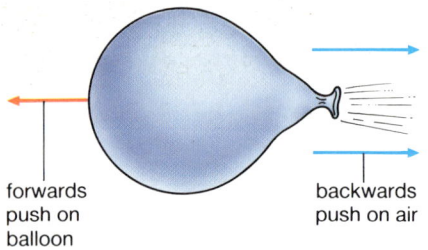

forwards push on balloon

backwards push on air

(b) Balloon has momentum to the left
Air has momentum to the right
Total momentum is still zero

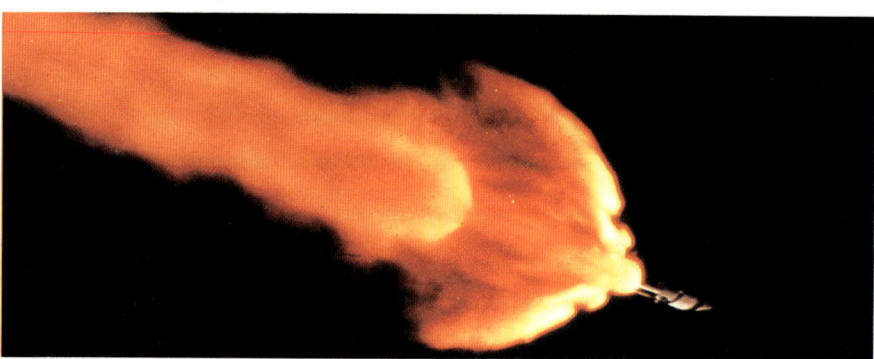

Saturn 1B at a height of 20 kilometres. The burnt fuel is driven away from the rocket with great force. In turn, the fuel gives an equal and opposite force to the rocket, which drives it into space

Figure 3 shows a simple design for a rocket. The rocket carries with it fuel, such as kerosene, and liquid oxygen (oxygen is needed for the fuel to burn). These are mixed and burnt in the combustion chamber. High pressure gases are then forced out backwards through the nozzle at speeds of about 2000 m/s.

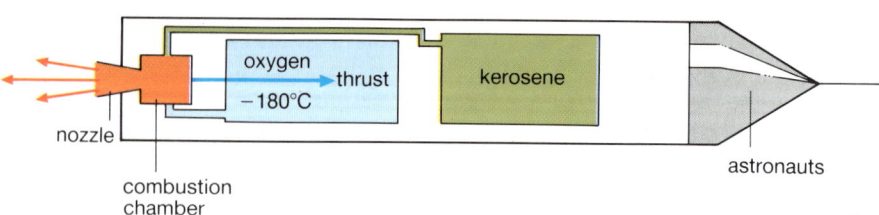

nozzle

oxygen −180°C

thrust

kerosene

astronauts

combustion chamber

Figure 3
A simple rocket

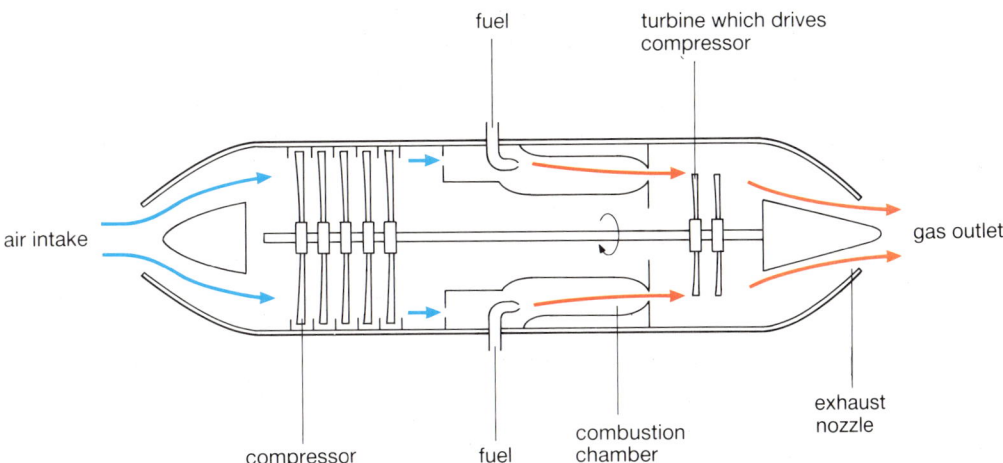

Figure 4
The jet turbine engine

Jets

Figure 4 shows a simplified diagram of the sort of **jet turbine engine** that is found on an aeroplane. At the front of the engine a compressor sucks in air, acting rather like a propellor, but the air is compressed and as a result it becomes very hot. Some of the heated air then goes through into the combustion chamber and is mixed with the fuel (kerosene). The fuel burns and causes a great increase in the pressure of the gases. The hot gases are then forced out at high speed through the exhaust nozzle. The escaping gases provide a forward thrust on the engine.

The engine needs to be started with an electric motor to rotate the compressor, but once the engine is running some of the energy from the exhaust gases is used to drive a turbine. The turbine is mounted on the same shaft as the compressor so air can now be sucked in without the help of an electric motor.

This is a large jet engine. You can clearly see the turbines

Questions

1 (a) Explain carefully how a rocket works.
(b) An aeroplane powered by jet engines cannot fly in space, but a rocket can. Why?
2 On the right you can see a velocity/ time graph for a firework rocket. The graph stops at the moment the fireworks stop burning. A positive velocity on this graph means the rocket is moving upwards.
(a) How can you tell from the graph that the acceleration of the rocket is increasing?
(b) Why does the acceleration rise?

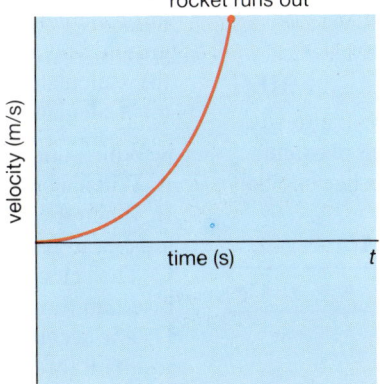

(Hint: What happens to the mass of the rocket as it burns?)
(c) Copy the graph and sketch how the velocity changes until the rocket hits the ground.
3 In the last unit you met the equation

$$F = \frac{mv - mu}{t}.$$

(In words: force equals the rate of change of momentum). Use this equation to calculate the thrust from a rocket that uses 2000 kg of fuel and oxygen each second, and ejects its exhaust gases at a speed of 1000 m/s.

SECTION C: *STUDY QUESTIONS*

1 The velocity against time graph represents the motion of a cyclist travelling along a straight, level road.

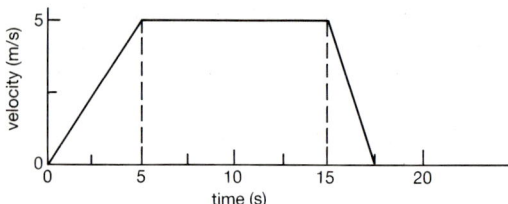

(i) For what distance did the cyclist travel at a constant speed?
A 25 m   **B** 50 m   **C** 68¾ m   **D** 75 m
E 118¾ m

(ii) During the journey the cyclist used the brakes to bring the cycle to a standstill. What acceleration did the brakes produce?
A $-2.0 \, \text{m/s}^2$ **B** $-0.5 \, \text{m/s}^2$ **C** 0   **D** $0.5 \, \text{m/s}^2$
E $2.0 \, \text{m/s}^2$

LEAG

2 Which graph of distance against time is for a body moving at a constant speed?

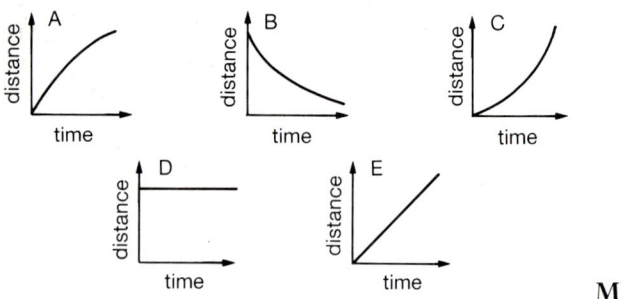

MEG

3 A sky diver, whose weight is 700 N, jumps from an aeroplane. After falling for some time he is travelling at a constant velocity. What is the force of air resistance at this time?
A less than 700 N
B exactly 700 N
C greater than 700 N
D exactly 700 kg
E greater than 700 kg

MEG

4 A freely-running 1 kg trolley pulled ticker-tape at a steady speed through a timer. At X on the tape, a lump of plasticine was dropped vertically on to the trolley at the dots became half as far apart.

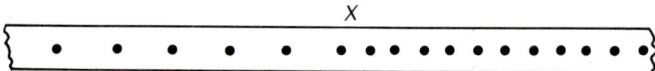

What was the likely mass of the plasticine?
A 0.25 kg   **B** 0.5 kg   **C** 1 kg   **D** 2 kg   **E** 4 kg

MEG

5 A small heavy object falls from an aircraft flying horizontally.

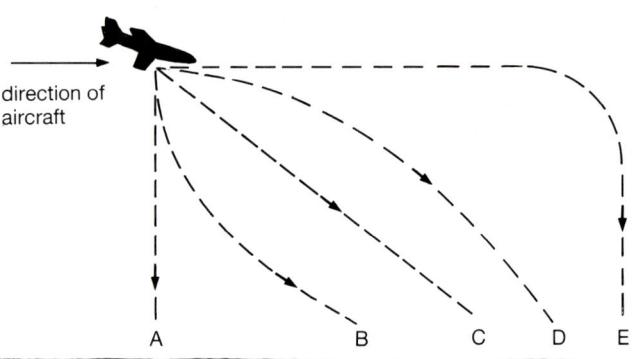

As seen from the ground, along which path, **A, B, C, D** or **E,** is the object most likely to fall?

MEG

6

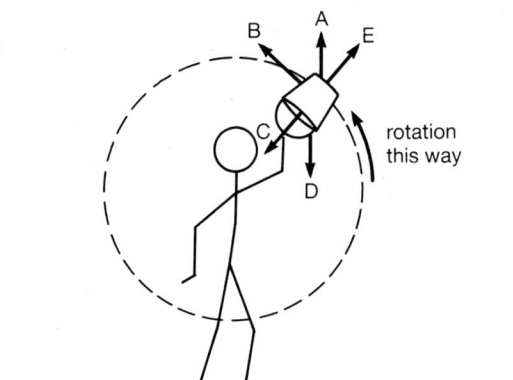

A man swings a bucket full of water in a vertical circle as shown in the diagram. It moves fast enough for the water to remain in the bucket. If he lets go of the handle, in which direction does the water move **immediately** afterwards?

MEG

7 Harry drops a stone off the edge of a cliff. It takes 2 s to reach the bottom.
(a) Calculate the speed of the stone just before it reaches the ground.
(b) Calculate the stone's average speed.
(c) Calculate the height of the cliff (y).

8 Which one of the following statements is always correct?
A When a body is at rest no forces can be acting on it at all.
B When there is no resultant force acting on a body, the body must be at rest.
C The velocity of a body is directly proportional to the resultant force acting on it.
D The acceleration of a body is directly proportional to the resultant force acting on it.
E The resultant force on a body is proportional to its mass.

NI

9 (a) A car engine is leaking oil. The oil drops hit the ground at regular time intervals, one every 2.0 seconds. The diagram below shows the pattern of the drops that the car leaves on part of its journey.

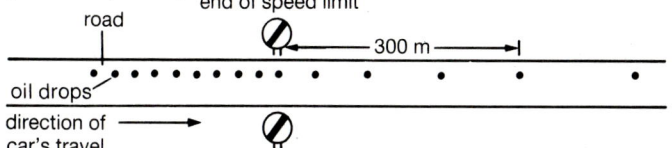

(i) What can you say about the speed of the car before it reaches the signs?

(ii) Calculate the distance between the drops on the road before it reaches the signs if the car is travelling at 10 m/s.

(iii) After the car passes the signs, what happens to the gaps between the drops of oil? What does this tell you about the motion of the car?

(iv) Further down the road it is found that the distance between the drops on the road has become 30 m. What is the speed of the car at this point?

(b) A front-wheel drive car is travelling at constant velocity. The forces acting on the car are shown in the diagram below. F is the push of the air on the car.

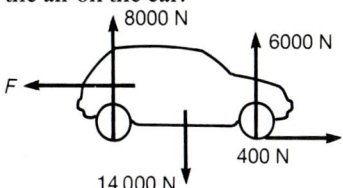

(i) Name the 400 N force to the right.

(ii) What is the value of F, the force to the left?

(iii) Taking the weight of 1 kg to be 10 N, calculate the mass of the car.

(iv) The force to the right is now increased. Describe and explain what effect this has on the speed of the car.

LEAG

10 (a) In testing the performance of an electric truck (milk float) the manufacturers made measurements of acceleration using the device shown below. The scale measures the resultant force of the springs acting on a mass M.

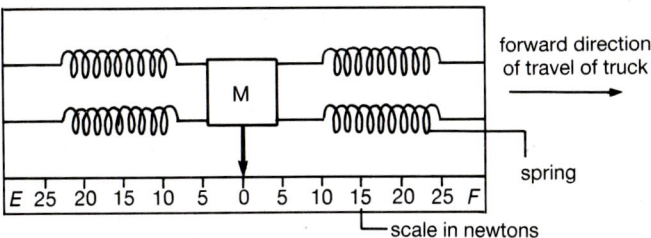

The device was clamped firmly and horizontally to the floor of the truck. When the truck was at rest the reading on the scale was zero as indicated in the diagram.

(i) What would the meter reading be when the truck had been travelling at a steady speed for some time? Explain your answer.

(ii) When the truck was accelerating forwards why was the scale reading towards end E?

(iii) Calculate the acceleration when a scale reading of 20 N towards end E was obtained. The mass of M was 2.5 kg.

(iv) What would be under test when readings were obtained towards end F?

(b) Suppose this device is now used vertically on the wall of a lift with end E towards the top as shown below.

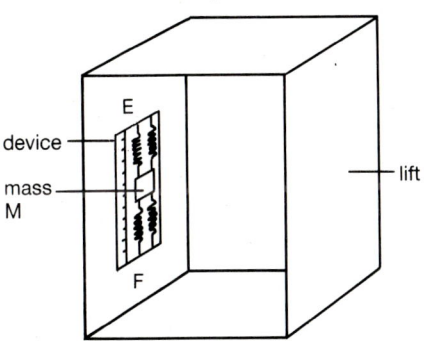

What can you decide about the motion of the lift if a reading of 20 N is observed towards end E? Give your answer (i) in words (ii) by adding relevant calculations. (The free fall acceleration may be taken as 10 m/s².)

LEAG

11 Mike has worked out a new way to measure his mass. He is in a trolley that accelerates down the slope when released. Use the data below to answer these questions.

- It takes 300 N to hold Mike stationary on the slope
- When released Mike travels 36 m in 6 s
- Mass of the trolley = 32 kg

(a) What is Mike's average speed over the 36 m?

(b) Assuming he accelerates at a steady rate, what is his final speed after travelling 36 m?

(c) Now work out his acceleration.

(d) Calculate Mike's mass.

STUDY QUESTIONS

12 The diagram below shows an alpha particle colliding head on with a stationary proton. Before the collision the alpha particle has a speed of 10^7 m/s. After the collision its speed is 0.6×10^7 m/s. The mass of the alpha particle is 4 times that of the proton. Calculate the speed of the proton.

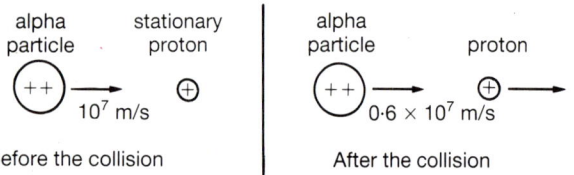

13 Archimedes' Principle states that 'if a body is wholly or partially immersed in a fluid then the upthrust is equal to the weight of fluid displaced'.

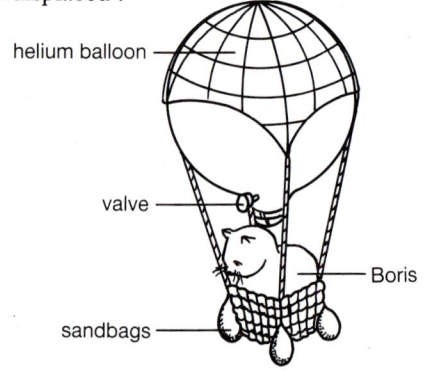

- Total mass of sandbags + Boris = 24 kg
- Volume of balloon = 30 m³
- Density of air at sea level ≏ 1.2 kg/m³
- Density of helium in balloon ≏ 0.2 kg/m³

(a) What is the force due to gravity on the balloon? (Include the helium).
(b) What upthrust is there on the balloon?
(c) What is Boris's initial acceleration upwards?
(d) Assuming this acceleration is constant how far will he have travelled after 2 seconds?
The graph shows how the density of air varies with height above sea level.

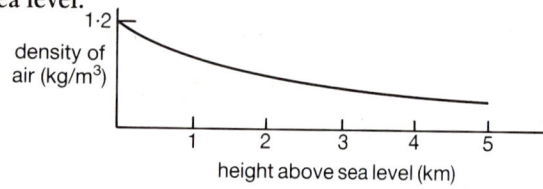

(e) Deduce from the graph how high Boris will rise.
(f) How could he go higher?
(g) How could he get down again?

14 Ravi and Sarah are very good friends. But unfortunately they live 10 km apart. However, at 7 pm every evening they set out to meet each other; Ravi walks at 6 km/h and Sarah walks at 4 km/h.

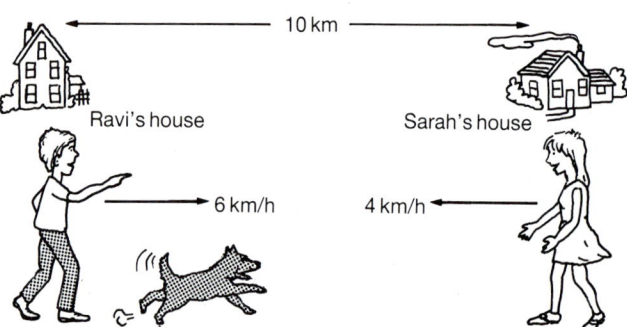

Ravi has a dog called Rover who always goes out with him. Rover thinks they must be bonkers to walk so far every night. However, he likes a good run. So as soon as he leaves Ravi's house he runs at 12 km/h until he meets Sarah. As soon as he meets Sarah he turns round until he meets Ravi. Rover runs backwards and forwards between Ravi and Sarah until they meet.
How far has Rover run when they meet?

15 When a bomber releases a load of bombs they fall freely under the action of gravity. The table below gives positions of the bombs at 2 s intervals.

(a) Plot a graph of the height of the bombs above the ground (y-axis) against the distance travelled horizontally (x-axis).
(b) How fast is the plane travelling?
(c) Use the graph to predict the position of the bombs after (i) 6 s (ii) 20 s.
(d) The bomber now flies on a mission at a height of 1500 m at the same speed. How far away from the target must the bombs be released if they are going to hit? Use your graph to help you answer this.
(e) Another bomber flies at a speed of 200 m/s. Draw a second graph, using the same axes as before, to show the position of bombs that fall from this plane.

	Time (s)										
	0	2	4	6	8	10	12	14	16	18	20
height above the ground (m)	2000	1980	1920		1680	1500	1280	1020	720	380	
distance travelled horizontally (m)	0	600	1200		2400	3000	3600	4200	4800	5400	

SECTION D
Energy and Power

The large industries that we rely on to make our lives easier and more comfortable often produce large amounts of pollution and use a great deal of energy. Huge power stations are needed to provide electricity to power them, and these use up valuable deposits of coal and oil. Pollution is the price that we have to pay

1 What is Work?

This horse must do work in order to jump the fence

Figure 1

Tony works in a supermarket. His job is to fill up shelves when they are empty. When Tony lifts up tins to put on the shelves he is doing some work. The amount of work Tony does depends on how far he lifts the tins and how heavy they are.

We define work like this:

$$\text{Work done} = \text{force} \times \text{distance} = F \times d$$

Work is measured in **joules** (J). 1 joule of work is done when a force of 1 newton moves something through a distance of 1 metre, in the direction of the applied force.

$$1\,\text{J} = 1\,\text{N} \times 1\,\text{m}$$

Example. How much work does Tony do when he lifts a tin with a weight of 20 N through a height of 0.5 m?

$$W = F \times d$$
$$= 20\,\text{N} \times 0.5\,\text{m}$$
$$= 10\,\text{J}$$

Tony does the same amount of work when he lifts a tin with a weight of 10 N through 1 m.

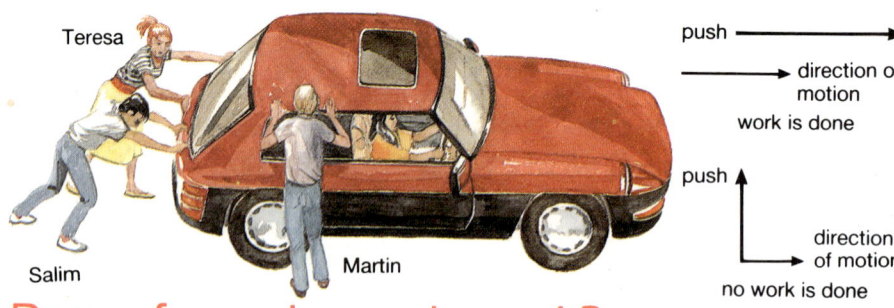

Teresa Salim Martin

push →
→ direction of motion
work is done

push ↑
→ direction of motion
no work is done

Does a force always do work?

Does a force always do work? The answer is no! In Figure 1 Martin is helping Salim and Teresa to give the car a push start. Teresa and Salim are pushing from behind; Martin is pushing from the side. Teresa and Salim are doing some work because they are pushing in the right direction to get the car moving. Martin is doing nothing useful to get the car moving. Martin does no work because he is pushing at right angles to the direction of movement.

In Figure 2 Samantha is doing some weight training. She is holding two weights, but she is not lifting them. She becomes tired, because her muscles use energy, but she is not doing any work, because the weights are not moving. To do work you have to do something useful, like lifting a load or pushing a car.

Finding the fuel

When you want a job done you have to pay for it. This is because you have to buy fuel. This provides **energy** for the job to be done. The supermarket manager has to pay Tony to do his work. The most important thing that Tony buys with his money is food. Food gives him the energy to do his work.

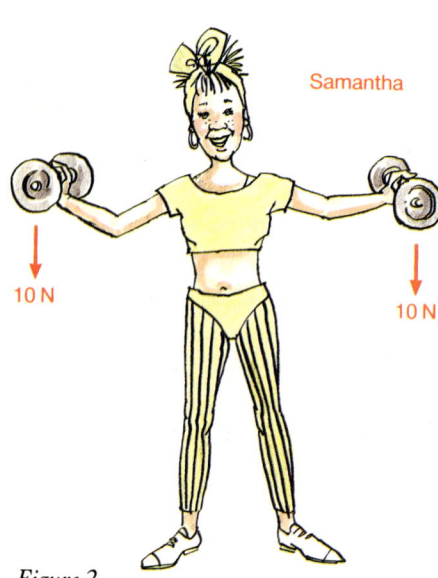

Samantha

10 N   10 N

Figure 2

Figure 3 shows a crane at work on a building site. The crane runs on diesel fuel. The table shows the amount of diesel used for some jobs. You can see that the amount of diesel used is proportional to the amount of work done lifting a load.

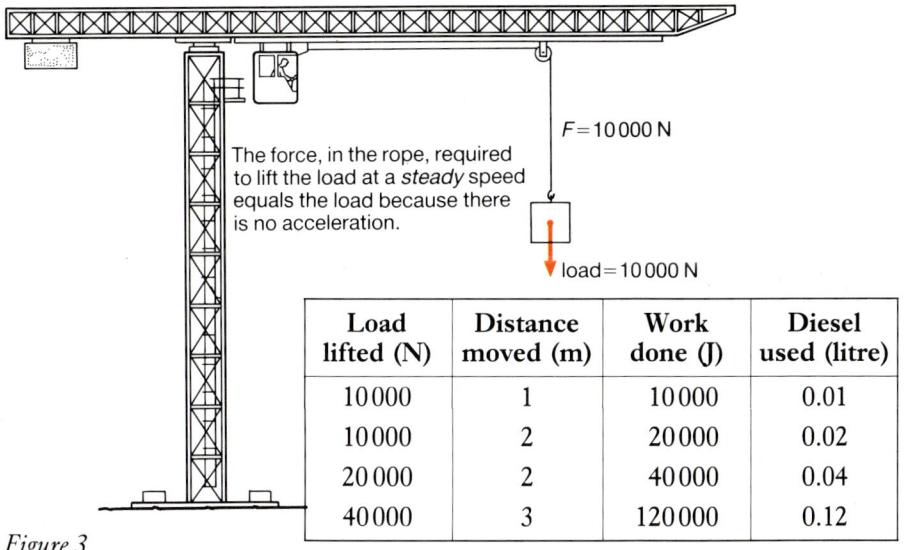

The force, in the rope, required to lift the load at a *steady* speed equals the load because there is no acceleration.

$F = 10\,000$ N

load = 10 000 N

Load lifted (N)	Distance moved (m)	Work done (J)	Diesel used (litre)
10 000	1	10 000	0.01
10 000	2	20 000	0.02
20 000	2	40 000	0.04
40 000	3	120 000	0.12

Figure 3

To do work a source of energy is needed. Rechargeable batteries provide the energy for this fork lift truck to do its work

Questions

1 The table below shows some more jobs done by the crane in Figure 3. Copy the table and fill in the missing values.

Load lifted (N)	Distance moved (m)	Work done (J)	Fuel used (litre)
5000	2		0.01
10 000	4		
	10	40 000	
6000			0.06
	5		0.1
25 000		90 000	

2 Joel is on the Moon in his spacesuit. His mass (and the suit) is 80 kg. The gravitational field strength on the Moon is 1.6 N/kg.
(a) What is Joel's weight?
(b) Joel now climbs 30 m up a ladder into his space craft. How much work does he do?

3 Mr Hendrix runs a passenger ferry service in the West Indies. He has three ships, which are all the same. Sometimes he has problems with the bottoms of the ships, when barnacles stick to them. This increases the drag on the ships, and they use more fuel than usual.
(a) Explain why a larger drag makes the ships use more fuel.
(b) The table below shows the amount of fuel used by Mr Hendrix's three ships, on recent journeys. Which one has barnacles on her bottom?

Boat	Journey	Distance (km)	Fuel used (litre)
Island Queen	Vieux Fort – Bridgetown	175	1050
Windward Beauty	Plymouth – Kingstown	220	1100
Caribbean Princess	Bridgetown – St. George's	255	1275

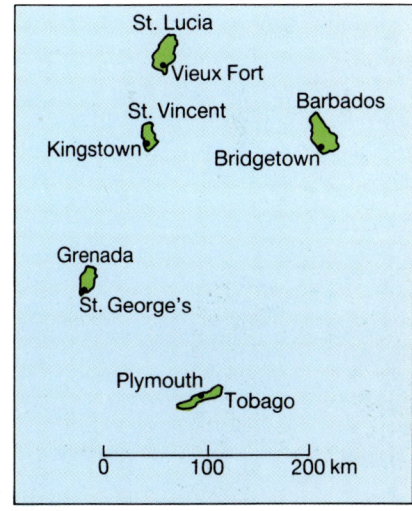

(c) Use the map and the data in the table to answer these questions: (i) how much fuel would the *Caribbean Princess* use to travel from Vieux Fort to Plymouth? (ii) how much fuel would the *Island Queen* have used for the same journey?
(d) The drag force on the *Caribbean Princess* is 50 000 N. How big is the drag on the *Island Queen*?

2 Energy

What is energy?

You have to put fuel into a machine to make it work. We say that the fuel has got some **energy**. For example, 1 litre of petrol has 40 million joules (MJ) of energy stored in it. This means that the greatest amount of work that we can get out of 1 litre of petrol is 40 MJ. Notice that work and energy are both measured in joules.

> Energy can be turned into work

Different types of energy

We cannot make energy and we cannot lose energy. This means that once we have got some energy we always have it. But energy often changes from one form to another. This is called the principle of conservation of energy. Several different types of energy are listed below:

- **Chemical energy**. Food has chemical energy stored in it, which is released by chemical reaction inside our bodies. Food provides energy to keep us warm and to do work. The stored energy in petrol is released when it is burnt inside a car's engine. (Burning is a chemical reaction.)
- **Gravitational potential energy**. When a rock is at the top of a hill, it has some stored energy. This sounds odd, because a rock at the top of a hill looks the same as a rock at the bottom of the hill. But when the rock is sent rolling down the hill, it can do some work. Water stored in a high-level dam helps us to produce electrical energy. The water has the potential to do work.
- **Kinetic energy** is the name given to the energy of motion. All moving objects have kinetic energy. A moving hammer has kinetic energy; it can do the job of knocking in a nail.
- **Heat energy**. When something is hot it possesses heat energy.
- **Strain energy**. If you have ever used a bow, you will know that you have to pull hard on the string before you can fire the arrow. You have done work to stretch the string. The bow is now strained, and it stores some energy. The strain energy is used to give the arrow kinetic energy.
- When a battery makes a current flow, **electrical energy** has been produced.
- You will also meet **nuclear**, **sound** and **light energy** later in the book.

In this fun-fair pirate ship, kinetic energy is changed into potential energy and then back into kinetic energy

Energy conversions

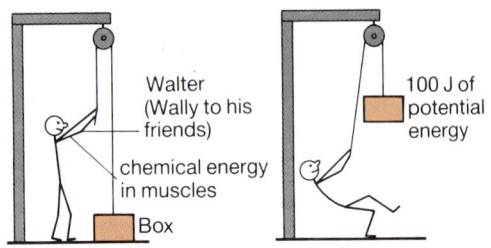
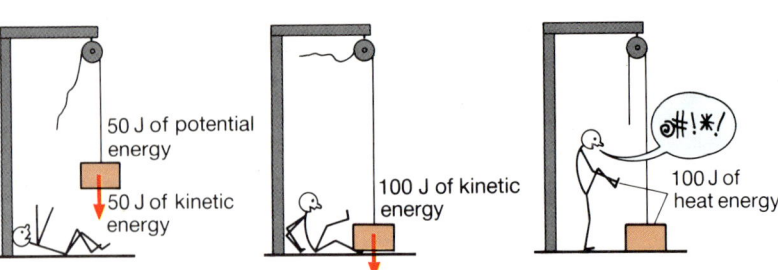

Figure 1

Chemical energy in Wally's muscles allows him to do work to pull up the box (Figure 1). This work gives the box potential energy. When Wally slips the potential energy turns into kinetic energy. When the box hits the ground the kinetic energy turns into heat energy.

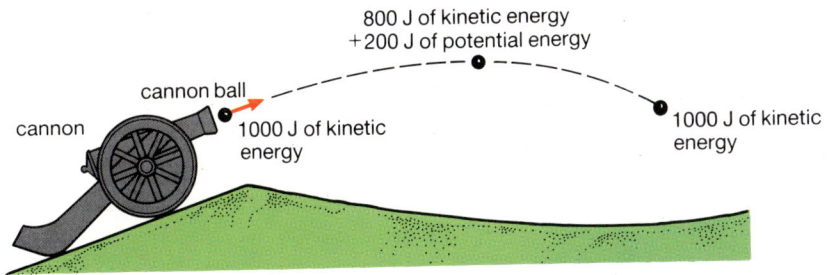

Figure 2

Small particles of rock called meteors, which are found in space, burn up in our atmosphere. They approach the Earth at about 40 000 kilometres per second. The friction between the meteors and the atmosphere turns the meteors' kinetic energy into heat and light energy. This photo was taken with a long exposure. Most of the trails are stars. Which two are the meteors?

In Figure 2 you can see how kinetic energy can be turned into potential energy, then back to kinetic energy again.

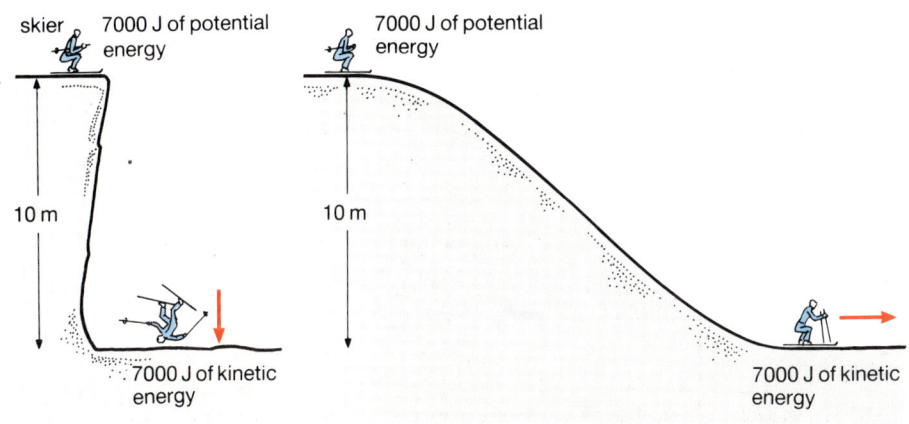

Figure 3

In Figure 3 a skier falls through a height of 10 m. He ends up with the same kinetic energy whether he falls straight down or accelerates more smoothly down the slope. In each case he loses the same potential energy.

Figure 4

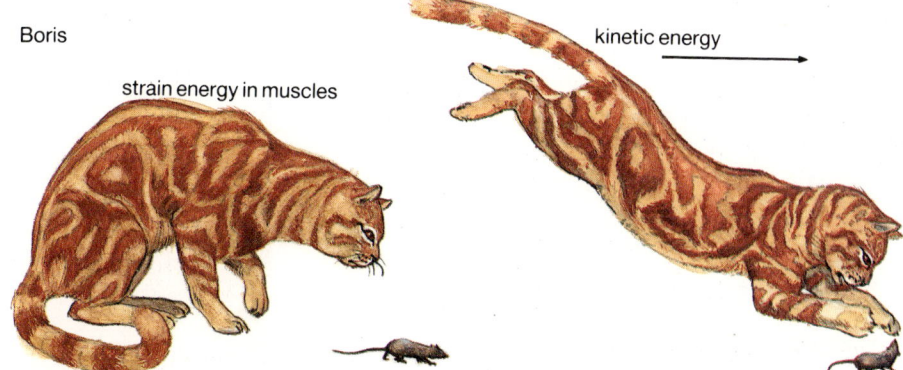

In Figure 4 Boris the cat is catching his supper. Chemical energy is used to create strain energy in Boris' muscles. This strain energy is converted to kinetic energy when Boris pounces on the mouse.

Questions

1 Judy pushes against a wall. Is she doing any work? Are her muscles using any energy?

2 A plane takes off in London and lands in New York. List and explain all the energy changes that happen during the flight.

3 Below, a ball falls to the ground.
(a) What is its kinetic energy when it hits the ground?
(b) What is its potential energy at B?

A ● potential energy = 50 J
   kinetic energy = 0

B ● kinetic energy = 30 J

C ● potential energy = 0

4 Paul kicks a stationary football. His foot is in contact with the ball over a distance of 0.1 m, and provides an average force of 500 N. How much kinetic energy does the ball have, just after it has been kicked?

3 Calculating the Energy

Noberto is lifting 110 kilograms. How much work does he do?

Potential energy

David is a weightlifter. When he lifts his weights he does some work. This work increases the gravitational potential energy (PE) of the weights. The mass on the bar is now *m*. How much work does David do when he lifts the bar a height *h*? (*m* is in kilograms and *h* is in metres.) The pull of gravity on the bar (its weight) is $m \times g$. We call the pull of gravity on each kilogram *g*; this is 10 N/kg.

$$\text{Work done} = F \times d$$
$$= mgh$$

> The work done is equal to the increase in potential energy.
>
> $$\text{Potential energy} = mgh$$

Kinetic energy

> The kinetic energy of a moving object is given by the formula:
>
> $$\text{kinetic energy} = \tfrac{1}{2}mv^2$$
>
> *m* is the mass of the object in kg, and *v* is its velocity in m/s.

This formula is very important for working out the stopping distance of a moving car. When cars are travelling very quickly, they need a large distance to stop in. The table below shows the stopping distances for a car travelling at different speeds. Next time you are travelling down the motorway at 40 m/s (90 mph) remember that your car needs about 120 m to stop in.

Speed (m/s)	Thinking distance (m)	Braking distance (m)	Total stopping distance (m)
10	6	6	12
20	12	24	36
30	18	54	72
40	24	96	120

Table 1. The faster the car travels, the greater the distance it needs to stop.

Can you estimate the total kinetic energy of the vehicles in this photograph?

When a driver sees a hazard, there is a small delay between taking his foot off the accelerator and putting it on the brake. In this time the car moves forward at its original speed. This is the *thinking distance*, which is proportional to the car's speed. When the brakes are applied, work is done by the braking force to take away the car's kinetic energy. Since the car's kinetic energy depends on v^2, the *braking distance* also depends on v^2. This means that when the car's speed doubles from 10 m/s to 20 m/s, the braking distance increases by a factor of 4.

Strain energy

When a spring is stretched it stores strain energy. This energy can be obtained from the spring when it is released, provided that it has not been stretched past its elastic limit.

Figure 1 shows a force/extension graph for a spring. How much energy is stored when the spring is stretched 0.1 m?

Energy stored = work done in stretching the spring

$$= \text{average force} \times \text{distance}$$

$$= 5\,\text{N} \times 0.1\,\text{m}$$

$$= 0.5\,\text{J}$$

When the spring is stretched the force pulling it changes, so we have to average the force.

Converting kinetic energy to potential energy

Sean throws a ball into the air with an upwards speed of 20 m/s. How high will it go? Using the principle that energy is conserved, we can say: the kinetic energy of the ball as it leaves Sean's arm, $\frac{1}{2}mv^2$, turns into potential energy, mgh, at its highest point.

So $\frac{1}{2}mv^2 = mgh$

$$h = \frac{v^2}{2g} = \frac{(20\,\text{m/s})^2}{20\,\text{m/s}} = 20\,\text{m}$$

Work is being done to increase the potential and kinetic energy of the Harrier jet

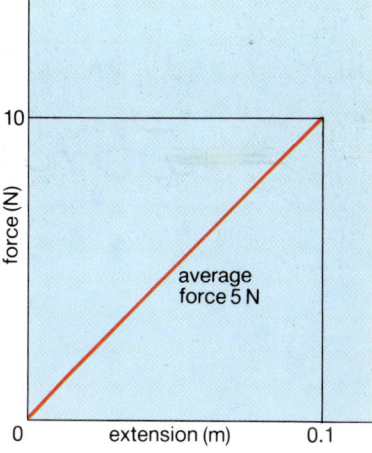

Figure 1

Questions

1 A charging rhinoceros moves at a speed of 15 m/s, and its mass is 1000 kg. What is its kinetic energy?

2 A car of mass 750 kg slows down from 30 m/s to 15 m/s over a distance of 50 m.
(a) What is its change in kinetic energy?
(b) What is the average braking force that acts on it?

3 This question is about the stopping distances in table 1.
(a) When the driver of the car sees a hazard, how long does it take him to start braking?
(b) In a built-up area the driver is travelling at 15 m/s. A child rushes out from behind an ice cream van 25 m in front of him. Use the data to decide whether he misses the child or not.

4 A catapult stores 10 J of strain energy when it is fully stretched. It is used to fire a marble of mass 0.02 kg straight up into the air.
(a) Calculate how high the marble rises.
(b) How fast is the marble moving when it is 30 m above the ground? (Ignore any effects due to air resistance.)

4 Power

The photographs on the left show Alice and David lifting some bricks. They each lift 20 bricks through a height of 5 m. This means that they do the same amount of work. However, David is large and powerful and he lifts all of his bricks in one go. Alice who is rather smaller, lifts her bricks one at a time. We use the word powerful to describe someone who can do the work quickly. **Power** is the rate of doing work or converting energy.

$$\text{Power} = \frac{\text{work done or energy converted}}{\text{time taken}}$$

From the equation above you can see that power is measured in J/s. But we give power its own unit called the **watt**. 1 watt is equal to a rate of working of 1 J/s. Power ratings can be very large; then we use the units of kW (kilowatt or 1000 W) and MW (megawatt or 1 000 000 W).

Body power

Alice measures her personal power output by running up a flight of steps. She takes 8.4 s to run up a flight of steps. Her mass is 14 kg. What power does she develop? She has done some work to lift her weight of 140 N through a height of 6 m.

$$\text{Work done} = \text{force} \times \text{distance}$$

$$= 140 \, \text{N} \times 6 \, \text{m} = 840 \, \text{J}$$

$$\text{power} = \frac{\text{work done}}{\text{time}}$$

$$= \frac{840 \, \text{J}}{8.4 \, \text{s}} = 100 \, \text{W}$$

Alice is converting energy at about the same rate as an electric light bulb.

David is carrying many more bricks than Alice and he is working about 20 times faster than her

The hind legs of this locust are extremely powerful. The insect takes off with a speed of 3 m/s. The jump is fast and occurs in a time of 25 milliseconds. The locust's mass is about 2.5 g. What power is generated? What is the power/mass ratio of a locust compared to a human?

Power of an express train

The photograph on the next page shows an inter-city express train moving along at a steady speed of 20 m/s. When the train moves at a constant speed, the driving force from the wheels is exactly balanced by opposing frictional forces. These opposing forces are caused by friction in the axles of the wheels and by wind resistance. So the train does work against these frictional forces.

How much power does the train have to produce when it is running at 20 m/s?

$$\text{Power} = \frac{\text{work done}}{\text{time}} = \frac{\text{force} \times \text{distance}}{\text{time}}$$

$$= F \times \frac{d}{t}$$

So the power developed is equal to the driving force × the distance travelled per second. But the distance travelled per second is the speed, v.

$$\boxed{\text{Power} = F \times v}$$

The resistive force on the train travelling at 20 m/s can be found from Figure 1.

Power = 8 kN × 20 m/s

$\qquad$ = 8000 N × 20 m/s

$\qquad$ = 160 000 W

or 160 kW

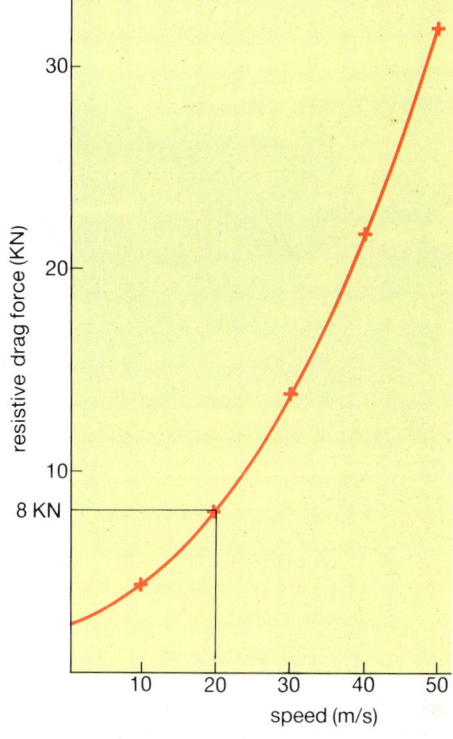

Figure 1
Graph to show the resistive force acting on a 125 train as the speed increases

This engine can provide 2 megawatts to drive the train at full speed

Questions

1 What is the unit of power?

2 David runs up the same flight of steps as Alice in 4 s. His mass is 100 kg. Roughly how many times more powerful is he than Alice? (Alice's power = 100 W)

3 On July 6th 1986, Leonid Taranenko lifted 265 kg above his head. In the last stages of his lift he raised the bar through 1 m in 2 s.
(a) How much work did he do?
(b) What power did he develop?

4 (a) Use the graph to estimate the resistive drag force on a 125 train when it is travelling at: (i) 40 m/s, (ii) 55 m/s.

(b) Show that the power that the engine produces to pull the train at 40 m/s is 880 kW.
(c) Calculate the power the engine produces when the train runs at 55 m/s.

5 (a) Use the formula $W = F \times d$ to calculate the work done against resistive forces when the train goes on a 200 km journey travelling at (i) 20 m/s (ii) 40 m/s. Give your answers in MJ.
(b) Now explain why, in 1973, when there was a temporary shortage of petrol, the government imposed a speed limit of 50 mph on all roads.

6 Here is part of an answer written by a student to explain the motion of a cricket ball. She has made some mistakes. Rewrite what she has written, explaining and correcting her errors.
  'When a batsman hits a cricket ball he gives it a certain force. As the ball rolls over the grass this force is used up and eventually the balls stops. All the power that the batsman used in hitting the ball ends up as heat.'

5 Machines

A car jack can multiply a force about 20 times

Force multipliers

Machines are clever devices that multiply forces for us. They help us to lift up loads that we are not strong enough to lift up directly. But do they multiply energy for us as well; do we get more energy out of the machine than we put in?

Figure 1 shows the principle of the **lever**, a simple machine. A 300 N downwards force acts 1 m to the left of the pivot. This force is balanced by a force of 100 N, that acts downwards at a distance of 3 m to the right of the pivot:

$$300\,\text{N} \times 1\,\text{m} = 100\,\text{N} \times 3\,\text{m}$$

So we can lift a load of 300 N by applying a force (effort) of only 100 N. The work done in lifting the load is

$$300\,\text{N} \times 0.5\,\text{m} = 150\,\text{J}$$

The work done by the person applying the effort is

$$100\,\text{N} \times 1.5\,\text{m} = 150\,\text{J}.$$

Energy is conserved, so we cannot get more out than we put in.

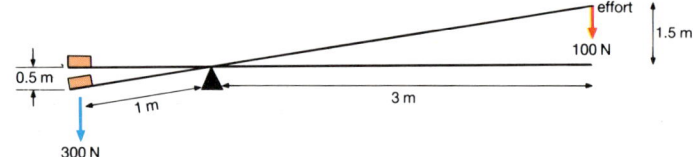

Figure 1
The principle of the lever

An **inclined plane** is another example of a common machine. Figure 2 shows a man rolling a barrel up to the top of the slope. The work he would do in lifting the barrel straight up through a height of 1 m is $1000\,\text{N} \times 1\,\text{m} = 1000\,\text{J}$. By rolling the barrel along the slope the effect of gravity is diluted. We can work out what force needs to be applied to the barrel.

Work $= F \times d$; work done = 1000 J; distance = 5 m.

$$1000\,\text{J} = F \times 5\,\text{m}$$

$$F = \frac{1000\,\text{J}}{5\,\text{m}} = 200\,\text{N}$$

This way the man will avoid serious damage to his back.

It is thought that the pyramids of Ancient Egypt were built by gangs of slaves pulling slabs of stone along a slope. Tree trunks were used as rollers.

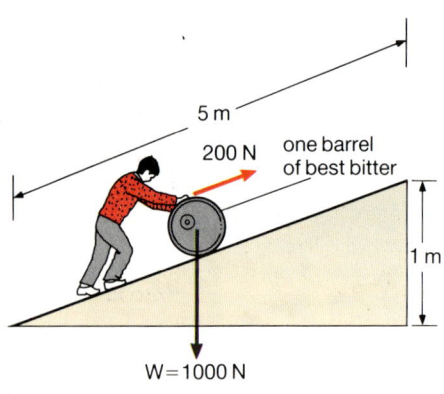

Figure 2
An inclined plane

Stonehenge consists of huge slabs of rock with masses of up to 50 tonnes. Can you suggest how the lintel stones (those across the top of the arches) were put into place?

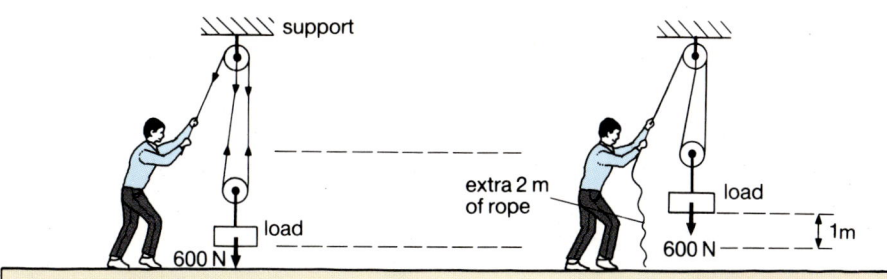

Figure 3
A pulley system

A **pulley system**, another simple machine, is shown in Figure 3. The load is 600 N and the lower pulley block is supported by two ropes. So if the tension in the rope is 300 N the load will be supported. The man can lift the load by applying a smaller effort, but he does not win from the energy point of view. If he lifts his load 1 m, each of the supporting ropes is shortened by 1 m, so he pulls the rope through 2 m. He applies half the force, but pulls through twice the distance, thereby doing the same work as if he had lifted the load straight up.

Gears

The machines that we have looked at so far increase forces; **gears** will increase (or decrease) couples (see page 25). Figure 4 shows two gear wheels in contact, the input shaft has to be turned twice to make the output shaft turn once. However, the couple from the output shaft is twice as big as that applied to the input shaft.

The bicycle is an example of a machine that is a distance multiplier. The force needed to push a bicycle along a road is small. So a cyclist applies a large force on the pedals, but moves his feet a small distance in one rotation of the pedals. This work is converted into a small force acting to push the bicycle forward, but the distance that the wheel rotates in one revolution is a lot more than the pedals' rotation. So the bicycle has increased the distance we have moved – that is the idea, to get there faster (Figure 5).

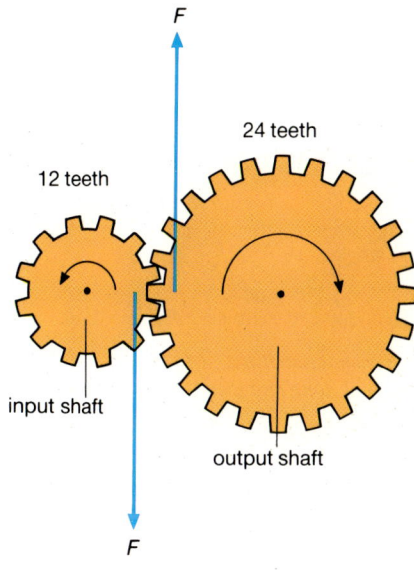

Figure 4
Where the gear teeth touch, equal and opposite forces act
Couple on input shaft = $F \times r$
Couple on output shaft = $2F \times r$

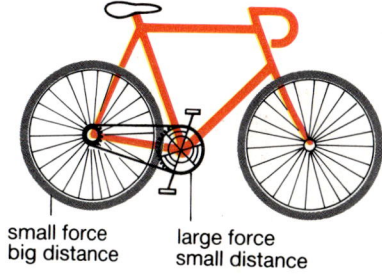

small force   large force
big distance   small distance

Figure 5
A distance multiplier

Questions

1 (a) Why is it useful to have machines to multiply forces for us?
(b) Can a machine save us energy? Explain your answer.

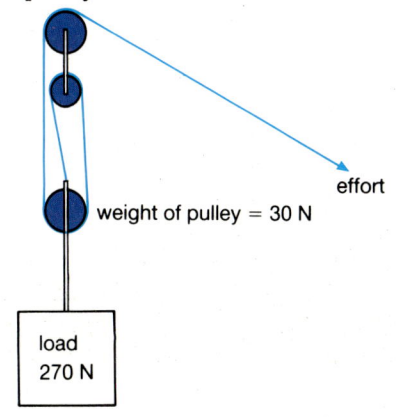

effort

weight of pulley = 30 N

load
270 N

2 The diagram on the left shows a pulley system. Calculate the smallest effort you need to apply to lift the load.
3 Soraya is pedalling her bicycle along level ground.
(a) How many times does the wheel turn for each rotation of the pedals?
(b) How far does the bicycle move for one rotation of the pedals?
(c) How much work does Soraya do in one rotation of her pedals?
(d) All the work Soraya does is used to push her along the road. Calculate F.
(e) Soraya turns her pedals twice per second. Calculate (i) the speed of the bicycle (ii) the power her legs produce.

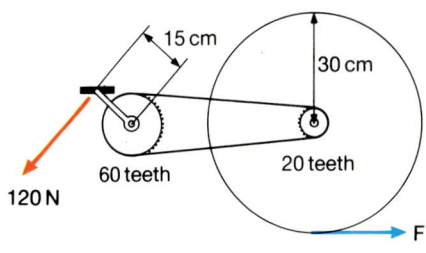

- distance moved by the pedal in one rotation = 0.94 m
- circumference of wheel = 1.88 m

6 Efficiency of Machines

Three hot machines. The heat that is lost has not done any work, and so these machines are inefficient

Figure 1

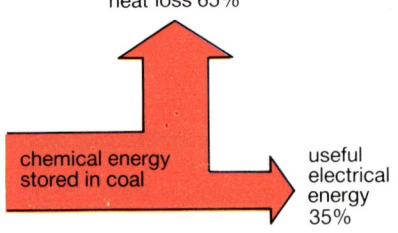

(a) How energy is used in a power station. The efficiency is 35%

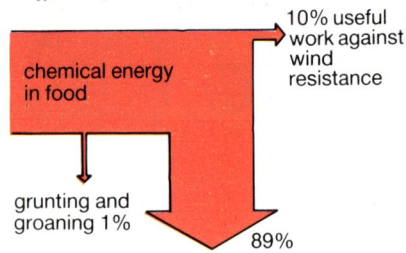

(b) How an athlete's energy is used. The efficiency is 10%

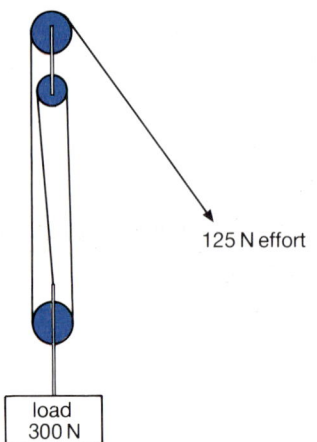

Figure 2

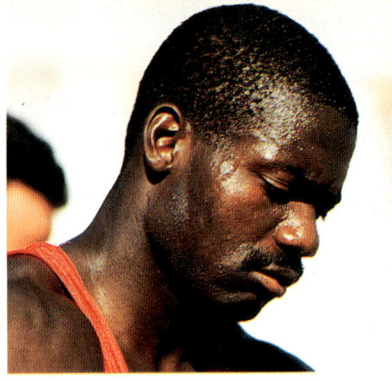

Unwanted heat

In the last unit we met some machines. It was assumed that we got as much work out of these machines as we put in. However, this is not usually true. Energy must be conserved, but it is often converted to a form that we do not want. The photographs in this unit are of three hot machines. The power station is producing electrical energy from chemical energy. The light bulb and athlete are producing light and kinetic energy, respectively, from their fuel. But in all cases heat energy is produced too. Figure 1 illustrates the principle for the power station and the athlete.

This leads us to the idea of **efficiency**, which is defined like this:

$$\text{Efficiency} = \frac{\text{useful energy (or work) out of machine}}{\text{energy (or work) put into a machine}}$$

$$or \text{ Efficiency} = \frac{\text{power out}}{\text{power in}}$$

Example. Figure 2 shows a pulley system being used to lift a load. What is its efficiency?
If the load moves 1 m, the effort rope will have to be pulled 3 m.

$$\text{Efficiency} = \frac{\text{work done on load}}{\text{work done by effort}}$$

$$= \frac{300\,\text{N} \times 1\,\text{m}}{125\,\text{N} \times 3\,\text{m}}$$

$$= 0.8 \text{ or } 80\%$$

We can express efficiency either as a fraction or a percentage.
The main causes for unwanted energy losses in the pulley system are
- we are lifting the lower pulley block
- there are frictional forces on the pulley axles which produce heat.

Human efficiency

We get our energy from food by the process of **respiration**. During respiration, food reacts with oxygen in our bodies forming carbon dioxide and water. We obtain the oxygen when we breathe in, and when we breathe out we get rid of the carbon dioxide formed during respiration.

food + oxygen → carbon dioxide + water + energy
(containing carbon
and hydrogen)

Figure 3
Philip doing pull-ups

When we take exercise we breathe faster. This means we get more oxygen to provide us with the mechanical energy. But the process is inefficient and our muscles make lots of heat too.

Figure 3 shows Philip doing pull-ups in the gym. While he does this his body uses 30 000 J of energy every minute. In one minute Philip does 20 pull-ups, before collapsing exhausted. Each pull-up was through a height of 0.5 m, his weight is 750 N; what is his efficiency?

$$\text{Work out} = 20 \times 750\,\text{N} \times 0.5\,\text{m}$$

$$= 7500\,\text{J}$$

$$\text{Efficiency} = \frac{\text{work out}}{\text{work in}}$$

$$= \frac{7500\,\text{J}}{30\,000\,\text{J}}$$

$$= 0.25 \text{ or } 25\%$$

Table 1 shows the efficiencies of two power stations. You can see that the efficiencies have improved over the years. But will we ever be able to do better than 37%? Surely we ought to be able to make a power station nearly 100% efficient? The answer is no. We are producing electricity (an ordered form of energy) from heat, which is associated with the disordered movement of molecules. Making order is far harder than making disorder, so the production of electricity will always be inefficient.

Station	Date	Efficiency
Battersea A	1933	16%
Drax	1975	37%

Table 1. Data from CEGB

Questions

1 Why do you get out of breath if you run too fast?

2 A 60 W electric light bulb uses 60 J of electrical energy each second. It produces 2 W of light. Calculate its efficiency. What happens to the other 58 W?

3 You have just got a job on a building site, as a bricklayer's mate. It is your job to carry bricks up a ladder 10 m high. In three hours you have to carry up 1000 bricks. Each brick weighs 40 N.

(a) How much work will you do in three hours?

(b) To do this work, your body will have to produce about four times as much energy. Explain why.

(c) Use the table in the next column to choose a breakfast that gives you enough energy to do the work.

Food	Energy value (kJ)
1 Weetabix	800
1 slice of bacon	700
1 egg	400
1 sausage	600
beans	150
1 slice of toast, butter and marmalade	600

4 The diagram (right) shows a wheel and axle. A load of 600 N is lifted a distance of 1 m by an effort of 200 N.

(a) How far does the effort force move to lift the load up 1 m? (Hint: look at the radius of each part of the machine.)

(b) How much work is done on the load?

(c) How much work is done by the effort?

(d) What is the efficiency of the machine?

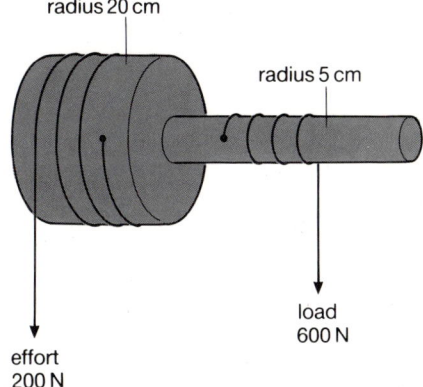

radius 20 cm

radius 5 cm

effort
200 N

load
600 N

7 Electrical Energy Production

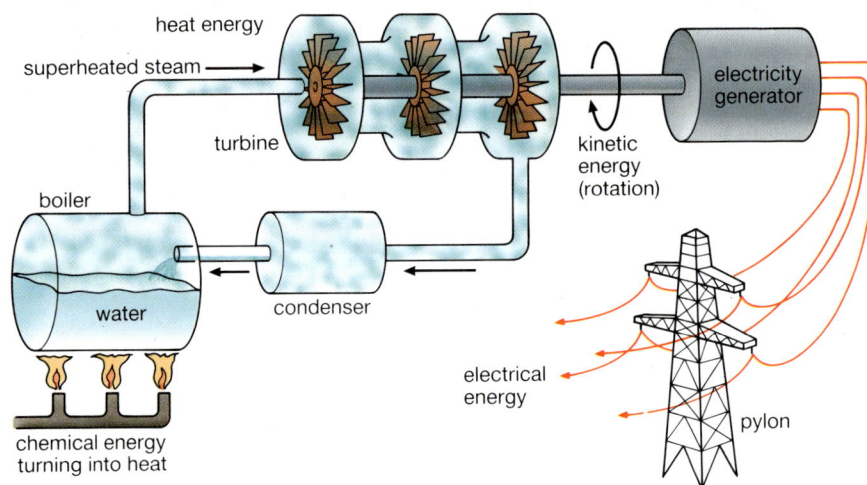

Figure 1
The principle of a power station

This is a 660 megawatt turbine inside the
coal-fired Drax power station. Superheated
steam is used to drive the turbine

Figure 1 shows the principle behind the production of electrical energy in a
power station. Most of our power stations use coal as their source of energy.
When coal is burnt its stored chemical energy is released as heat energy. This
heat energy boils water at high pressure to make superheated steam at
temperatures of about 700°C. The kinetic energy in the superheated steam is
used to drive **turbines**. These are connected to the electricity generator by large
coils rotating inside a strong magnetic field.

You can see in Figure 2 a chart showing how energy is used in this process.
There is a lot of heat lost in the power station. The electrical energy itself is also
converted to heat energy in factories and houses.

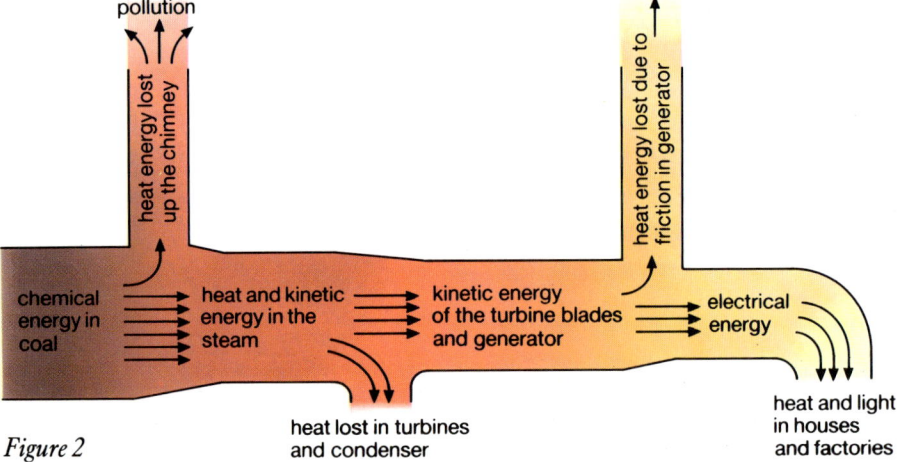

Figure 2

There are two problems that arise from the production of electricity that worry
a lot of people.

- **Pollution**. Burning coal makes the gases carbon dioxide and sulphur dioxide.
These pollute the atmosphere. When sulphur dioxide dissolves in water, an
acidic solution is formed containing sulphurous acid.

water + sulphur dioxide → sulphurous acid

So when sulphur dioxide gets into rain, the rain becomes acidic. We call this
acid rain. Acid rain damages stonework in buildings. It is thought that acid

rain is also killing trees in Scandinavia. It is likely that sulphur dioxide produced in Britain is blown across to Scandinavia by the prevailing north-easterly winds.

Heat energy is always produced when we make electrical energy. People worry that we will warm the Earth up, as we increase our use of electrical energy. This warming is caused by (1) the extra heat from power stations, factories and homes; (2) extra carbon dioxide in the atmosphere trapping heat in, so that the Earth's temperature will rise slowly. An **increase in the Earth's temperature** could have very serious consequences. An increase of 1°C or 2°C to the Earth's average temperature would probably melt a large amount of ice from the polar ice caps. Although we get great benefits from electrical energy, we have to consider its effect on the environment. If we are not careful we will damage the world that we live in.

The turbine blades at the Drax power station

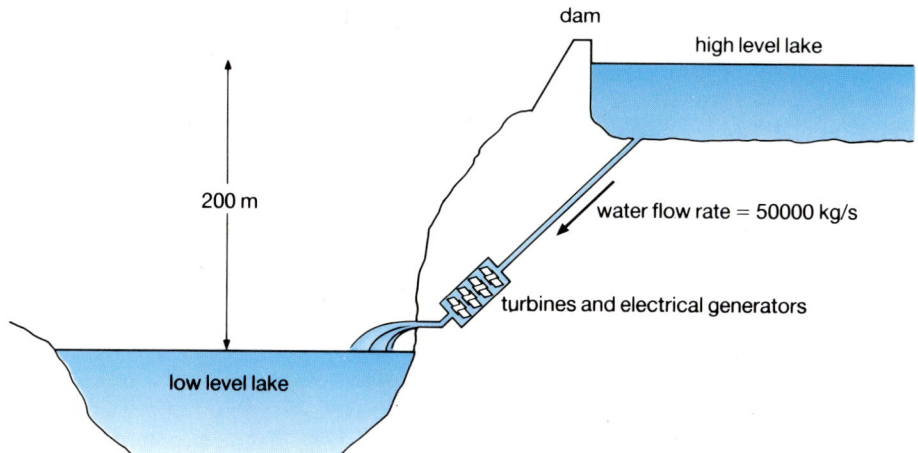

Figure 3
A pumped storage power station

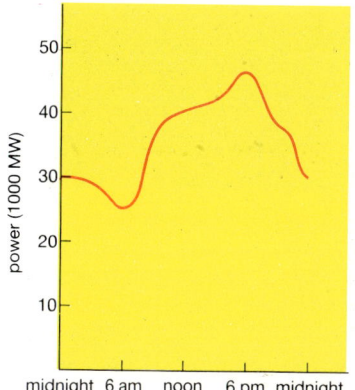

Figure 4
The typical use of electrical energy on a winter's day in Britain

Questions

1 (a) What is smog? Why is it usually found in large cities?
(b) Why do large cities have smokeless zones?
2 (a) What is acid rain?
(b) How is acid rain formed?
(c) What effects does acid rain have on the environment?
(d) What would you do to reduce the problems caused by acid rain?
3 Figure 3 (above) shows the layout of a pumped storage power station. Water from the high level lake produces electrical energy by flowing through the turbine generators. These are placed just above the low level lake. When

there is a low demand for electricity, the generators are driven in reverse to pump water back into the high level lake. This means there will be enough water to generate electricity again, when demand is high.
(a) Why is this sort of power station useful to the Central Electricity Generating Board?
(b) Would this pumped storage power station pollute the atmosphere when it generates electricity?
(c) Where does the energy come from to pump the water back up the hill again?

(d) What energy changes occur as water flows from the high level lake to the low level lake?
(e) Calculate the loss of potential energy of the water in 1 second. (Use the information in Figure 3.)
(f) The turbines are 60% efficient. Calculate the electrical power output of this station. Give your answer in MW.
(g) For the next two questions, look at Figure 4. At what times of day will the pumped storage power station be producing electricity?
(h) Explain the shape of the graph in Figure 4. Why does it reach a maximum at about 6 pm?

8 The World's Energy Resources

A large city like Los Angeles might use $10^{14}J$ of energy in one day. You need to burn 100 000 tonnes of coal to provide this amount of energy

Fossil fuels

At the moment the world faces an energy crisis. This may seem surprising; you have learnt that energy cannot be lost. The problem is that we are burning fuels like coal, oil and gas. These fuels produce electricity, warm our houses and provide energy for transport. The end product of these fuels is heat energy. We cannot recapture the heat and turn it back into coal or oil. These fuels are known as **non-renewable energy sources**. Once we have burnt them they have gone forever.

Coal, oil and gas are known as *fossil fuels*. They are the remains of plants and animals that lived some hundreds of millions of years ago. Supplies of these fuels are limited. We use coal mainly for the production of electrical energy. If we go on using coal at its present rate, it will last us for about another 300 years.

We get petrol from oil. So this energy source is vital for running cars and aeroplanes. At our present rate of use, oil will last us about 60 years. Gas will last for about the same time. Twenty years ago Britain started to drill for oil in the North Sea. This supply of oil is going to run out, in the 1990s. So this crisis is not something that is going to happen a long way ahead. It is going to happen in your lifetime. It is important that we use our fossil fuels carefully, and we must also look for other sources of energy.

Other energy sources

- **Nuclear power** is being used more and more to generate electricity. By the year 2000 France will generate 95% of its electrical energy using nuclear power. The nuclear fuel that is used is uranium. This is also a non-renewable energy source. But nuclear power could provide our energy for a few thousand years. Nuclear power also worries people because of the radioactive waste that is produced. This is discussed further in Section L.

Figure 1
How long will our fossil fuels last?

Some further sources of energy are described below; these are **renewable energy sources**. In these cases we extract energy from the environment and these energy sources will always be available to us.

- **Biomass** is the name we use when talking about plants in the sense that we can extract energy from them. The most common way of extracting energy from biomass is to cut down a tree and burn it. Trees are a renewable energy source provided forests are looked after properly. Unfortunately, in Africa, Asia and South America trees are being cut down far faster than they are being replanted. It is thought that cutting down large forests has changed our climate by reducing rainfall in some parts of the world.

  In some countries vegetable oils are already in use to drive farm machinery. So the world's biomass may be used in future to power our cars.

- **Tidal power.** Figure 2 shows a map of the Severn estuary, which is a suitable position for a tidal barrage. The idea is that water flows in through the sluice gates at A at high tide. At low tide water flows out through the turbogenerators at B. We are using the potential energy of the water to generate electrical energy. If the Severn estuary barrage is built it will produce about 7000 MW of power.

- **Hydroelectric power** is widely used in Scandinavia, where a lot of water flows down the mountains from melted snow. The principle is the same as tidal power: the potential energy of the water is used to generate electrical energy.

- **Wind power.** Energy from the wind can also be used to generate electricity. The photograph opposite shows a wind farm in Scotland.

- **Geothermal power.** Energy can be obtained from a hot spring. When the water in the spring boils, the steam formed can be used to drive electrical turbines. In China, warm water (50°C) is pumped directly into factories and houses to provide central heating.

- **Solar power** can be used directly to warm up water in panels in the roofs of houses. Energy from the sun can also be used to generate electricity using photocells. A lot of us now use solar-powered calculators.

A small wind turbine on Scoraig on the west coast of Scotland. Scoraig has no mains gas or electricity and its dozen or so houses are entirely dependent on wind turbines and storage batteries

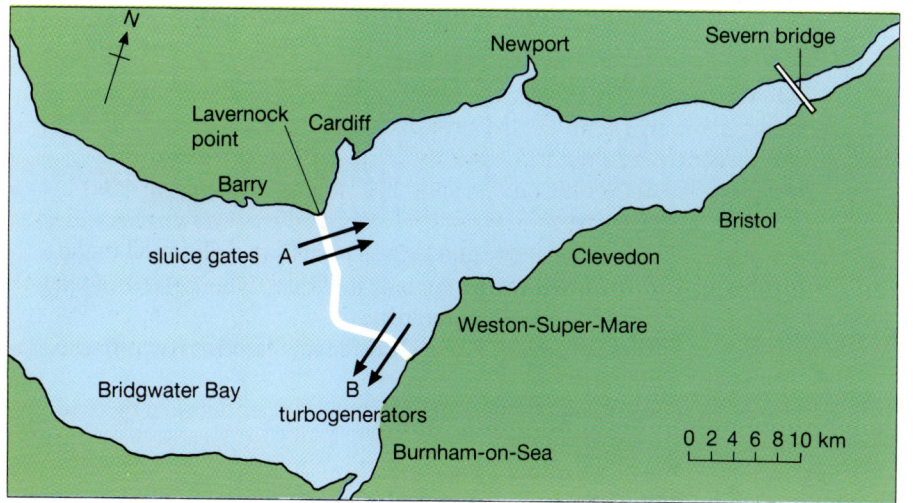

Figure 2
The Severn Barrage. A report from the Severn Barrage Committee suggested that it was technically feasible to build the barrage. At the time of the report (1981) the cost was estimated at £7.5 billion. The barrage would be able to produce 7000 MW of power. This amounts to about 6% of the national need

Questions

1 Explain the difference between a renewable and a non-renewable fuel.
2 Why are air fares likely to be very expensive in 50 years' time?
3 Why is petrol wasted if a car accelerates or brakes rapidly?
4 Discuss whether wood is a renewable energy source.
5 Which of the following countries do you think are high users of energy, and which are low users: USA, Norway, India, Britain, Thailand, Ethiopia? Is there a connection between wealth and a country's use of energy?

SECTION D: *STUDY QUESTIONS*

1 Each statement describes a force acting. Which force is causing work to be done?
A The attraction between a magnet and a nail which is hanging from it.
B The pull of a moving railway engine on its coaches.
C The push of a person's feet when standing on the floor.
D The tension in an elastic band wrapped around a parcel.
E The weight of a book at rest on a table.

MEG

2 Which energy change involves frictional forces?
A Chemical energy to heat energy.
B Chemical energy to kinetic energy.
C Heat energy to sound energy.
D Kinetic energy to heat energy.
E Potential energy to sound energy.

MEG

3 Five machines used energy from fuel to do useful work, as shown in the table. Which of the machines wasted the most energy?

Machine	Energy used (J)	Useful work done (J)
A	500	250
B	1 000	250
C	1 000	500
D	5 000	4 500
E	10 000	9 500

MEG

4 A lady weighing 400 N climbs a vertical ladder. How much work has she done after climbing 2 m?
A 0.5 J   B 200 J   C 402 J   D 800 J   E 8000 J

LEAG

5 An electric motor raises a mass of 0.2 kg, at a constant speed, through a vertical distance of 3.0 m in 2 s. If the acceleration of free fall is 10 m/s^2, the power, in W, developed by the motor in raising the load is
A 0.3   B 1.2   C 3.0   D 6.0   E 12.0

NI

6 An electric motor is used to lift a load of 500 N through a vertical height of 4 m.
(a) Calculate the work done on the load.
(b) If the load is raised 4 m in a time of 10 s, calculate the power output of the motor.
(c) Explain what is meant by the efficiency of the motor.
(d) If the efficiency of the motor is 40%, calculate the power input to the motor.
(e) Give one reason why the efficiency of the motor is less than 100%

NEA

7 A stone is thrown upwards. When it reaches *P* which of the following has the greatest value for the stone?
A Its acceleration.
B Its kinetic energy.
C Its potential energy.
D Its velocity.
E Its weight.

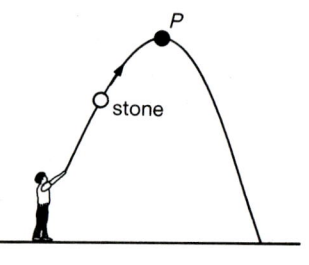

MEG

8 *Voyager 1* was launched on September 5th, 1977. After a journey of nearly two years it reached Jupiter. On January 30th 1979 *Voyager 1* was 35.1 million km away from Jupiter. The data below show how *Voyager* approached Jupiter during the next month until passing close by Jupiter on March 5th. The times recorded are from 00.00 h GMT on January 30th 1979.

	Distance from Jupiter's centre R (millions of km)	Velocity v (m/s)	Time from 00.00 h GMT Jan 30th 1979 days	hours	Date
1	35.1 0	10 900	0	00	Jan 30th
1A	34.7 1	10 904	0	10	
2	28.0 8	11 000	7	10	Feb 6th
2A	27.6 8	11 006	7	20	
3	21.0 6	11 135	14	18	Feb 13th
3A	20.6 6	11 145	15	4	
4	1.10 4	11 402	21	23	Feb 20th
4A	13.8 8	11 411	22	3	
5	7.0 2	12 165	28	21	Feb 27th
5A	6.9 3	12 184	28	23	
6	3.5 1	13 564	32	1	Mar 3rd
6A	3.4 6	13 601	32	2	
7	2.8 1	14 212	32	15	Mar 3rd
7A	2.7 6	14 271	32	16	

During all of the approach to Jupiter, the motors on *Voyager* were turned off. So the only force acting on *Voyager* was the gravitational pull of Jupiter.
(a) Explain why *Voyager*'s speed increased as it approached Jupiter.
(b) Work out the acceleration of *Voyager* between these two pairs of points: (i) 1 and 1A (ii) 7 and 7A. Express your answer in m/s^2. (Hint: 1 hour = 3600 s)
(c) Now work out the force acting on *Voyager* at points 1 and 7. *Voyager*'s mass is about 2000 kg. Explain why the force changed.
(d) Work out *Voyager*'s increase in kinetic energy as it moved from point 2 to point 6. Where did this increase in kinetic energy come from?

9 A staircase at an underground station has 105 steps, each one 0.17 m high. A boy, weight 400 N, runs up these stairs in 20 seconds.

(a) How much work does he do climbing the stairs?

(b) What is his average power output while climbing the stairs?

(c) The energy content of 100 g of cornflakes is 1465 kJ. The boy eats 30 g of cornflakes for breakfast. If he could use all the energy from this 30 g of cornflakes to climb the stairs how many times would he be able to climb them?

(d) Explain why the cornflakes alone will not provide the boy with enough energy to climb the stairs the number of times you have just calculated.

MEG

10 (a) The diagram shows a windmill with a rotor 100 m across. This could be used to generate electricity.

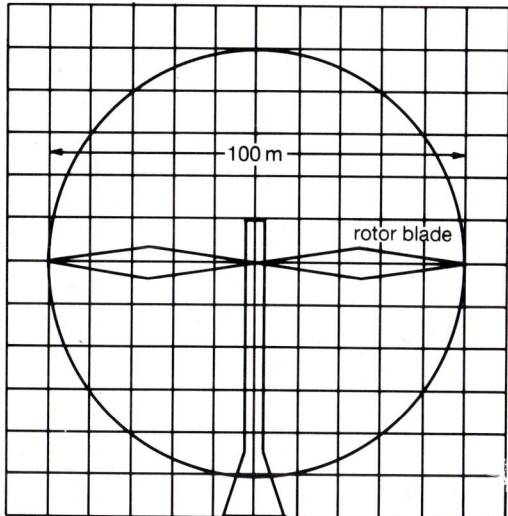

(i) Estimate the total area of the rotor blades.

(ii) The wind has a velocity of 10 m/s. The force it can cause on a surface of 1 m² is 90 N. How much work can this wind do on 1 m² in 1 s?

(Work = force × distance)

(iii) Wind is used to turn the rotor blade. The effective work done on 1 m² of the blade is 50% of that calculated in (ii). Find how much energy is transferred to the rotor in 1 s.

(iv) Calculate the maximum electrical power in watts which would be generated by this windmill.

(v) A conventional power station generates about 900 megawatts of electric power (1 MW = 1 000 000 W). How many of these windmills would be needed to replace a power station?

(b) In unit D8 you saw where a barrier could be built across the estuary of the River Severn. This would make a lake with a surface area of about 200 km² (200 million m²).

The diagram below shows that the sea level could change by 9 m between low and high tide. But the level in the lake would only change by 5 m.

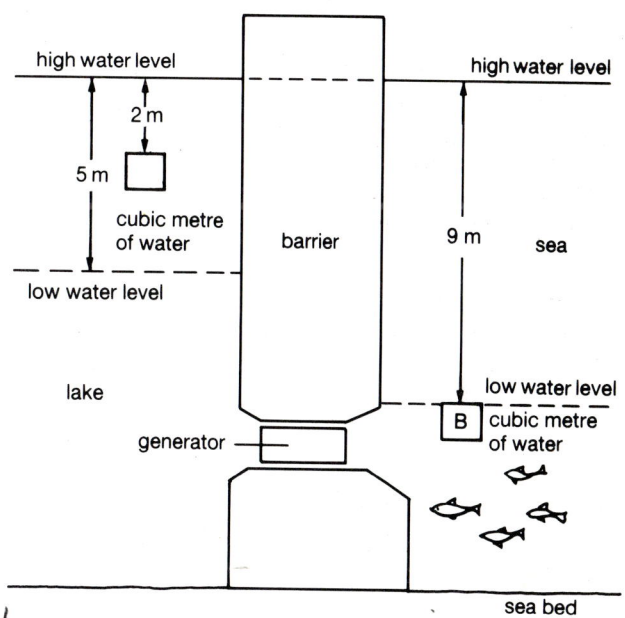

(i) What kind of energy is given up when a cubic metre of water falls from position A to position B?

(ii) A cubic metre of water has a mass of 1000 kg. The acceleration due to gravity is 10 m/s².

Calculate the force of gravity on a cubic metre of water. Calculate the work that this can do as it falls from A to B.

(Force = mass × acceleration)

(Work = force × distance)

(iii) How many cubic metres of water could flow out of the lake between high and low tide?

(iv) Use you answers to parts (ii) and (iii) to find how much energy could be obtained from the tide. Assume that position "A" is the average position of a cubic metre of water between high and low levels in the lake.

(v) The time between high and low tide is approximately 20 000 seconds (about six hours). Use this figure to estimate the power available from the dam. Give your answers in megawatts.

(c) Explain the advantages and disadvantages of the windmill and the Severn Barrier as sources of power.

LEAG

11 A piece of elastic used in a catapult has a force-extension graph as shown. When the elastic was stretched to the maximum extension shown on the graph, the energy stored in the elastic was 25 J.

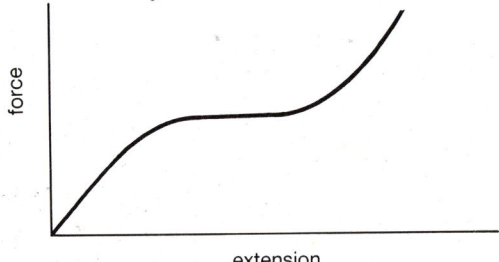

(a) Does the elastic obey Hooke's Law over the range of values shown? Justify your answer.
(b) When the stone was released from the catapult, 80% of the energy stored in the elastic was given to the stone as kinetic energy. The stone, which had a mass of 100 g, took 0.2 s to leave the catapult. What was the maximum speed of the stone? Explain your result.
(c) The original piece of elastic was replaced by a piece twice as long.
(i) Copy the original graph. Then sketch as accurately as you can the force-extension graph that you would expect to obtain for this new piece of elastic.
(ii) Explain the shape of the graph you have drawn.

MEG

12 The diagram below shows a trolley on a horizontal table. The trolley is connected to a mass M by a string. When let go, the trolley runs along the table but is stopped when it hits the pulley. (Forces of friction on the trolley and pulley are so small that they can be ignored.)

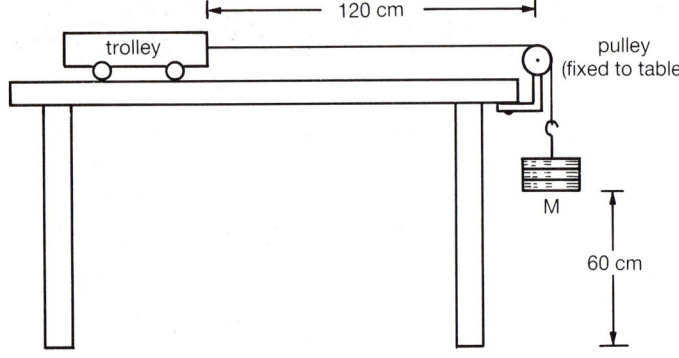

(a) Sketch the shape of a graph to show how the speed of the trolley changes as times goes by.
(b) Describe what changes, if any, occur in each of the following during the entire motion of the trolley.
(i) The gravitational potential energy of the trolley
(ii) The gravitational potential energy of mass M
(iii) The kinetic energy of the trolley.

MEG

13 The diagram shows a motor that is used to lift up a pile driver. The driver is dropped from a height of 10 m on top of the pile. The driver and pile each have a mass of 500 kg.

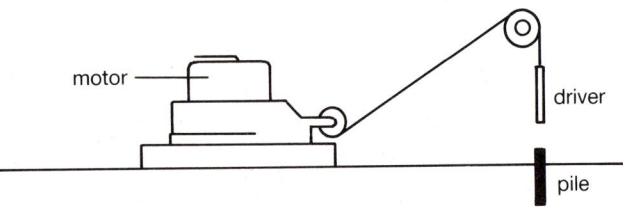

(a) How much potential energy does the driver have when it is 10 m above the pile?
(b) Calculate the speed of the driver just before it hits the pile.
(c) The driver and pile stick together on impact. Now calculate their speed immediately after impact. (Hint: use the principle of conservation of momentum.)
(d) Calculate how much kinetic energy is lost on impact. What has happened to this energy?
(e) The driver and pile come to rest after a time of 0.1 s. Calculate the average force that the ground exerts on the pile during this time.
(f) Calculate how far the pile goes into the ground.

14 Boris has caught Mikhail the mouse and has set him to work in the tread mill. When Mikhail runs he can just lift Boris at a speed of 0.03 m/s.
(a) How fast must Mikhail be running relative to the treadmill?
(b) What force (*F*) does Mikhail exert on the treadmill? Assume the machine is 100% efficient.

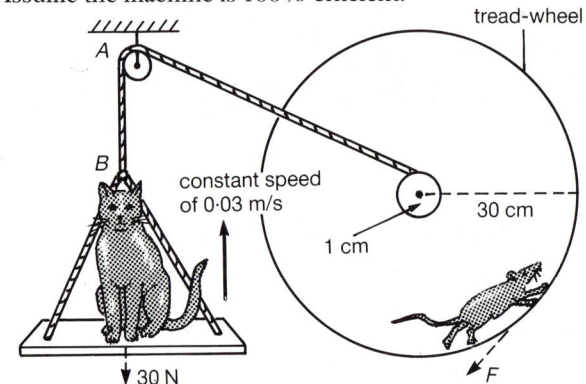

15 Which one of the following statements about power stations is NOT true?
A Hydro-electric power stations use water to drive turbines.
B In a power station, turbines drive generators.
C In power stations, generators produce electricity.
D Electricity from nuclear power stations differs from the electricity from a coal-fired power station.
E Some energy is wasted as heat in the power station.

MEG

SECTION E
Matter

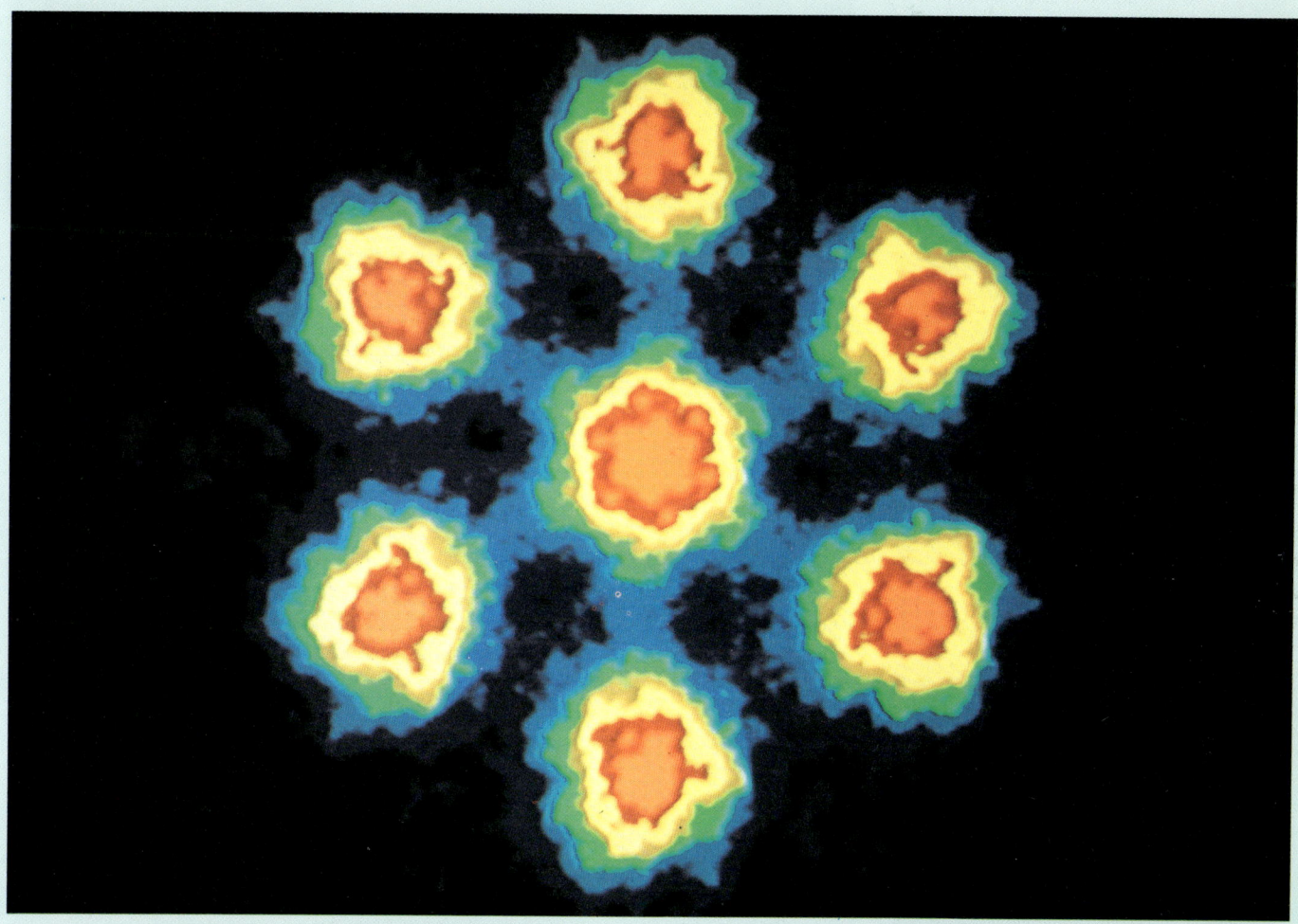

A false colour photograph of a uranyl microcrystal taken with a scanning electron microscope. The image shows the uranium atoms arranged in a perfect hexagonal shape around a central atom. Each atom is spaced about 0.000 000 000 32 metres from its neighbour, which means that the magnification in this photograph is about 100 million times!

1 Small Particles

This male moth has large antennae for detecting small particles of the female moth's scent

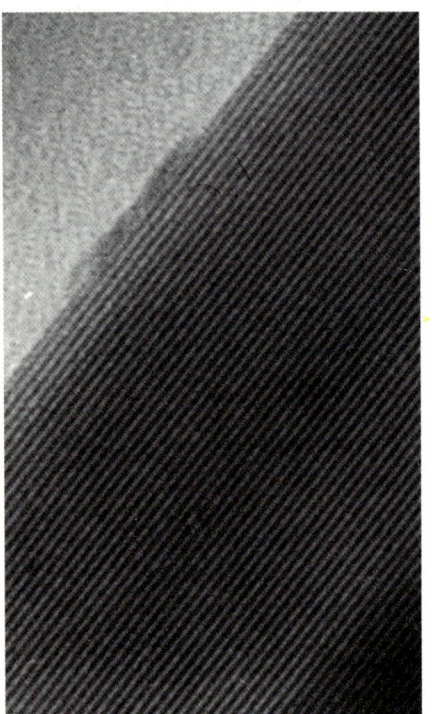

This is an electron microscope photograph of a manganese compound. The dark lines are rows of manganese atoms neatly arranged in the crystal

Everything we touch, swallow and breathe is made out of tiny particles. The smallest particles are **atoms**. There are only about 100 different types of atom. Materials that are made from only one type of atom are called **elements**. For example, aluminium contains only aluminium atoms. Some other common elements are oxygen, hydrogen, nitrogen and carbon.

The small particles in lots of materials are **molecules**. Atoms combine chemically to make molecules. For example, a water molecule is made up of two hydrogen atoms and one oxygen atom, while a carbon dioxide molecule contains one carbon and two oxygen atoms.

Atoms and molecules are far too small for us to see directly. But the photograph shows you some molecules seen through a powerful electron microscope. Electron microscopes show us that oil molecules are about 0.0000001 cm (10^{-7} cm) long. That means that 10 000 000 molecules put end to end would be about 1 cm long. (The diagrams on page 5 help you to see how small molecules are).

Discovering molecules

The idea that things were made from atoms and molecules was thought of over a hundred years ago, long before electron microscopes were invented. Below are some examples which suggest that matter is made up of small particles. **Matter** is the name that we use to describe all solids, liquids and gases.

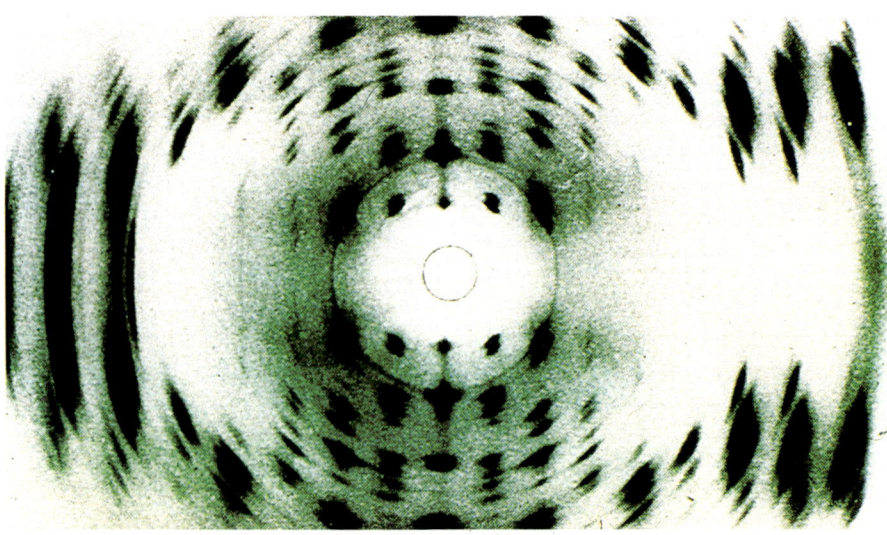

This is an X-ray diffraction photograph of DNA. The dark spots are caused by X-rays being diffracted by DNA molecules. A photograph like this one played a vital part in the unravelling of the complex structure of the DNA molecule

- The photograph opposite shows a male Emperor moth. He uses his very large antennae to detect the scent of a female. A male Emperor moth can be attracted by the scent of a female at a distance as far as 10 km. The female produces only a small quantity of scent. This suggests that tiny particles of her scent must spread out through the air.

- There is a simple experiment that you can do for yourself, which shows that liquids are also made up of very small particles. You can take a small drop of blue ink and put it into a glass of water (see page 91). If you stir the water it will turn a very pale blue. The small particles in the ink have now been spread further apart.

- Growing crystals. Figure 1 shows how you can grow a crystal. You dissolve some copper sulphate in water to make a strong solution. Then a small crystal of copper sulphate is placed in the solution. During a week or so, this crystal grows very slowly into a larger one. This can be explained by saying that the solution contains very small particles of copper sulphate. These particles stick to the crystal so that it grows larger.

Growing a crystal is a bit like making a neat pile of oranges. As another orange is added the pile grows. You can see that the shape of a pyramid of oranges is the same as the shape of an alum crystal.

When oranges are piled together in neat rows, they form a pyramid shape

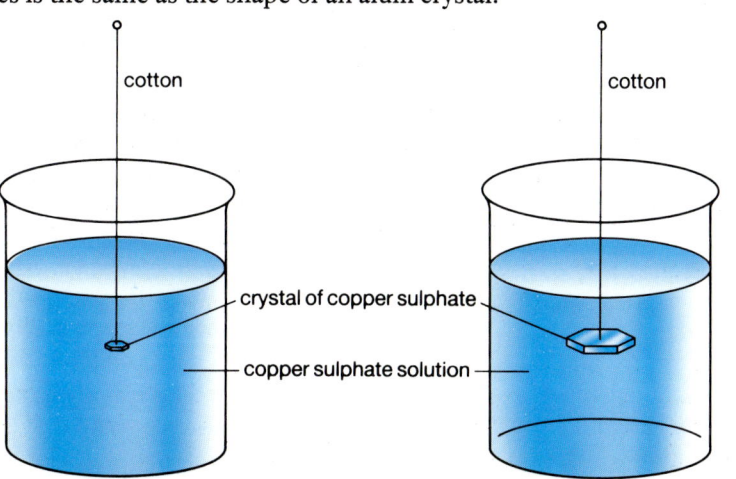

Figure 1
(a) A small crystal of copper sulphate is placed in a copper sulphate solution

(b) A week later a large crystal has grown

cotton

cotton

crystal of copper sulphate

copper sulphate solution

The shape of this alum crystal suggests that the molecules in it are also stacked neatly in rows

Questions

1 When you cook a curry the smell spreads right through the house. Why does this suggest that the curry powder is made up of small particles?

2 Why do we think that molecules in crystals are packed into neat rows?

3 In the Straits of Hormuz in the Persian Gulf, an oil tanker is holed by machine-gun fire and develops a small leak. The tanker spills $10\,m^3$ of oil. The diagram opposite shows how the oil spreads out over the sea's surface.

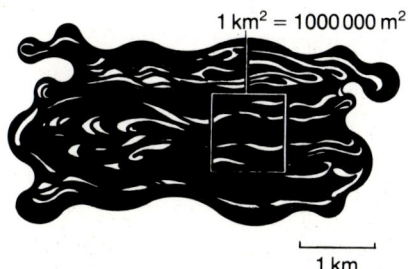

$1\,km^2 = 1\,000\,000\,m^2$

1 km

(a) Estimate the area of the spill in km^2.
(b) What is the area of the spill in m^2?
(c) Calculate the oil spill's thickness.

(d) Does the thickness of the oil spill tell you the size of an oil molecule? Explain your answer.

4 The South American golden dart-poison frog produces the world's most lethal poison, batrachotoxin. Only 0.002 g of this substance is enough to kill an average adult (mass 70 kg).
(a) Work out the lethal dose of this poison per gram of human flesh.
(b) Why do you think that this poison is made up of small particles?

2 Molecules on the Move

Diffusion

In the last unit you read how a male Emperor moth can detect a female moth's scent at a great distance. This supports the idea that the scent is made up of molecules and also the idea that the molecules in a gas are moving. Here are some simple experiments to show you more about the movement of molecules. In Figure 1 you can see a long tube with all the air pumped out of it. It is closed with a tap. On the other side of the tap is attached a small capsule of bromine inside some rubber tubing. The capsule is broken and then the tap opened. As soon as the tap is opened the bromine vapour fills the long tube. This tells us that the bromine molecules are moving quickly. They are actually moving at a speed of about 200 m/s.

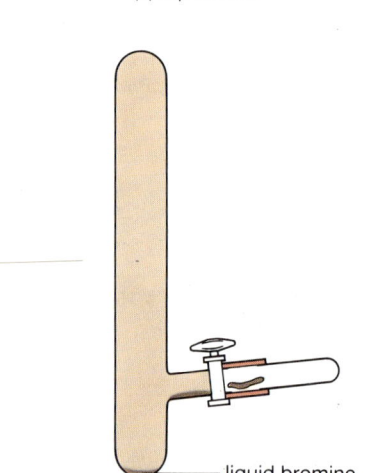

Figure 1
Bromine vapour filling a vacuum

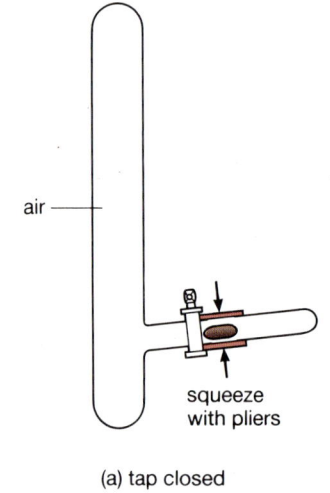

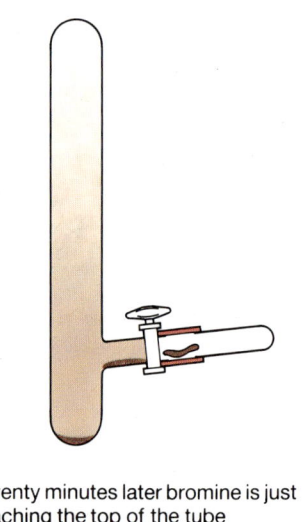

(a) tap closed

(b) twenty minutes later bromine is just reaching the top of the tube

Figure 2
The bromine molecules bump into the air molecules, and are slowed down

Figure 2 shows what happens when the experiment is repeated with air inside. This time when the tap is opened the bromine does not fill the tube quickly. After about 20 minutes the bottom half of the tube is coloured dark brown, but the top is only light brown. So although the bromine molecules travel very quickly it takes a long time for them to reach the top.

The reason for this is that the air molecules are also moving quickly. The air molecules get in the way of the bromine molecules. When two molecules bump into each other they will change direction. The bromine molecules keep changing direction and so take a long time to reach the top.

> This process of one substance spreading through another is called **diffusion**.

Diffusion also occurs in liquids, but only very slowly in solids. Diffusion is very important for all living things. When animals have eaten a meal, food is digested and diffuses into the blood. Blood then carries the food all round the body. Plants need nitrogen, potassium, phosphorus and other elements. These diffuse through the soil to the plants' roots.

The essential nutrients for a plant diffuse through the soil to its roots. The roots are split up into many branches that spread out so that the plant can absorb as much of the nutrients as possible

Figure 3
Place some ink in a glass of water. After an hour or so the water is a light blue colour. Diffusion works more slowly in liquids, which suggests that molecules move more slowly in liquids than in gases.

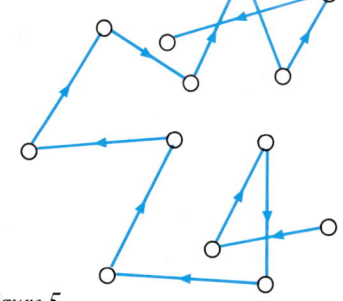

Figure 4

Brownian motion

Robert Brown, a botanist, looked through his microscope at some grains of pollen which were in water. He noticed that the grains of pollen were moving around randomly. They jiggled from side to side.

You can see the same sort of **Brownian motion** if you look through a microscope at smoke particles. Figure 4 shows a typical experimental arrangement. A small puff of smoke is put into a smoke cell under the microscope. Light from a small bulb shines on to the cell. The smoke particles show up as tiny specks of light through the microscope. Figure 5 shows the sort of path a smoke particle follows.

The smoke particles move randomly because they are always being knocked by air molecules. These air molecules are too small to see under the microscope, but they are moving so quickly that they can deflect a much larger smoke particle.

Figure 5
The path of a smoke particle as seen through the microscope

Questions

1 The diagram shows two bromine molecules moving in a jar full of air. Both molecules reach the top after a while.
(a) Copy the diagram and complete the paths of A and B, to show how they reached the top.
(b) Explain why bromine molecules move in this way.
(c) Further up the jar is molecule C. Will it reach the top before A and B?

2 When molecules are warmed up they travel faster. Explain what effect a high temperature would have on
(a) the motion of smoke particles in a smoke cell.
(b) the rate at which gases diffuse.
3 Below, you can see some data collected in a series of experiments on diffusion. An approximate time for different gases to diffuse through a distance of 10 cm is recorded. The temperature of the gases in all experiments was 20°C. The purpose of the experiments was to study how the

rate of diffusion of a molecule depends on its mass.
(a) Plot a graph of the time of diffusion (*y*-axis) against the relative mass of the molecule (*x*-axis).
(b) Ammonia has a relative molecular mass of 17. Use your graph to predict how long, on average, ammonia molecules will take to diffuse 10 cm through air.
(c) What conclusion can you draw about the relationship between speed and mass of a gas molecule?

Substance	Relative molecular mass	Time to diffuse 10 cm through air (s)
Hydrogen	2	30
Carbon dioxide	44	160
Chlorine	71	200
Bromine	160	300
Iodine vapour	254	380

3 The Kinetic Theory of Matter

Water appears in different forms . . . Ice

. . . Water

When something is moving it has **kinetic energy**. In the last unit you met the idea that molecules are always on the move.

This idea is called the **kinetic theory**. The most important points in the theory are:

- Every kind of material is made of small particles (molecules or atoms).
- The sizes of particles are different for different materials.
- The particles themselves are very hard. They cannot be squashed or stretched, but the distance between particles can change.
- The particles are always moving. The higher the temperature of a substance, the faster its particles move.
- At the same temperature all particles have the same energy. So heavy particles will move slowly and lighter particles will move quickly.

The last point helps to explain why hydrogen can diffuse through air much more quickly than bromine.

Solids, liquids and gases

Ice, water and steam are three different states of the same material. We call these three states solid, liquid and gas. You will now see how kinetic theory helps us to understand them.

- **Solid**. In a solid the particles are packed into rows just like apples or oranges stacked in shops (Figure 1(a)). The particles can never move out of their rows, but they vibrate around their fixed positions. As the temperature of a solid is raised the particles vibrate more and the material expands. This means that the distance between particles has increased a little.

  In a solid the particles are very close to each other and held in position by very strong forces. This makes it very difficult to change the shape of a solid by squashing it or pulling it.

- **Liquid**. In a liquid the particles are still very close to each other, which means that liquids are also very difficult to compress (Figure 1(b)). The particles can now move around from place to place. A liquid can change shape to fit into any container, but its volume will remain constant at a given temperature. As a liquid warms it also expands a little.

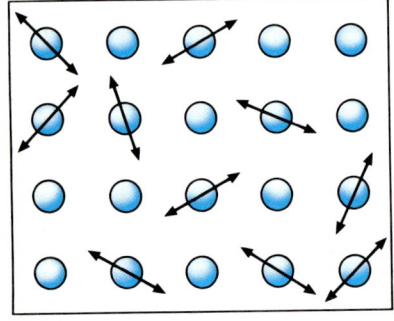

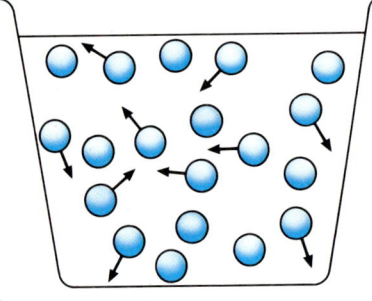

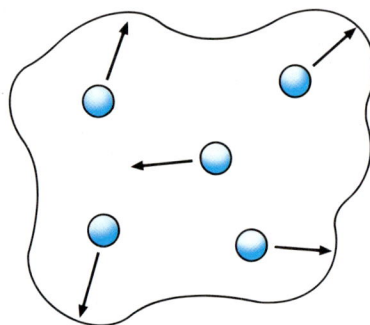

Figure 1

(a) *Particles in a solid are arranged in neat rows. The particles can vibrate but do not leave their positions in the rows*

(b) *Particles in a liquid are close together, but free to move around*

(c) *Gas molecules are free to move into any available space. They move quickly and there are large distances between them*

- **Gas**. In a gas the particles are separated by big distances (Figure 1(c)). The forces between gas particles are very small. It is easy to compress a gas because there is so much space between the particles. The gas particles are in a constant state of rapid random motion. This makes a gas expand to fill any available space.

Gas pressure

A gas in a container exerts a pressure on the container walls. This is because the gas particles are always hitting the walls. The pressure that the gas causes depends on:
- The number of collisions that the particles have with the container walls per second.
- How hard the particles hit the walls.

The pressure inside a container of gas can be increased in three ways:
- Putting more molecules in the container. The number of collisions that the molecules make with the walls each second is now larger.
- Making the volume of the container smaller. The same number of molecules make more collisions with the walls, because they travel less distance between collisions.
- Heating the container. The molecules travel faster and they now hit the container walls harder and more often.

. . . Vapour and steam. Geysers are formed when water deep in the Earth is heated under pressure. The water finds a route to the surface and boils to form a jet sometimes as much as 70 metres high

Questions

1 Use the kinetic theory to help you explain the following:
(a) Some solids, like metals, are very stiff and difficult to bend.
(b) A liquid can be poured.
(c) If a gas pipe has a small hole in it, gas will escape from the pipe and you will be able to smell it a few metres away from the leak.

2 The data in the table below shows how the pressure in a gas cylinder varies with the mass of the cylinder.

Mass of cylinder (kg)	Pressure in the cylinder (Pa × 10^5)
8.7	1.5
9.0	2.7
9.6	5.0
10.2	7.3

(a) Plot a graph of pressure (y-axis) against mass (x-axis).
(b) Use the graph to determine the mass of the empty gas cylinder.

(c) The greatest pressure the cylinder can take safely is 20×10^5 Pa. What will the mass of the cylinder be then?

3 Diagram *(i)* shows a box with two molecules in it. One molecule, X, moves parallel to the line BC. The molecule bounces backwards and forwards and hits the pink wall 100 times a second. The other molecule bounces backwards and forwards and hits the blue wall 100 times a second. The volume of the box is now halved by pushing in the pink wall as shown in diagram *(ii)*.
(a) How many times a second does X hit the pink wall now?
(b) How many times a second does Y hit the blue wall now?
(c) (i) Suppose the box is filled with a lot of molecules. When the volume is halved as shown in diagram *(ii)* the pressure acting on the pink wall of the box doubles. Explain why. (ii) What happens to the pressure acting on the blue wall?

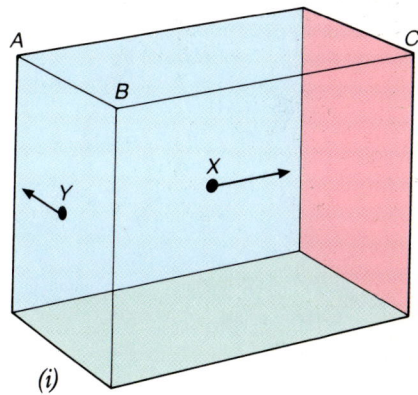

(i)

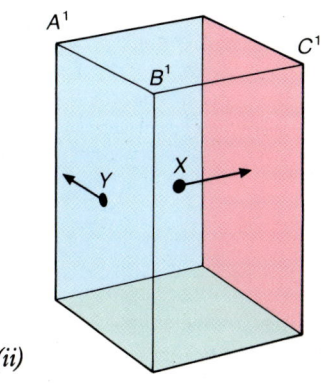

(ii)

4 Melting and Evaporation

When water freezes it expands, sometimes causing pipes to burst

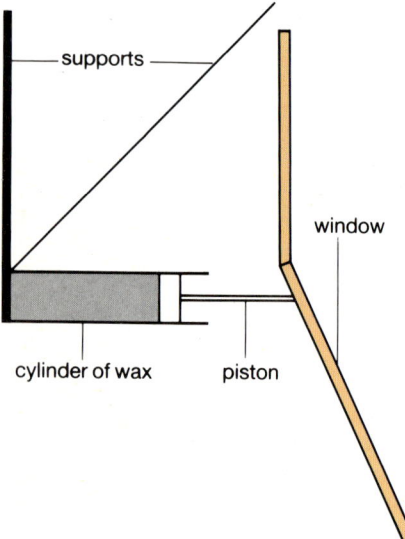

Figure 2
Automatic window opener

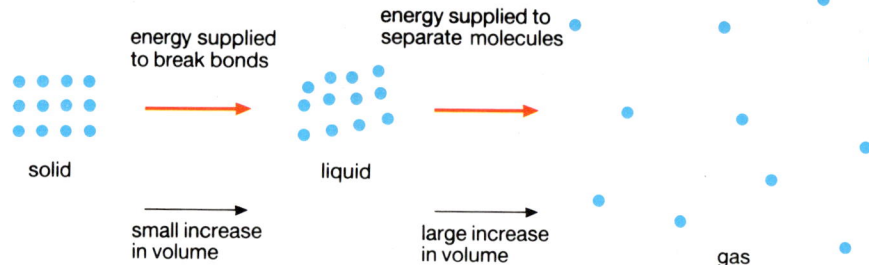

Figure 1

Changing solid to liquid

When a solid is heated the molecules inside it vibrate more and more quickly. If enough heat is supplied the molecules will break away from their fixed positions and start to move around. The solid has melted. The temperature at which this change happens is called the **melting point** of the material (Figure 1).

Most materials expand when they melt, because molecules are a little further apart in the liquid state. Figure 2 shows how we can use this in a greenhouse. A cylinder is filled with wax that melts at about 25°C. When air in the greenhouse reaches this temperature the wax melts and expands. This pushes the window open. This automatic window opener prevents damage to crops due to overheating.

Water is an unusual substance because it expands when it freezes (see Figure 3 Water is at its most dense at 4°C, so when a pond freezes over there is a layer of dense water at 4°C at the bottom. Ice is less dense than water which is why we see ice on the surface of ponds.

Changing the melting point

The expansion of water on freezing can be a nuisance. It causes pipes to burst in houses or in cars. To prevent freezing in car cooling systems, drivers add antifreeze. This lowers the melting point of water. When salt is added to water the melting point is lowered. This is why salt is put onto our roads in winter.

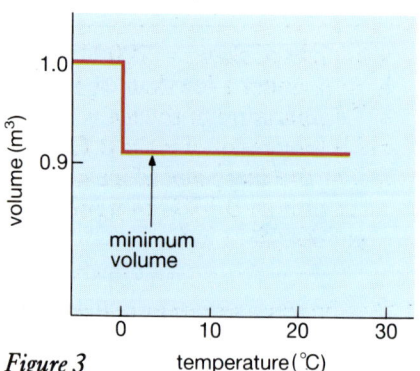

Figure 3
Volume change on freezing for water

This road gritter is used in icy conditions to make roads less slippery

The melting point of ice can also be lowered a little by applying very large pressures to it. An extra pressure equal to that of the atmosphere will lower the melting point of ice by about 0.01°C.

Evaporation

As a liquid warms up, the average speed of the molecules in it gets larger. However, not all of the molecules in the liquid will be travelling at the same speed. Figure 4 shows that some will be travelling slowly and some more quickly. Some molecules near to the surface of the liquid have enough energy to escape. They evaporate from the liquid to form a vapour. Evaporation from the surface of a liquid can happen at any temperature. But as the temperature gets higher more molecules have enough energy to escape. So evaporation happens at a faster rate.

Eventually the temperature rises to the **boiling point** of the liquid. At this point, evaporation also happens inside the liquid. Bubbles of vapour form inside the liquid and rise to the surface. Heat applied to a boiling liquid gives the molecules enough energy to evaporate.

Evaporation causes cooling. This is because it is the faster (hotter) molecules that escape, leaving behind the slower (colder) molecules. The evaporation of sweat from your skin keeps you cool on a hot day. But the evaporation of water from us can also be very dangerous. You can lose heat from your body very rapidly in wet and windy conditions.

This wine cooler has been soaked in water before the wine bottle was placed in it. Evaporation of the water through the porous pot takes heat away from the wine and keeps it cool

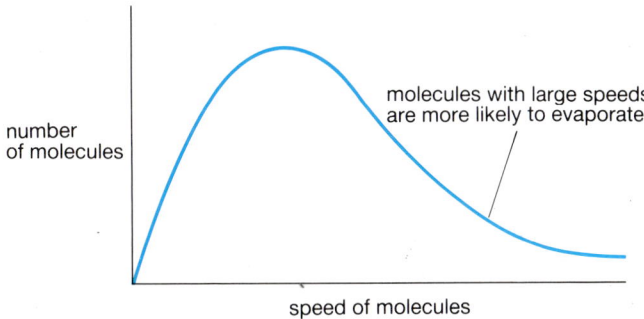

number of molecules

molecules with large speeds are more likely to evaporate

speed of molecules

Figure 4
The molecules in a liquid do not all travel at the same speed

With the lid in place this coffee would stay hot longer. Why is this?

Questions

1 (a) Why does washing dry faster on warm, windy days?
(b) Why do you hang out washing, rather than leaving it in a pile to dry?
(c) Why does the amount of water in the atmosphere (humidity) affect the drying rate?

2 Use Figure 3 to help you sketch a graph of the way in which the density of water changes as it freezes.

3 When you go ice skating a layer of water forms between the bottom of your skate and the ice. This layer of water is important for reducing friction, so that you can skate quickly. Two theories have been suggested to explain how this water forms:

I When pressure is applied to ice its melting point is lowered. So the pressure from an ice skate will melt some ice.

II When the skate moves over the ice, heat is generated due to frictional forces.

Use the following data to help you decide which theory is more likely.

- Weight of skater = 700 N
- Area under 1 ice skate = 10^{-3} m^2
- Temperature of the ice = −5°C
- Melting point of ice = 0°C
- The melting point of ice will be lowered by 0.01°C by a pressure of 100 000 N/m^2

4 Explain why a saucepan will come to the boil more quickly if it is covered by a lid.

5 Physics in the Kitchen

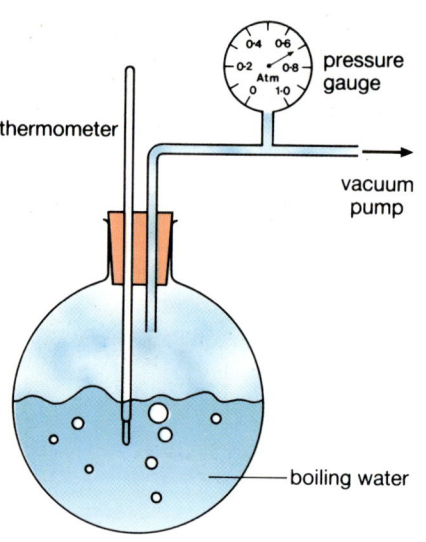

Figure 1
Experiment to see how pressure changes the boiling point of water

Changing boiling points

When water boils, bubbles of water vapour form inside the liquid. The pressure inside these bubbles is equal to the pressure of the air above the water. So when the air pressure changes the water boils at a different temperature. If the air pressure is greater than 1 atmosphere, water boils above 100°C; if the pressure is lower than 1 atmosphere, water boils below 100°C.

Figure 1 shows an experiment that you can do to see how pressure changes the boiling point of water. You put some water in the flask and heat it until it is boiling at 100°C. Then you stop heating and use a pump to reduce the pressure. As the pressure gets less the water keeps boiling at lower and lower temperatures.

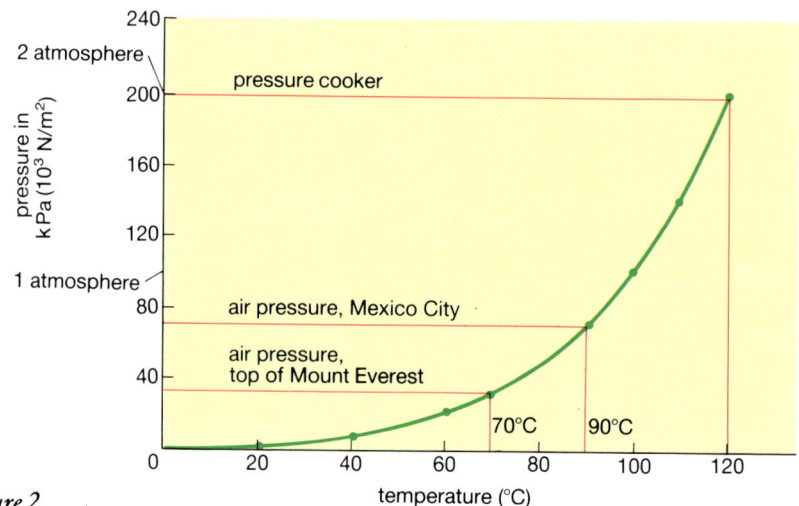

Figure 2
Graph showing how the boiling point of water changes with pressure

Figure 2 shows a graph of how the boiling point of water changes with pressure. You would notice this effect if you lived somewhere like Mexico City which is 3000 m above sea level. The air pressure there is only about 0.7 atmospheres, and water boils at 90°C. At the top of Mount Everest, 9000 m above sea level, a kettle would boil at only 70°C.

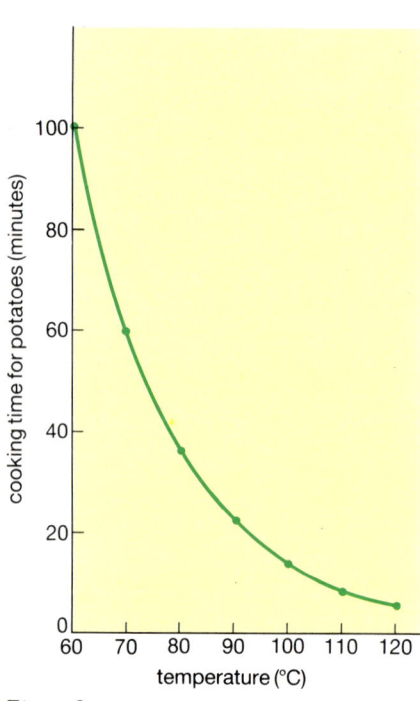

Figure 3
Graph to show how the cooking time for potatoes decreases as the temperature is raised

As the air pressure is lower up a mountain, water boils at a lower temperature and food takes longer to cook

In a pressure cooker the air pressure is increased to about 2 atmospheres. The cooker has an airtight lid except for a small hole at the top. Weights are put on top of this hole so that the air pressure inside must be greater than atmospheric pressure before steam will escape. The advantage of cooking at high pressure is that the boiling point of water is raised and cooking times are considerably reduced (Figure 3, Figure 4).

The boiling point of water can also be raised by adding some impurities. The water in a car radiator boils at a temperature above 100°C partly because of the antifreeze that has been put in it but also because it is pressurised. Salt water also has a slightly higher boiling point than fresh water.

These vegetables can be cooked very quickly in the pressure cooker

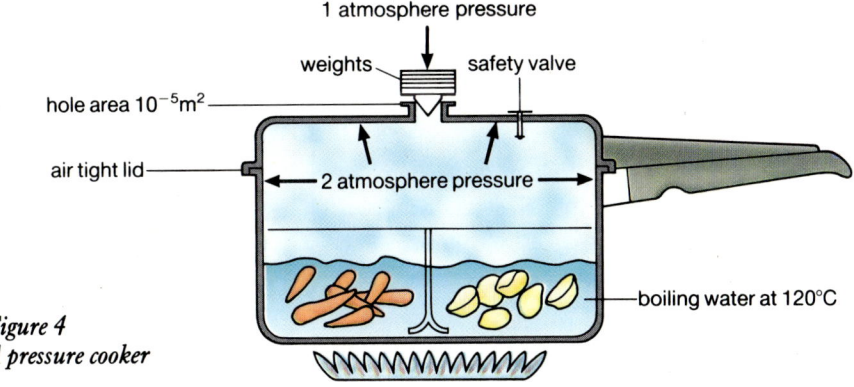

Figure 4
A pressure cooker

Refrigeration

Figure 5 shows a convincing demonstration of how evaporation can cause cooling. Ether has a low boiling point which means that it is fairly easy to evaporate. When air is blown through ether it evaporates very quickly. The ether molecules carry away heat energy with them and rapid cooling occurs. The water surrounding the ether soon turns to ice. The refrigerator in your kitchen works using a similar idea.

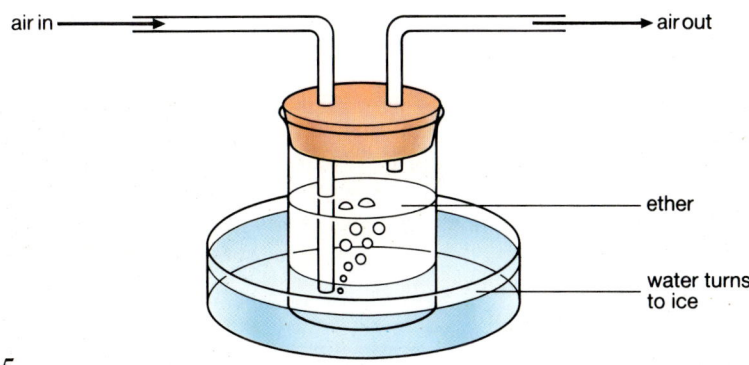

Figure 5
Cooling by evaporation

Figure 6 on the next page shows the important points of a refrigerator. The cooling in a refrigerator is done by chemicals called freons, which boil at low temperatures (about −30°C). The cooling occurs in the ice box where boiling freons evaporate. The freon gas then reaches the compressor which squashes the gas to a high pressure. The compression also warms the gas. The gas turns back into liquid freon as it cools.

The cooling fins on the back of a fridge. Warm freon is pumped through the tubes which pass through the fins. As it circulates, the freon loses heat and cools down

On the back of your refrigerator you will see cooling fins which cool the liquid back to room temperature. Then the freon is allowed to expand. When the pressure is low, the freon starts to boil and its temperature falls. The freon now passes through the ice box to start the cooling cycle again.

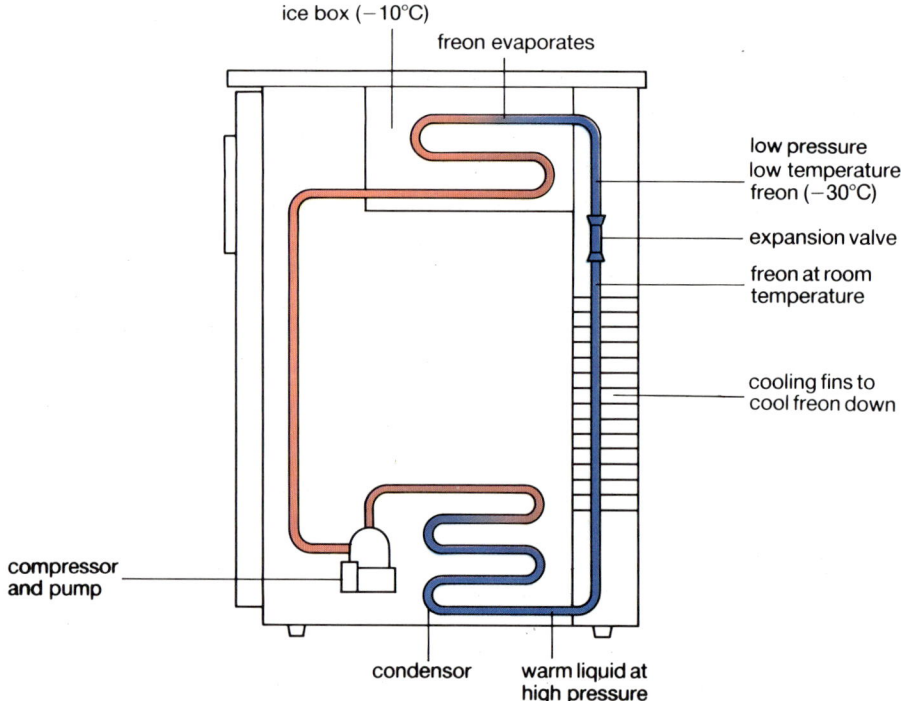

Figure 6
A refrigerator

Questions

1 When salt is added to water, its boiling point is increased. You often add salt to water when you cook vegetables. Does this make the vegetables cook more slowly or more quickly?

2 (a) The area of the hole in the top of the pressure cooker in Figure 4 is 10^{-5} m. What weight must be put on top of the hole to keep the pressure inside the cooker at 2 atmospheres (2×10^5 N/m^2)? Remember the pressure outside is 1 atmosphere.
(b) What is the pressure inside the cooker when a weight of 0.5 N is used?

3 This question is about the time taken to cook your potatoes as you climb Mount Everest. The higher you get the longer it takes.

(a) The table below shows you how the air pressure drops as you climb up the mountain. Copy the table and then use Figure 2 to fill in the column of boiling points.
(b) Use Figure 3 to fill in the column for cooking times.
(c) Now plot a graph to show how your cooking times (y-axis) change with your height above sea level (x-axis).
(d) Use your graph to predict cooking times at (i) 5000 m (ii) 8500 m.
(e) Would your cooking times have changed so much if you used a pressure cooker to cook your potatoes? Explain your answer carefully.

Height above sea level (m)	Air pressure kPa (10 N/m^2)	Boiling point of water (°C)	Cooking time for potatoes (minutes)
0	100	100	14
2000	79		
4000	62		
6000	47		
8000	36		

6 Internal Combustion Engines

Petrol engines

If you put your finger over the end of a bicycle pump, and pump the cylinder up and down a few times, you find that the air inside the cylinder gets hot. As the piston moves it collides with moving molecules. The molecules rebound off the moving piston at a greater speed. This means the gas gets hotter. The pressure inside the bicycle pump cylinder increases for two reasons. First the air has been compressed, and then this compression warms the air up. If you compress the air slowly the temperature will stay the same because heat will escape through the sides of the cylinder (Figure 1).

Figure 2 shows how the idea is used in a **four-stroke petrol engine**.

(1) On the **intake** stroke a petrol vapour and air mixture is fed into the cylinder.

(2) On the **compression** stroke both the inlet and exhaust valves are closed. The piston moves up rapidly to compress the air/petrol mixture to about $\frac{1}{10}$ of its original volume; the *compression ratio* is 10:1. The pressure inside the cylinder will now be about 20 atmospheres (2 MPa) and the temperature about 350°C.

(3) The **working** stroke. Once the gas has been compressed the sparking plug produces a spark which starts the fuel burning. The temperature in the cylinder now rises to about 1000°C. The pressure now rises to about 50 atmospheres (5 MPa). The higher pressure now forces the piston back. Although energy is used to squash the gas during the compression stroke, the working stroke gives out far more energy.

(4) In the **exhaust** stroke, the exhaust valve opens to allow the high pressure gases to escape.

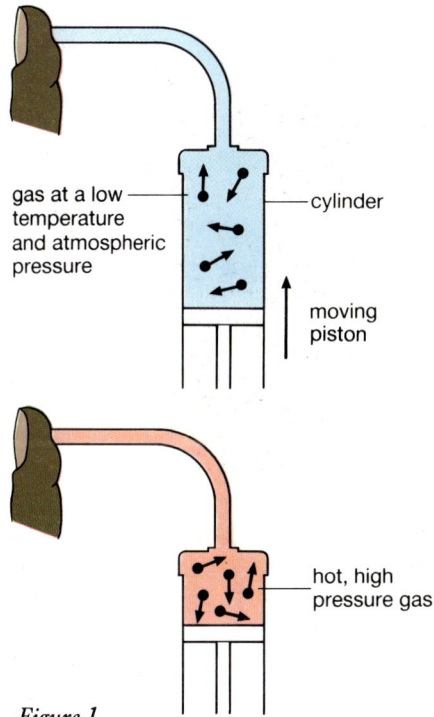

Figure 1
Heating air up in a bicycle pump

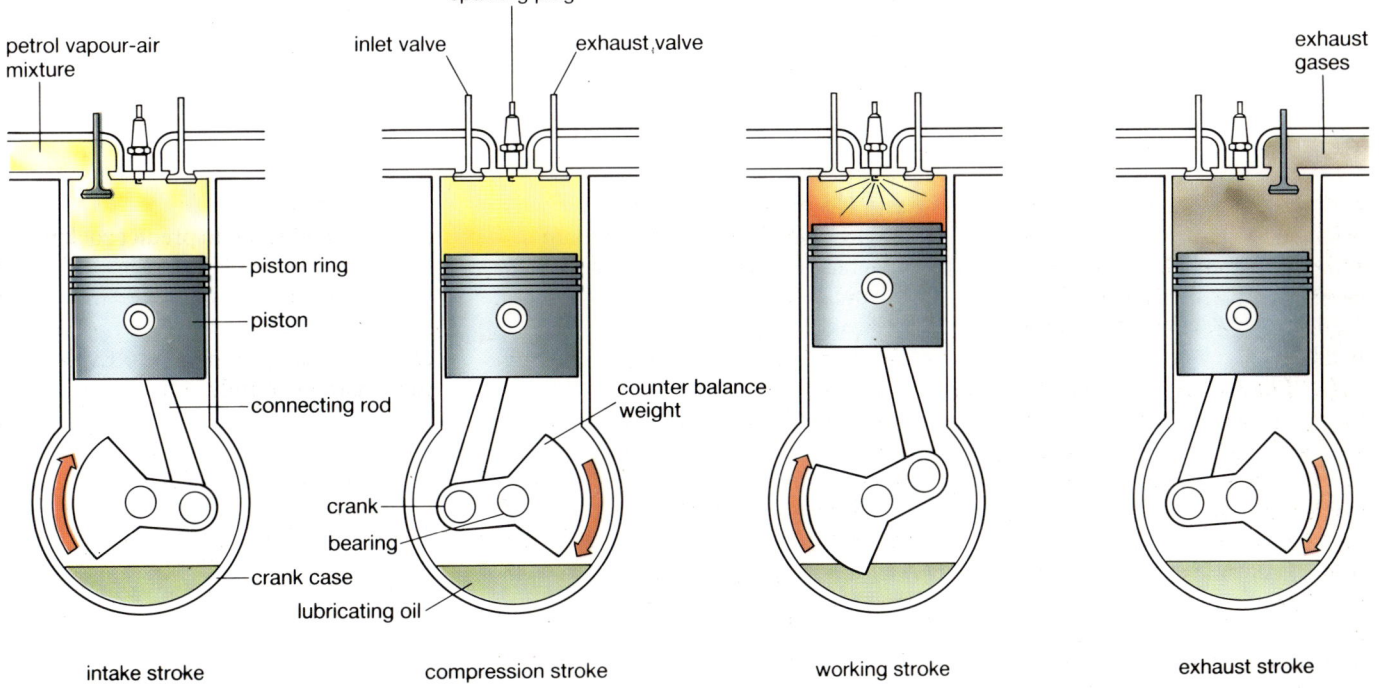

Figure 2
The four-stroke petrol engine

Most cars have four **cylinders** using a four-stroke cycle. The cylinders are arranged so that they take turns in performing one of the four strokes (Figure 3). At any instant, one of the cylinders will be producing power. A car with only one cylinder would produce a very jerky ride.

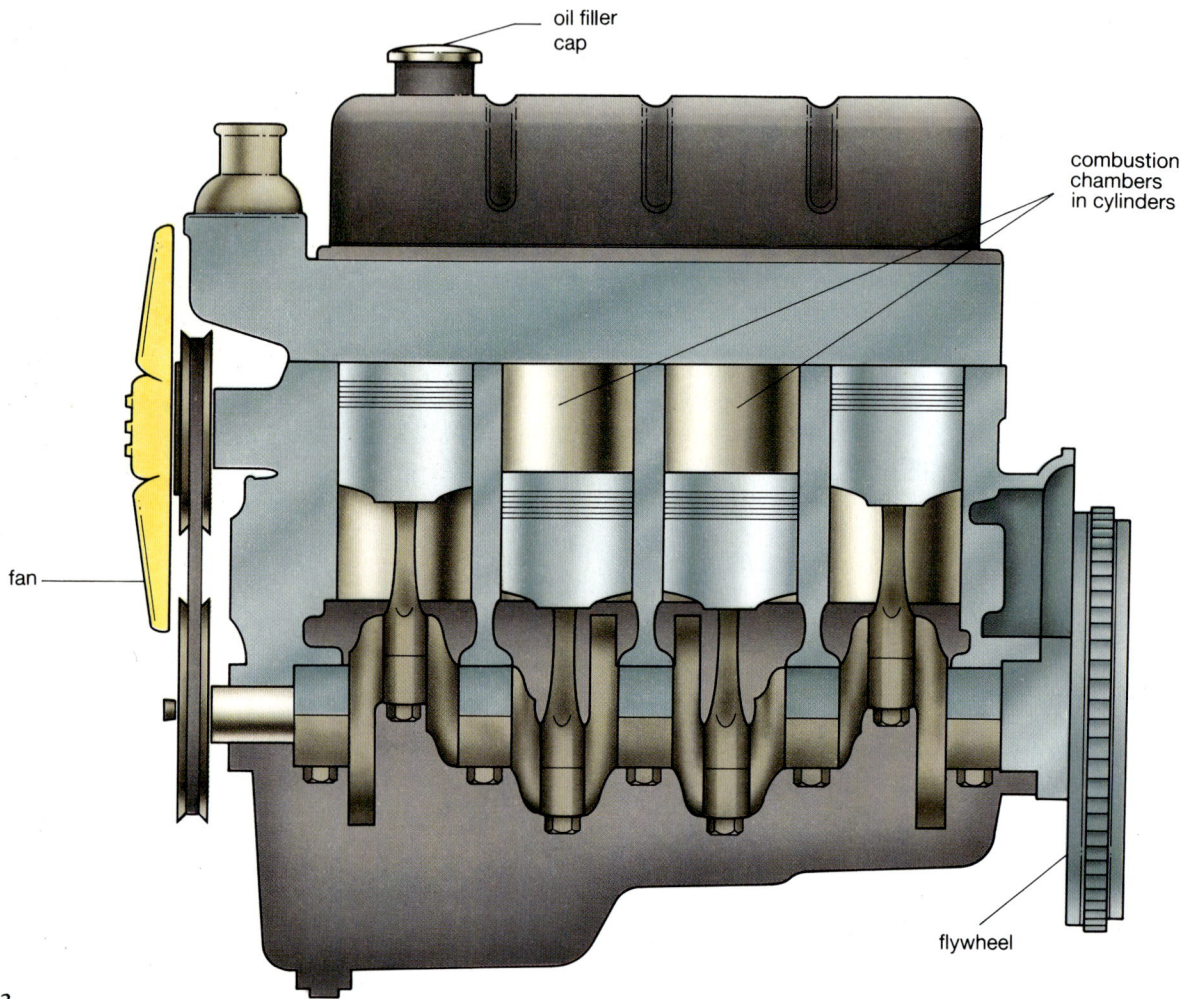

oil filler cap

combustion chambers in cylinders

fan

flywheel

Figure 3
A car engine cut away to show its four cylinders

Diesel engines

A four-stroke diesel engine works in a similar way to the petrol engine you have just read about. The differences are these:
- During the intake stroke only air is taken into the cylinder.
- The compression ratio of diesel engines is higher than that of petrol engines. At the end of the compression stroke the air has been squashed to about $\frac{1}{20}$ of its original volume. At this point the temperature of the air is about 700°C, and its pressure is about 35 atmospheres (3.5 MPa).
- The diesel engine has no sparking plug. At the end of the compression stroke fuel is forced into the cylinder under high pressure. The temperature of the air is so hot that the fuel burns as soon as it has mixed into the air. When the fuel burns, the temperature inside the cylinder reaches about 2500°C and the pressure is about 100 atmospheres (10 MPa)
- Such an engine using diesel oil has a higher **efficiency** than a petrol engine.

Figure 4 shows a simple two-stroke diesel engine, which could be used in a lawn mower. In this engine the intake, compression and exhaust strokes are all combined into one upward movement of the piston.

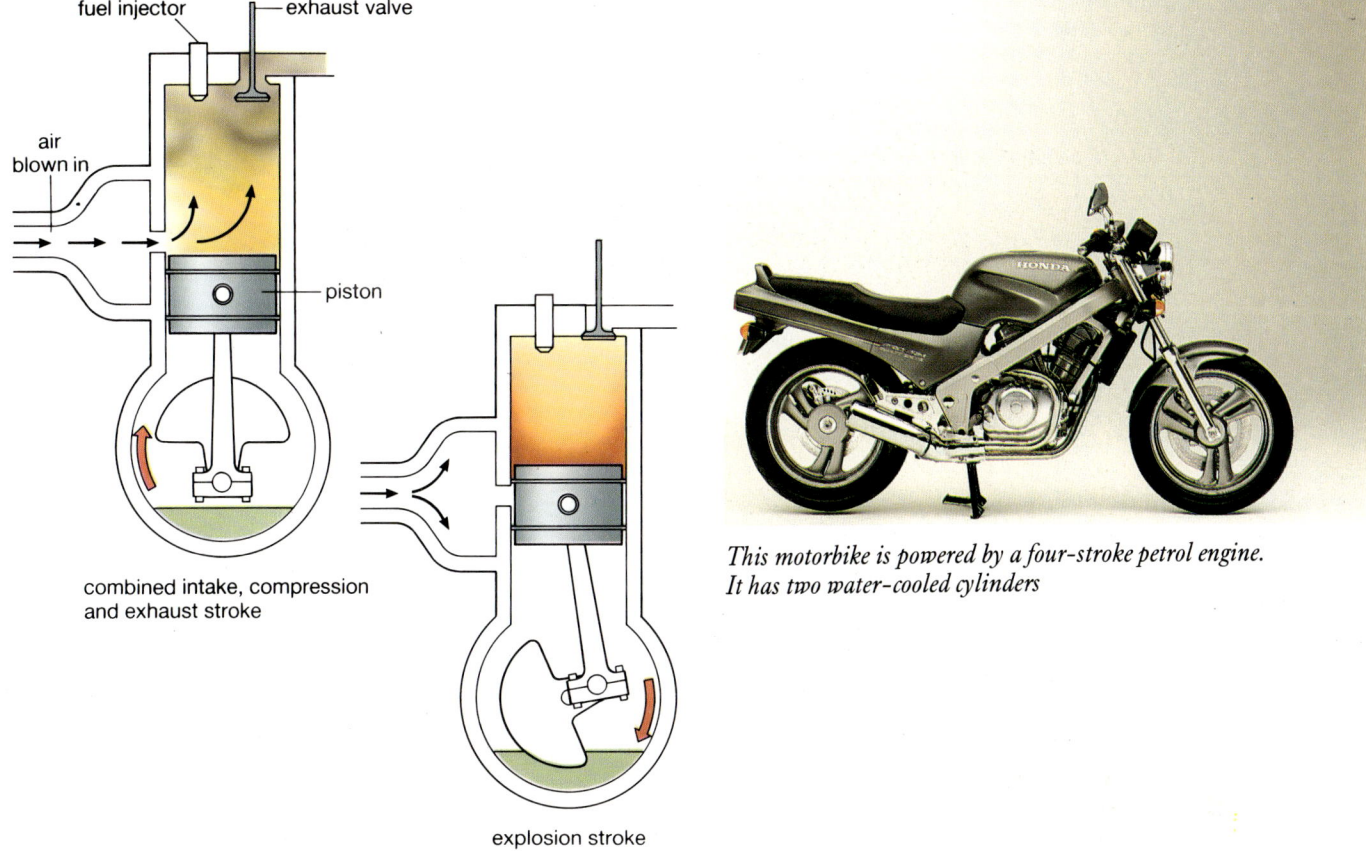

fuel injector — exhaust valve

air blown in

piston

combined intake, compression and exhaust stroke

explosion stroke

This motorbike is powered by a four-stroke petrol engine. It has two water-cooled cylinders

Figure 4
The two-stroke diesel engine

Questions

1 What is meant by the term 'compression ratio'?

2 At the beginning of the working stroke in a petrol engine the pressure in a cylinder is 5 MPa (5×10^6 N/m^2). The area of the piston is 0.008 m^2.
(a) Calculate the force acting on the piston.
(b) Explain why cylinders and pistons in a petrol engine have to be well built. Why do they have to be even stronger in a diesel engine?

3 Why does a diesel engine not have a sparking plug?

4 Why do most cars have four cylinders?

5 (a) You were told in the text that diesel engines have higher efficiencies than petrol engines. What does that mean?
(b) Use the data opposite to work out the cost of fuel to drive: (i) a petrol-driven car 10 000 km, (ii) a diesel-driven car 10 000 km.
(c) How far do you have to drive a diesel-fuelled car before you have recovered the extra cost of its engine?
(d) If you were buying a car which sort of engine would you choose?

- cost of diesel engine £1000
- cost of petrol engine £700
- 1 litre of petrol will take the car 10 km
- 1 litre of diesel oil will take the car 14 km
- petrol costs 40p per litre
- diesel costs 35p per litre
- cars with diesel engines are noisier and more sluggish than petrol driven cars

7 The Gas Laws

Pressure (atmospheres)	Volume of air (l)	$\frac{1}{V}(l^{-1})$
0.5	2.0	0.5
1.0	1.0	1.0
2.0	0.5	2.0
4.0	0.25	4.0
10.0	0.1	10.0

Table 1

Boyle's Law

Figure 1 shows a cylinder and piston. The pressure of the air in the cylinder can be changed by moving the piston up and down. Provided the piston is moved slowly the temperature of the gas stays the same. This apparatus can be used to see how the pressure, P, changes as the volume, V, of the air is increased or decreased.

Table 1 shows the sort of results you can get if you do the experiment. When the gas is compressed to half its volume the pressure is doubled. We say that the pressure of the gas is *inversely proportional* to its volume:

$$P \propto \frac{1}{V}$$

The table of results also shows that when you multiply together the pressure and volume, you always get the same number. (What is it?)

When the mass of a gas stays the same, and its temperature does not change:

$$\boxed{P \times V = \text{constant} \qquad \text{This is **Boyle's law**}}$$

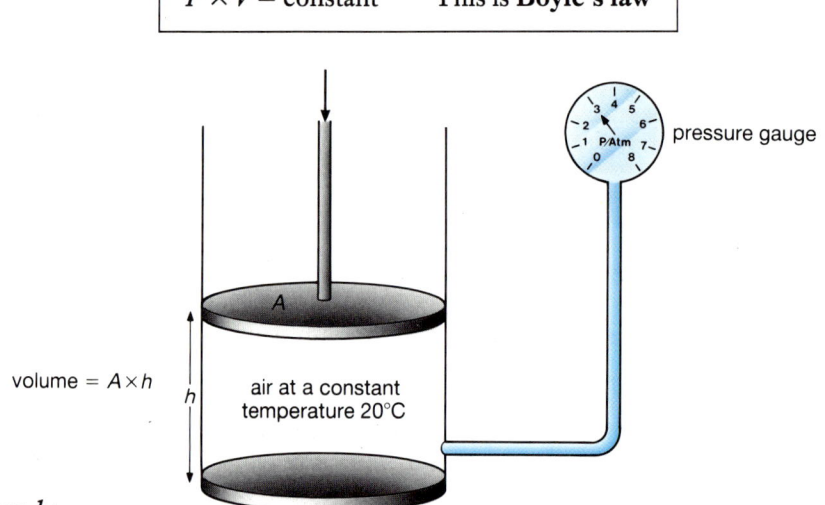

volume = $A \times h$

Figure 1
Testing a piston and cylinder

Law of pressures

Figure 3 shows you the same apparatus, placed in a large tank of water. This time the volume of the gas is kept constant. The pressure is measured at a lot of different temperatures.

The results of such an experiment are shown in Figure 4. As the temperature, T, is raised from 0°C to 100°C the pressure rises steadily from 1 to about 1.4 atmospheres. We can draw a straight line through these results. If this line is extended below 0°C you can predict what will happen to the pressure at lower temperatures. You can see that the pressure will be about 0.5 atmospheres at −136°C. At a temperature of −273°C the pressure due to the gas is nothing.

A gas exerts a pressure on the walls of its container because of its moving molecules. If a gas no longer exerts a pressure it is because the molecules have stopped moving. This means that −273°C is the lowest possible temperature. At that temperature molecules have stopped moving so we cannot cool them any further. We call −273°C **absolute zero**.

Figure 2
As the volume of the gas is halved, the pressure doubles

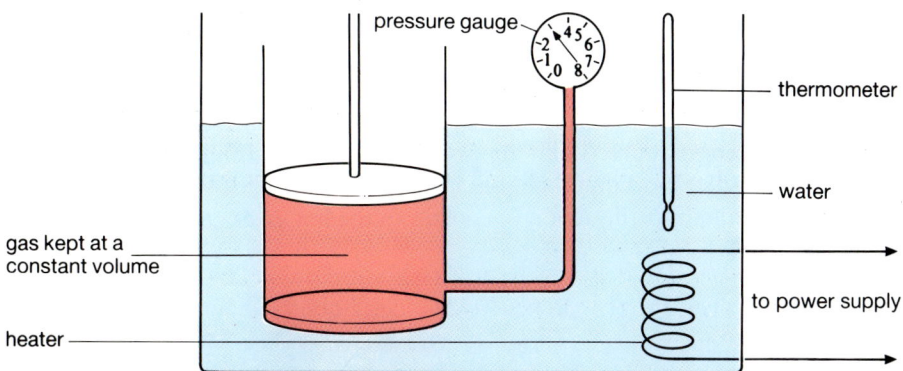

Figure 3

Absolute temperatures are measured in *degrees Kelvin*, or K. Absolute zero is 0 K and 0°C is 273 K. The size of 1 degree Kelvin is the same size as 1°C. Notice that no degree sign is used with degree**s** Kelvin. We write 77 K, *not* 77°K.

Figure 4 shows that the pressure of a gas is proportional to its absolute temperature, provided its volume stays the same:

$$P \propto T \qquad \text{This is the } \textbf{law of pressures}$$

Charles' Law

The apparatus shown in Figure 3 can be used for a third experiment. The pressure of the gas is now kept constant. The gas is allowed to expand when it is heated.

Experimental results show that the volume of the gas is proportional to its absolute temperature, provided its pressure remains the same:

$$V \propto T \qquad \text{This is } \textbf{Charles' law.}$$

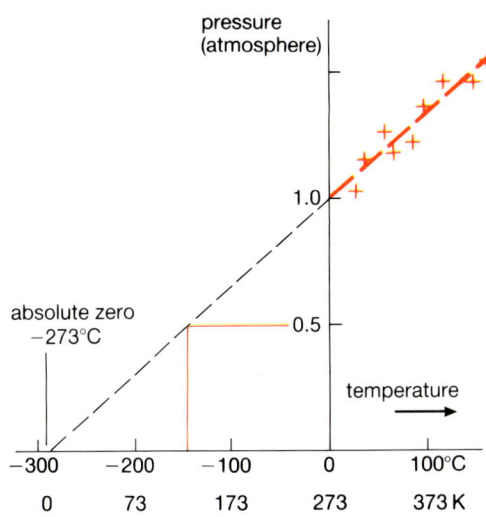

Figure 4
Results of an experiment using the apparatus of Figure 3

Questions

1 This question refers to the experiment shown in Figure 1.
(a) Use Table 1 or Figure 2 to work out the volume of the air when the pressure was: (i) 3 atmospheres, (ii) 8 atmospheres
(b) Use the data to plot a graph of the pressure, P, against the reciprocal of the volume, $1/V$.
(c) A student took a measurement of the volume and pressure and recorded 0.16 litres and 7.0 atmospheres. Plot this point on your graph. The student made an error in measuring the pressure. Suggest what the value should have been.

2 This question refers to the experiment described in Figure 3.
(a) Use the graph in Figure 4 to predict the pressure in the cylinder at:
(i) a temperature of 150 K,
(ii) a temperature of 150°C.
(b) What will the pressure in the cylinder be at 600 K?

3 The diagram (right) shows a gas storage vessel of volume 160 000 m³. The pressure in the main gas grid pipelines is about 800 kPa. In the storage vessel the pressure is just greater than 100 kPa (atmospheric pressure). Calculate the volume of gas from the grid pipelines required to fill the gas storage vessel.

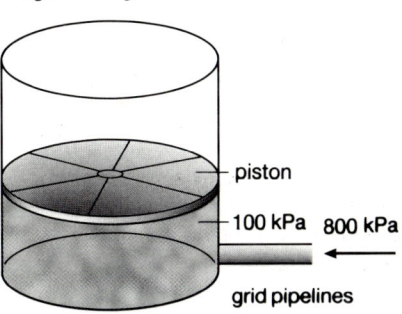

8 Liquid Gases

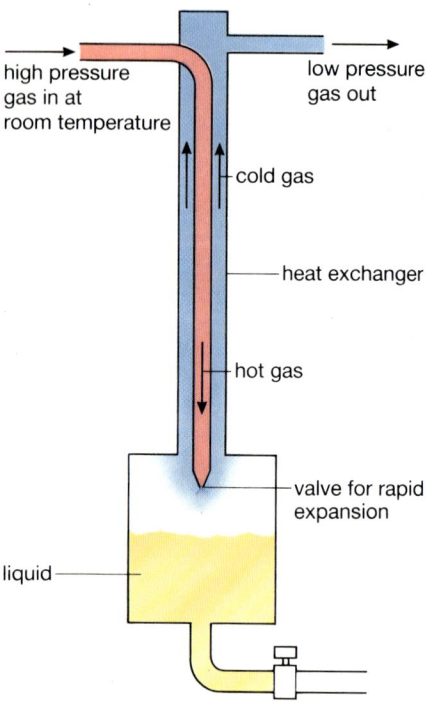

Figure 1
The principle of a gas liquefier

You learnt earlier that when gases in a cylinder are compressed rapidly, the temperature of the gases rises. If a gas is expanded very rapidly it cools down. This idea is used to turn gases like oxygen and nitrogen into liquids. Figure 1 shows how a **liquefier** works. First the gas is compressed and allowed to cool back to room temperature. Then the gas expands rapidly through a valve; it cools down so much that it becomes a liquid. A heat exchanger allows the cold gases, flowing out through the liquefier, to cool down the hot gases coming into the liquefier.

A spray of liquid nitrogen is used to freeze these cakes as they pass through the freezing tunnel on a conveyor belt

Sometimes liquefied gases are used to cool things down. However, the main reason for liquefying nitrogen and oxygen is that large amounts of them can be transported in a small volume. Oxygen becomes a liquid at −183°C; 1 litre of liquid oxygen at that temperature turns into 600 litres of gas at room temperature.

Figure 2 shows how the volume change occurs. When the liquid turns to gas there is a large expansion, and 1 litre of liquid oxygen turns to 200 litres of gas. But as the gas warms up, at constant pressure, there is a further increase in volume. In warming from 90 K (−183°C) to 270 K (−3°C) the gas expands from 200 litres to 600 litres. Liquid gases are stored or transported in special thermos or vacuum flasks (see page 125), which provide the best means of keeping something cold. A tanker for transporting liquefied gases has to be very strongly built to make sure that damage cannot occur to the vacuum flask.

Use of gases

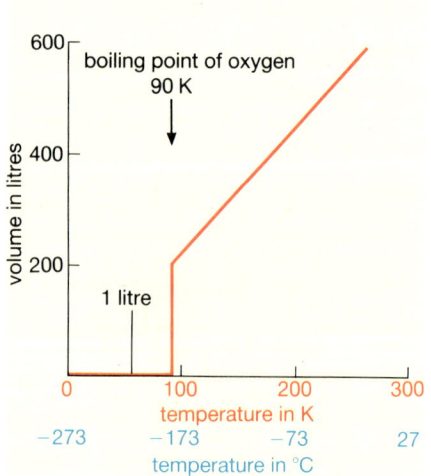

Figure 2
The expansion of oxygen when the pressure is kept constant at 1 atmosphere

- **Oxygen.** One of the most important uses of oxygen is in hospitals. When patients are seriously ill it is sometimes necessary for them to breath air with extra oxygen in it. Liquid oxygen is stored in a vacuum-insulated tank and then oxygen as a gas is distributed around the hospital, through pipes, to the patient's bedside. Liquid oxygen is also stored in readiness for use in important industrial processes such as the production of iron and steel.

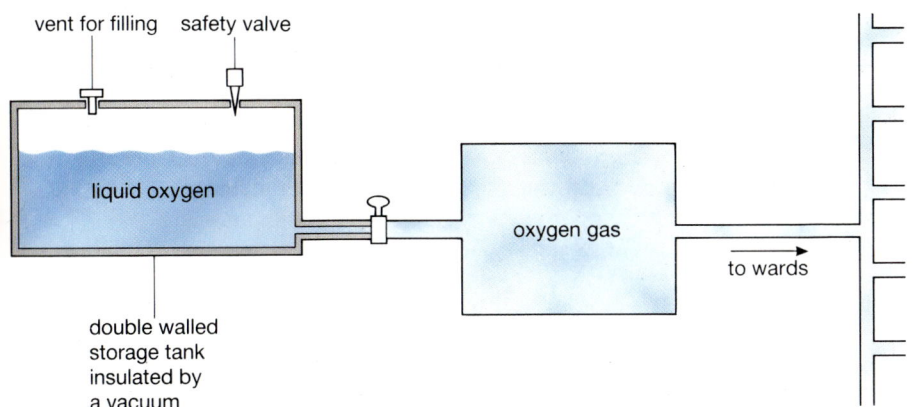

Figure 3
Oxygen storage and distribution for a hospital

Gas	Boiling point	
	°C	K
Oxygen	−183	90
Argon	−186	87
Nitrogen	−196	77
Neon	−246	27
Hydrogen	−253	20
Helium	−269	4

Table 1 Boiling points of 'liquid' gases.

- **Nitrogen.** Liquid nitrogen is widely used for rapid freezing. The frozen peas you buy have probably been frozen in liquid nitrogen. When something is frozen in liquid nitrogen it freezes so quickly that its structure is well-preserved. This is why hospitals use liquid nitrogen for freezing and preserving blood.

   Nitrogen in gas form also has many uses. Nitrogen is a gas that does not burn. It is used as a 'blanket' to prevent fires. When a cargo of oil is loaded into a tanker, the oil is covered with nitrogen, so if a spark occurs a major explosion is prevented. Nitrogen also keeps food fresh. You all know how nasty stale bread is. It is oxygen that makes things go stale. Foods like crisps and cereals are kept fresh by putting nitrogen into the packets.

- **Helium** boils at 4 K. Helium is usually only of use in research laboratories. Some materials such as niobium and tin become superconducting at liquid helium temperatures. A **superconductor** has no electrical resistance and can carry a large current without getting hot. Superconductors can be used for making very strong magnetic fields. Research is being done now to try to find materials that are superconducting at room temperature.

Questions

1 Explain why gases do not obey Charles' law at temperatures near absolute zero.

2 You are the administrator for a large hospital. You are faced with the problem of supplying oxygen to the intensive therapy unit (ITU).
(a) Use the data on the clipboard below to calculate the maximum volume of oxygen that patients might breathe in one day.
(b) Use the information in Figure 2 to calculate the volume of liquid oxygen needed to supply this amount of oxygen gas to the patients.
(c) Suggest what size of storage vessel would be suitable for the hospital's liquid oxygen. A delivery of oxygen is made only once a week.

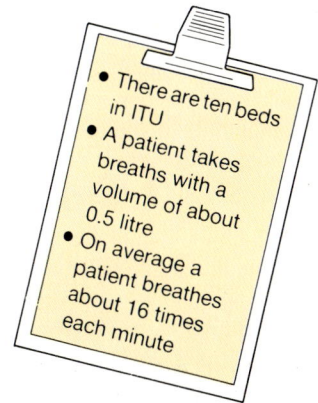

- There are ten beds in ITU
- A patient takes breaths with a volume of about 0.5 litre
- On average a patient breathes about 16 times each minute

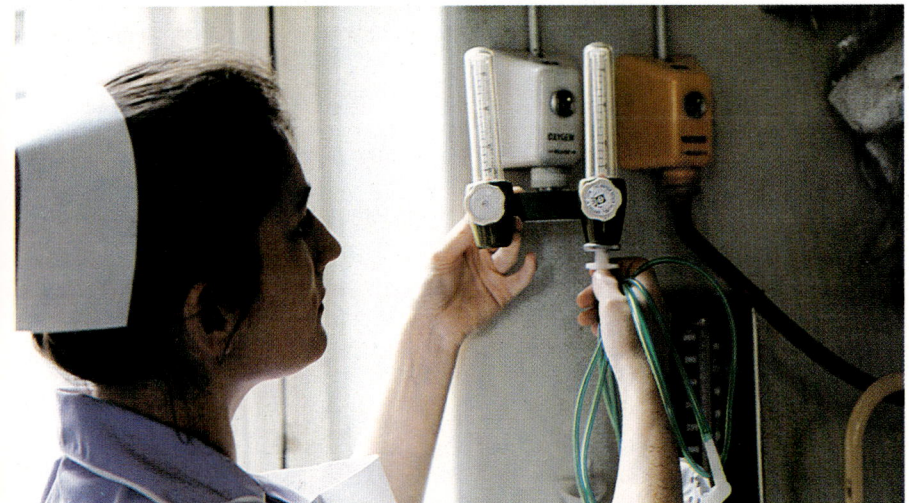

A nurse attaches a double regulator to the wall-mounted oxygen supply in a hospital ward. In this way oxygen from one outlet can be given to two patients

SECTION E: STUDY QUESTIONS

1 Smoke particles in air are seen through a microscope. The smoke particles move and make frequent changes in direction because they
A repel each other.
B attract each other.
C are able to move themselves.
D are colliding with each other.
E are colliding with invisible air particles.

MEG

2 (a) In this diagram the bottom circle represents the molecules in the liquid. Copy and complete the top circle to represent the molecules in the gas.

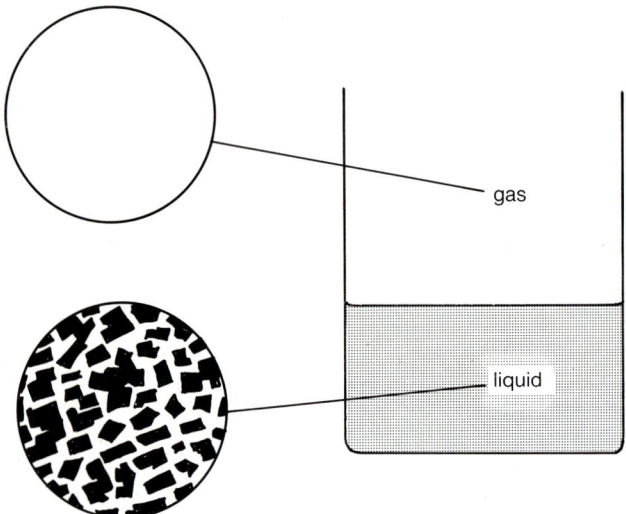

(b) With reference to the diagram explain why gases can be compressed easily but liquids cannot.
(c) Brownian motion is usually shown in a school laboratory by looking at smoke specks under a microscope.
(i) Describe the motion of the smoke specks.
(ii) What causes this motion?

LEAG

3 A drop of oil, of volume 0.4 mm³, was placed on a water surface and spread out to a maximum area of 1200 mm². Assuming the oil patch was one molecule thick, what is the best estimate of the diameter of an oil molecule, in mm?
A 0.00004   B 0.00008   C 0.000012
D 0.00033   E 0.00048

NEA

4 Which one of the following will cause the boiling point of water to rise?
A Reducing the pressure over the surface of the water.
B Dissolving some salt in the water.
C Boiling the water faster by increasing the supply of heat.

D Using an immersion heater instead of an external supply of heat.
E Boiling the water at a greater height above sea level.

NEA

5 The Kinetic Theory assumes that gases are made up of rapidly moving molecules. Which of the following facts does **not** support the idea that gas molecules are moving?
A A tiny smoke particle undergoes 'Brownian Motion' when in a gas.
B Gases exert a pressure on the walls of their containers.
C Gases have lower densities than solids and liquids.
D When placed together, two gases will diffuse into each other.
E When released into an empty container, a gas will rapidly fill it.

MEG

6 The two strokes of a four-stroke engine cycle during which only **one** value is open are
A induction and compression.
B compression and exhaust.
C induction and exhaust.
D compression and power.

SEG

7 When a drop of ether is put onto the skin, the skin feels cold. This is because
A liquids like ether are poor conductors of heat.
B ether is a good conductor of heat.
C the ether is an anaesthetic.
D the ether evaporates by taking heat from the skin.
E heat cannot reach the skin through the ether.

NEA

8 A sealed can contains air. If the can is heated the pressure of the air inside the can increases. Which of the following statements best describes why the pressure increases?
A When heated, air molecules expand and press harder on the can.
B More air molecules are created and so more molecules press on the can.
C When heated, air molecules slow down and collide less frequently with the can.
D Air molecules are attracted to the heated sides of the can.
E When heated, air molecules move faster and collide with the can more often.

LEAG

9 There are practically no attractive forces between the molecules in
A a solid.
B a liquid.
C a gas.
D both a liquid and a gas.
E both a solid and a gas.

NEA

10 The diagram shows Boris in a diving bell. He is just about to explore the depths of the black lagoon. As the bell is lowered into the sea, water rises to fill the bell.
(a) Explain why water starts to fill the bell as it is lowered.
(b) the pressure below the surface of the sea is given by the formula: $P = (100\,000 + 10\,000\,d)$ N/m^2; d is the depth below the surface in metres. What is the pressure of air in the diving bell at a depth of 10 m?
(c) How deep can the bell go, before Boris gets his feet wet?

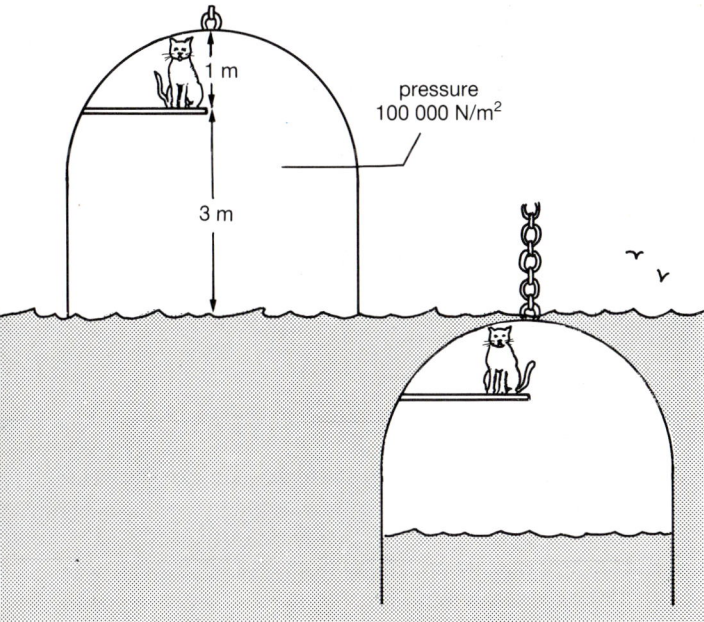

pressure
100 000 N/m^2

1 m

3 m

11 A petrol engine repeats the same sequence of four strokes continuously. Each stroke involves the following parts of the engine: *inlet valve*, *outlet valve*, and the *piston*. The *spark plug* produces the spark needed to ignite (set fire to) the fuel. The diagrams on page 99 show the sequence of the four strokes. Use them to help you answer the questions which follow.

(a) Copy and complete the following table to show the sequence of events during the four strokes.

Stroke number	1	2	3	4
Direction piston is moving (*up* or *down*)	down			up
Inlet valve position (*open* or *closed*)		closed	closed	
Outlet valve position (*open* or *closed*)		closed	closed	

(b) (i) In which stroke does air and petrol *enter* the cylinder?
(ii) In which stroke do the exhaust gases *leave* the cylinder?
(c) (i) Copy this flow chart. Complete the boxes to show the main types of energy transferred in the engine.

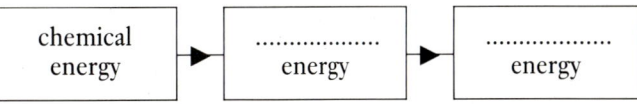

chemical energy → .... energy → .... energy

(ii) Suggest what type of energy provides the spark.
(d) (i) Explain why air is needed as part of the fuel mixture.
(ii) Explain why it is important that the fuel and air are well mixed.
(e) (i) Petrol is a mixture of compounds. Most of these compounds contain the elements carbon and hydrogen and are called hydrocarbons. Suggest two substances which are likely to be produced when a hydrocarbon burns in oxygen.
(ii) What process in living organisms is similar to this reaction?

(f) A diesel engine has the same moving parts as a petrol engine but differs in the way it works. Some of these differences are given in this table.

	Type of engine	
	Petrol	Diesel
fuel	petrol	diesel
greatest pressure just before fuel ignites	about 900 KPa	about 2200 KPa
spark needed	yes	no
efficiency	24%	40%

(i) In both types of engine the temperature of the gas mixture in the cylinder increases as the piston moves upwards with both valves closed. Use your understanding of particles to explain why this increase in temperature happens.
(ii) In which type of engine will this increase in temperature be greater?
(iii) Suggest and explain one important difference between the general structure of the two engines.
(iv) Explain what is meant by saying that the petrol engine is *24% efficient.*
(v) Explain what happens to the remaining 76% of the energy.

SEG

12 Which substance is a liquid at room temperature?

substance	melting point (°C)	boiling point (°C)
A	−218	−183
B	−39	357
C	44	280
D	119	444
E	1038	2336

MEG

13 The diagram shows some bubbles rising in a glass of a fizzy drink. Can you explain why the bubbles get larger as they rise to the surface?

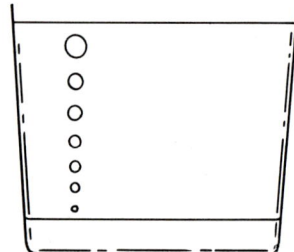

14 The picture below shows part of a chemical works in which a tanker is being filled with a chemical. The chemical is normally a gas. It is invisible, has no smell and is heavier than air. It is a non-ionic compound which is non-flammable and non-toxic.

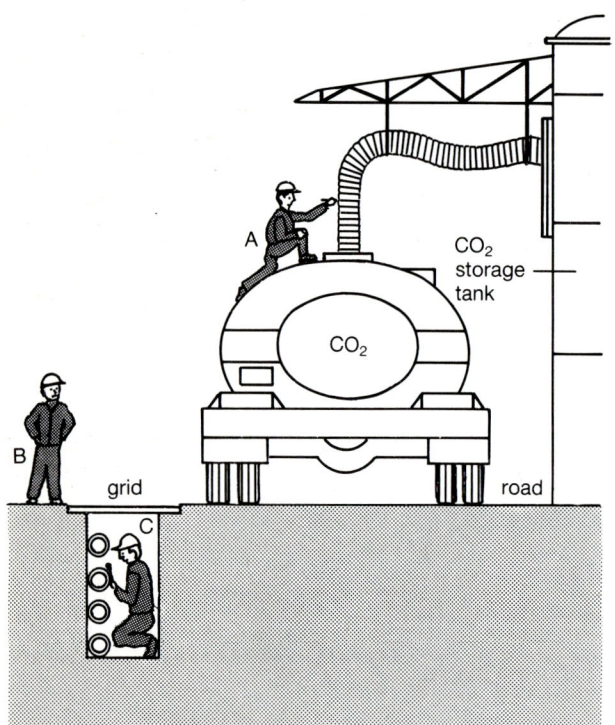

(a) What is the *name* of the chemical?
(b) If there is a leak of the chemical, which worker will be *most* affected? Give **two** reasons for your answer.
(c) The same chemical can be used in fire extinguishers designed to put out fires involving live electrical equipment.
(i) Suggest and explain **one** reason for this.
(ii) Explain why water is unsuitable for this type of fire.
(d) Although the chemical in the tank is normally a gas, it is stored and transported as a liquid. This is because each kilogram of the chemical takes up far less space as a liquid.
(i) Explain why the liquid chemical takes up less space.
(ii) Explain why it may be an advantage to store and transport the chemical as a liquid.
(iii) Suggest and explain one method by which the gas might be turned into a liquid.
(e) Suggest **two** safety features that you would make part of the tanker's design. Give reasons for your choice.

SEG

15 A sealed test tube contains air at atmospheric pressure and room temperature. When it is surrounded by ice, which of the following statements about the air in the test tube is **not** correct?
A The pressure becomes lower.
B The average kinetic energy of the molecules becomes lower.
C The average momentum of the molecules becomes less.
D The molecules collide less often with the test tube.
E The average number of molecules in each cm^3 becomes less.

MEG

16 (a) Explain in terms of molecules, why a gas exerts a pressure on the walls of its container.
(b) The air inside an old syrup tin is heated with a bunsen burner. The lid is pressed firmly on. Explain why the pressure in the tin increases.
(c) If the tin is heated enough the lid flies off. Explain why.
(d) The experiment is now repeated, but with a small amount of water placed in the tin. Explain why the lid flies off at a lower temperature.

17 In the diagram below, you can see the planet Mars in an elliptical orbit around the Sun. At its closest (position A) Mars is about 200 million km away from the Sun. At its most distant (position B) Mars is about 250 million km away from the Sun.

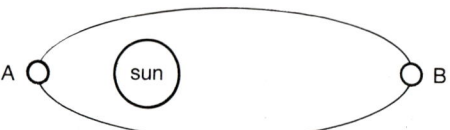

Mars has a thin atmosphere which is mostly carbon dioxide. The pressure of the atmosphere changes from winter to summer. The table below shows some details of the Martian climate. The data was taken by the Viking Lander, near to the equator in 1977. Can you explain why the pressure of the Martian atmosphere changes so much?

	Average daily temperature (°C)	Average pressure (N/m^2)
Position B (winter)	−73	50
Position A (summer)	−13	65

SECTION F

Heat

Solar furnaces like this one at Odeillo in France, are power stations that can produce up to 10 megawatts of electricity using energy from the Sun. To find out how they work, turn to page 124

1 Hot and Cold

The Namib desert: daytime temperatures of 58°C have been recorded here in midsummer. However, at night, temperatures can reach freezing point. This range of temperature makes the desert a very hostile environment for living things

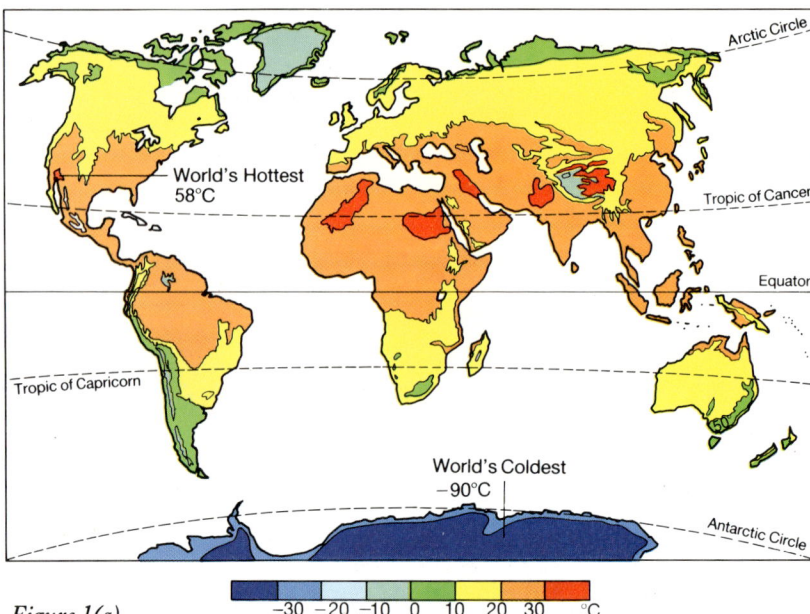

Figure 1(a)
Mean temperature in July

The Antarctic: during the continuous darkness of mid-winter, the temperature here can be as low as −87°C

Everybody knows the meanings of the words hot and cold. If you pick up a hot plate you burn your fingers. On a cold winter's day when you go outside without gloves, your fingers freeze. When you get too hot or too cold life becomes very unpleasant. Your body works normally at a temperature of about 37°C. You will become very ill if your body temperature rises as high as 40°C or falls as low as 34°C.

We are only comfortable wearing light clothing when the air temperature is around 10°C to 20°C. The maps in Figures 1a) and b) show the world's average temperatures in January and July. As you can see there are many places where it is too hot or too cold. In winter the average temperature in Verkhoyansk (Siberia) is below −40°C; the lowest temperature recorded there is about −70°C. People only survive the winter by staying indoors for months at a time. If you lived in Ethiopia you would face a different problem. The average yearly temperature in some parts of Ethiopia is about 35°C. In July the midday temperature can reach 50°C. Under such scorching conditions there is little water and food is hard to grow. Millions of Ethiopians have died over the past few years through shortage of food and water.

Life can also be uncomfortable in other places. In Bombay, you might prefer to live in an air-conditioned flat. In Madrid, where midday temperatures reach 45°C in the summer, banks open from 9.00 am−12 noon and then again from 4.30 pm−6.00 pm. Most Spaniards have a siesta in the middle of the day, simply because it is too hot to be working outside.

Heat and temperature

It is a very common mistake to confuse the two words heat and temperature. When we talk about **heat** we mean the heat energy that we have had to put into a material to warm it up. Heat energy is measured in **joules**. The word **temperature** is used to describe how hot something is. Temperatures are usually measured in **degrees celsius** (°C).

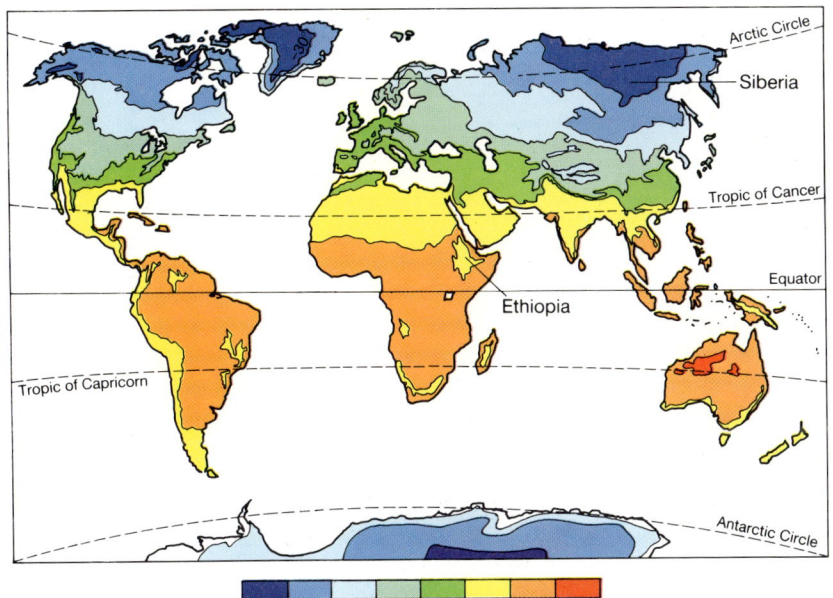

Figure 1(b)
Mean temperature in January

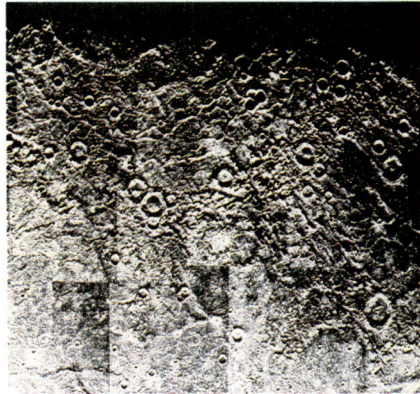

The solar system also has extremes of temperature. The daytime temperature on Mercury is about 330°C — hot enough to melt lead!

On Uranus, the daytime temperature is a cool −216°C

Peter is a plumber and he uses a soldering iron which has a mass of 0.6 kg. Elsie is an electrician and she uses a soldering iron which has a mass of 0.1 kg. They each plug their irons in and wait for them to warm up. After one minute Peter's iron has been given 20 000 J of heat and it has reached a temperature of 50°C. In that time Elsie's iron has been given 10 000 J of heat and it has reached a temperature of 100°C. Peter's iron has more heat energy than Elsie's but Elsie's iron is at a higher temperature.

Figure 2 shows the molecules in the two irons after they have been warmed up. The length of the arrows in the diagrams shows how fast the molecules are moving. The temperature is a measure of the energy of *each* molecule. The heat energy that you have to supply to warm an iron up is a measure of the energies of *all* of the molecules.

Questions

1 Use the maps of the world to help you answer the following questions.
(a) Why do very few people live in:
(i) Greenland, (ii) the centre of Australia?
(b) Which part of the world has the greatest difference in average temperatures, between January and July?
(c) What other factors, besides temperature, affect how many people live in a country?
2 Which will cause the worse burn: (i) a pan of hot water at 70°C falling on your foot, (ii) a spark from a bonfire at about 500°C landing on your hand? Explain your answer.
3 Criticise this statement: 'The heat of the inside of the Sun is about 15 million degrees celsius'.

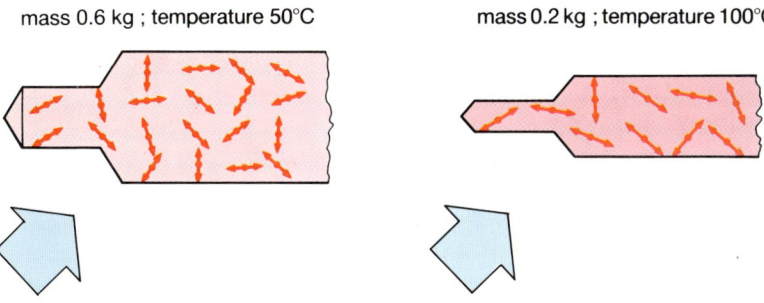

mass 0.6 kg ; temperature 50°C   mass 0.2 kg ; temperature 100°C

Figure 2

(a) 20 000 J of heat energy warmed up Peter's soldering iron to 50°C

(b) 10 000 J of heat energy warmed up Elsie's soldering iron to 100°C

2 Expansion of Solids

In the photograph above, you can see how the engineers who built the Humber Bridge overcame the problem of the bridge expanding in hot weather. When the bridge expands the plate on the left slides under the plate on the right, so that the surface of the bridge stays flat.

The molecules in solids and liquids are always vibrating. If the temperature falls, molecules vibrate less and they move closer together (Figure 1(a)). Then a substance will **contract**. When the temperature rises, molecules vibrate more and push each other apart. This causes substances to **expand** (Figure 1(b)). If you try to stop something expanding then very large forces may result. Sometimes these forces cause us serious problems but at other times they may be useful.

Expansion causing problems

In Saudi Arabia the temperature at midday in July can rise as high as 40°C. But on a winter's night in January, the temperature can drop below 0°C. These large temperature changes cause problems with oil pipelines. A long straight length of pipe would soon buckle under the forces caused by expansion. The problem is solved by putting a series of expansion loops into the line. These loops make the line more flexible, so that it is free to expand or contract as the temperature changes. Figure 2 shows how the expansion of overhead power lines is tackled.

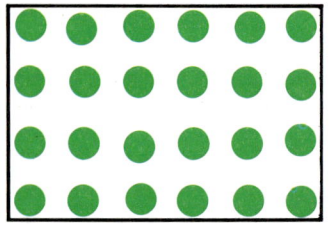

Figure 1
(a) Cold solid. Atoms vibrate slowly

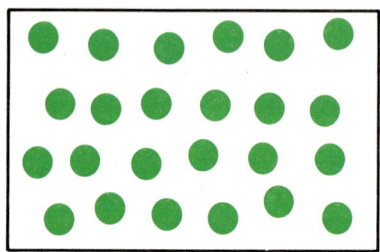

(b) Hot solid. Atoms vibrate quickly and push each other apart. Expansion results

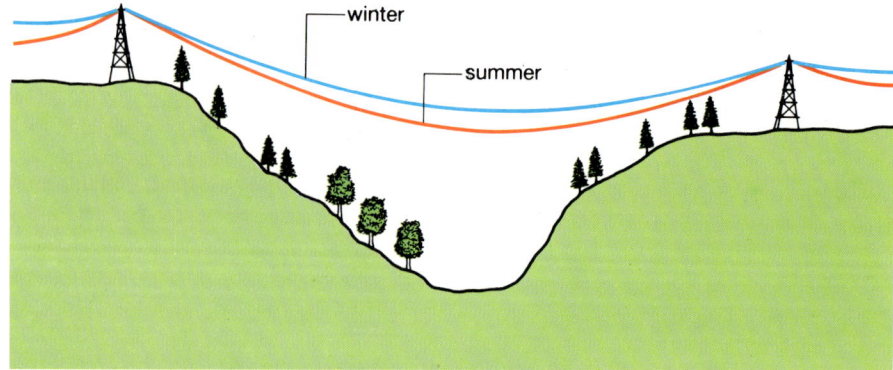

Figure 2
Engineers must allow for the expansion of overhead power lines. In summer the lines will be longer and they will sag more

The forces due to expansion have also caused British Rail some trouble. A few years ago the steel rails used to be laid in lengths of about 20 m. It was necessary to leave a gap between each rail to allow for expansion on hot days. On some lines you can hear the clickety-click sound as your coach rattles over the gaps.

Nowadays, most rails are laid in long lengths and welded together to form continuous tracks of several kilometres. The advantage of this is that the passengers get a smoother journey and the train suffers less wear. However, to avoid problems with expansion the rails have to be fixed to heavy concrete sleepers that can resist large forces.

Another way to avoid trouble is to stretch the rails as they are laid. This means that on a cold day the rails will be under tension. The rails will be compressed only on a very hot day. This can cause buckling. The rails can stand quite large tension (stretching) forces.

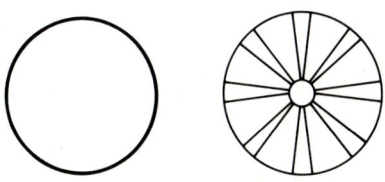
the tyre is too small when cold

the tyre is too big when hot

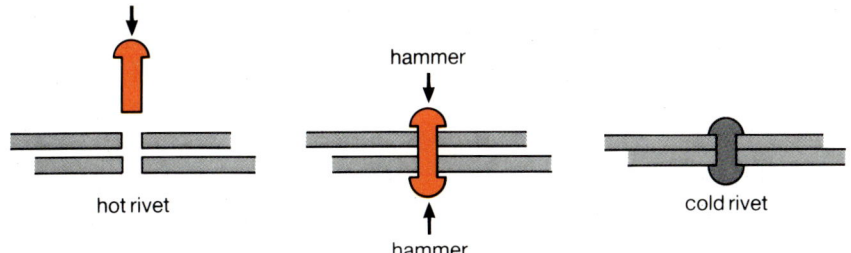

hot rivet

hammer

hammer

cold rivet

Figure 3
Expansion is useful in riveting. A hot rivet is hammered to join two plates together. When the rivet cools it contracts and pulls the two plates together tightly

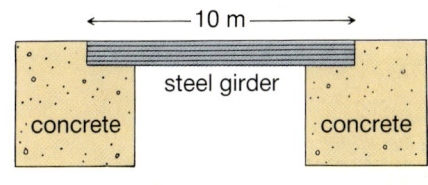
the tyre is a tight fit

Figure 4
British Rail engineers have problems with expansion, but they can also use it to their advantage. The steel tyre to be fitted to a wheel is too small when it is cold. By heating it up it will expand to fit around the wheel and will be a tight fit when cooled down again

Questions

1 (a) Explain why you sometimes find gaps between railway lines.
(b) Modern railway lines are laid in long stretches without gaps. What problems does that cause?
(c) What do the words tension and compression mean?
(d) Explain why it is safe to put a rail under tension but not under compression.
(e) Is it better to lay railway lines on a hot or a cold day? Explain your answer.
2 Look at the diagram of a bridge, and read the data in the box below it. The bridge was built in the middle of a cold Russian winter when the temperature was −20°C. The ends of the main steel girders were fixed in cement and no room was left for expansion.
(a) By how much does a steel rod of length 10 m expand when it warms up by (i) 1°C, (ii) 40°C?
(b) Explain why the girders in the bridge are under compression in a hot summer when the temperature is 20°C.
(c) Calculate the force required to compress the bridge girder by 1 mm.
(d) Now calculate the force compressing the girder in the summer when the temperature is 20°C.
(e) Explain what might happen to the girder or the concrete in the summer.

←——— 10 m ———→

steel girder

concrete   concrete

- A rod of steel 1 m long expands by 0.01 mm when it warms up by 1°C.
- A steel rod 10 m long with a cross-sectional area of 1 cm² needs a force of 2000 N to compress it by 1 mm.
- The cross-sectional area of the girder in the bridge is 100 cm².

3 Putting Expansion to Use

Material	Increase in length (mm) for 1 m heated through 1°C
aluminium	0.025
brass	0.019
iron	0.012
steel	0.011
glass (common)	0.009
glass (Pyrex)	0.003
invar (an alloy of nickel and steel)	0.001

Table 1

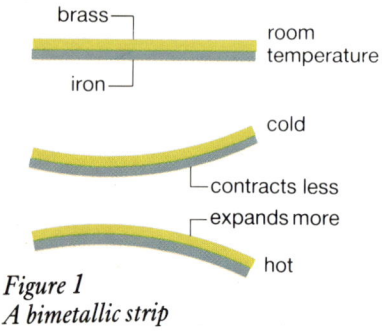

Figure 1
A bimetallic strip

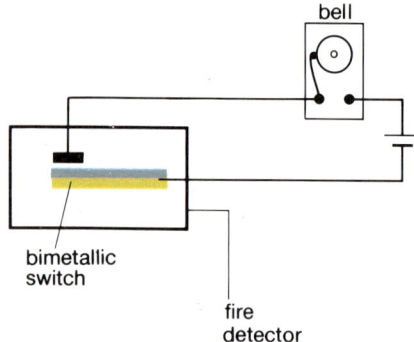

Figure 2
A switch for a fire alarm

Not all materials expand by the same amount when heated. Table 1 opposite shows you the increase in length of various materials when they are heated through 1°C. You can see that brass expands nearly twice as much as iron for the same temperature rise. Some materials such as invar and Pyrex glass hardly expand at all.

This difference in expansion between materials allows us to make a very useful device called a **bimetallic strip** (Figure 1). A strip is made out of two different metals such as iron and brass. The two metals are fixed together so that they cannot move separately. When the strip is cooled the brass contracts more than the iron so the strip bends upwards. When the strip is heated it bends the other way because the brass is now longer than the iron. This principle is used to make fire alarms and thermostats.

- **Fire alarm.** Figure 2 shows the principle behind a simple fire alarm. Inside the fire detector itself there is a bimetallic switch. When the bimetallic strip gets hot it bends upwards and completes an electrical circuit. A current now flows and makes a fire alarm sound.
- **Thermostat.** A bimetallic switch can also be used to keep the temperature of your clothes iron constant (Figure 3). When the iron is cold the switch is up and current flows to the heating element. However, when the iron is hot the bimetallic strip bends downwards and breaks the circuit. A control knob allows you to choose your iron's temperature; hot for cotton, cooler for nylon. The hotter you want the iron to be the more you turn the control knob downwards. This means that the bimetallic strip has to bend more before the circuit is broken, so it must be hotter.

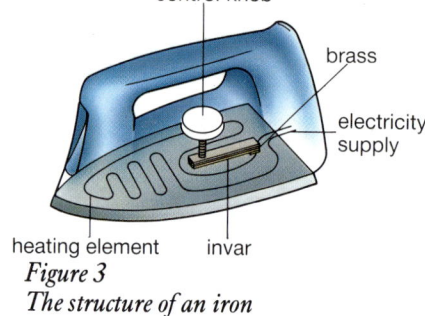

Figure 3
The structure of an iron

Thermometers

Liquids also expand when heated. We use the expansion of mercury to measure temperatures. You can see opposite a diagram of a **mercury thermometer**. The mercury is contained in a bulb at one end. When the bulb is warmed, the mercury expands into a thin capillary. The hotter the bulb is the further the mercury moves along the tube. We can read the temperature from the scale marked along the thermometer.

Fixing a scale

We make a temperature scale by choosing two fixed **reference points** (Figure 4). On the Celsius or centigrade scale the lower fixed point is the melting point of pure ice. The higher fixed point is the boiling point of pure water when the water is boiling under normal atmospheric pressure (760 mm Hg). The position of the mercury thread is marked when the thermometer is in the melting ice. This is defined as 0°C. The position of the thread is again marked when the thermometer is in the steam. This is defined as 100°C. The length between the

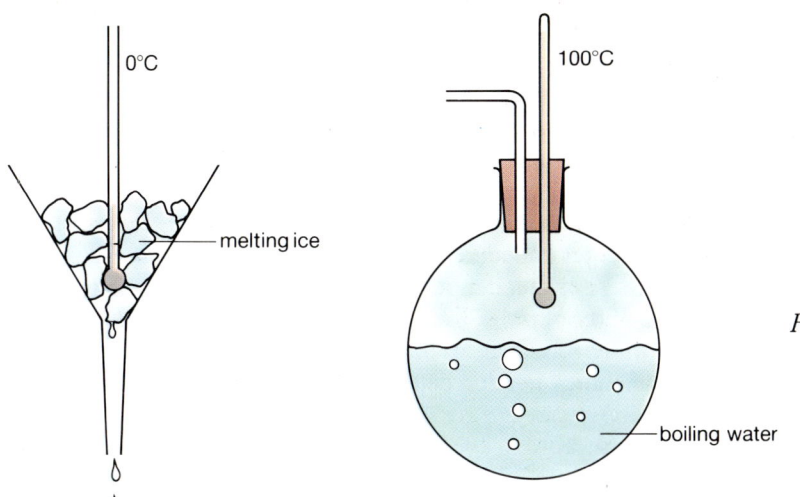

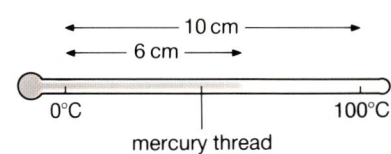

Figure 5

Figure 4
(a) The lower fixed point   *(b) The higher fixed point*

0°C mark and the 100°C mark can now be divided up. In Figure 5 the length of the scale between 0°C and 100°C is 10 cm; the mercury has moved 6 cm along the scale, so the temperature is 60°C.

The clinical thermometer

When the mercury in a thermometer cools it contracts, so the thread shrinks back down the thin tube. This is a nuisance if we want to take the temperature of a patient, because the temperature will always appear too low. To avoid this there is a small narrow kink in the tube. When the thermometer is taken out of the patient's mouth the mercury thread breaks at this constriction and we can then read the temperature (Figure 6).

Figure 6
Normal temperature for us is about 37°C This patient looks ill

Questions

1 Below is a diagram of a gas oven thermostat. When the oven is hot enough the valve controlling the gas supply closes. When the oven cools down the valve opens again.
(a) Explain how it works.
(b) How would you add a control knob to adjust the oven temperature?

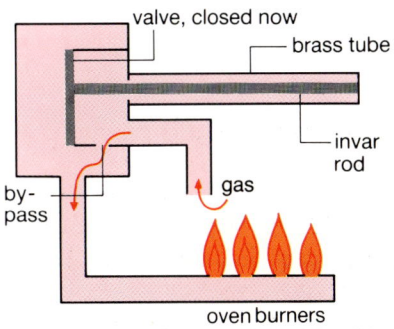

2 The graph below shows how the current flowing in the heating element of a clothes iron changes with time.
(a) Explain why it behaves as it does.
(b) Sketch a graph to show how the temperature of the iron changes with time.

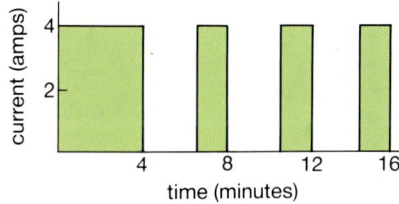

3 Which liquid would you use in a thermometer in the following places? Choose mercury or alcohol, and use the table in the next column to help you.
(a) The Antarctic.
(b) A desert.
(c) An oven.

	Mercury	Alcohol
boiling point (°C)	360	78
melting point (°C)	−37	−110

4 To make a thermometer which can show very small temperature changes it is necessary to have a large quantity of mercury in the bulb and a very narrow tube. A particular thermometer has a scale that reads 10°C per cm. How many °C per cm does it read if:
(a) the volume of mercury in the bulb is doubled
(b) the capillary tube is doubled in diameter?

4 Supplying Heat

Night storage heaters use concrete to store heat energy. The concrete is heated up at night when electricity is cheap and it then releases heat during the day.

When you want a cup of coffee you put some water in the kettle and switch it on. The water gets hotter because heat energy is being supplied which makes the molecules vibrate more rapidly. The hotter the water is, the faster the molecules move.

The amount of heat energy that must be supplied to warm up the kettle depends on two things:
● More energy is needed for a large temperature rise.
● More energy is needed if there is a lot of water in the kettle.
Warming 200 g of water from 20°C to 80°C needs 50 000 J of heat. Warming 200 g of wine from 20°C to 80°C (for making fondue) needs 40 000 J of heat. The amount of energy needed depends on the substance.

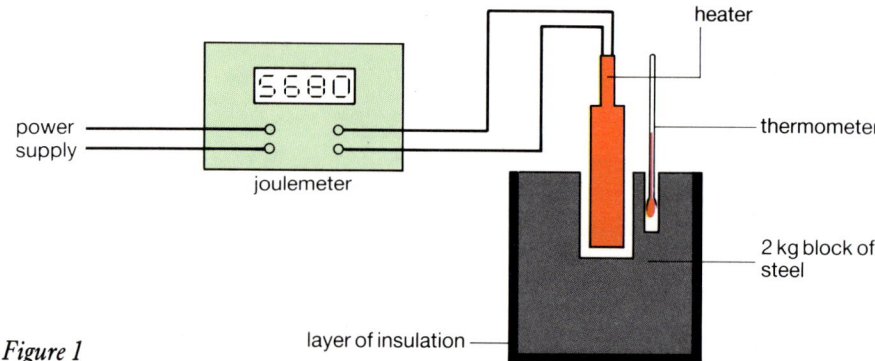

Figure 1
An experiment to measure the heat energy required to warm up a block of steel

Specific heat capacities

You can see in Figure 1 an experiment to measure the heat energy required to warm up a block of steel. A thermometer records the temperature rise, while a Joulemeter measures the number of joules that the heater has supplied. The block of steel is wrapped in thick insulating material to make sure that no heat can escape to the surrounding air. Figure 2 shows how the temperature of the block rises as heat energy is supplied.

The amount of heat required to warm 1 kg of a substance by 1°C is called its **specific heat capacity**. It is measured in units of **J/kg°C**. We can work out the specific heat capacity of steel like this. Figure 2 shows us that 18 kJ warmed the block from 20°C to 40°C.

$$18\,000\,\text{J warmed 2 kg by 20°C}$$

so $9000\,\text{J would warm 1 kg by 20°C}$

and $\dfrac{9000\,\text{J}}{20} = 450\,\text{J would warm 1 kg by 1°C}$

The specific heat capacity of steel is 450 J/kg°C. As you can see from table 1, different substances have very different specific heat capacities. In general you can work out the heat required to warm up a substance using the equation below.

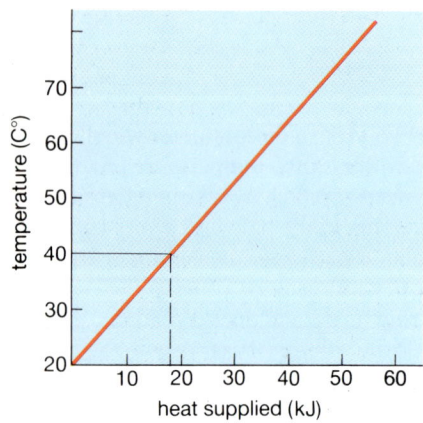

Figure 2
A graph to show how the temperature of the steel block rises as heat energy is supplied

---

Heat supplied = mass × specific heat capacity × temperature change

$$\text{joules} = \text{kg} \times \frac{\text{joules}}{\text{kg°C}} \times °C$$

Example. How much heat is required to heat up the tip of Elsie's soldering iron by 400°C? The tip of the soldering iron is made of copper and has a mass of 30 g.

$$\text{Heat supplied} = m \times s \times (T_2 - T_1)$$

$$= 0.03 \, \text{kg} \times 380 \, \frac{\text{J}}{\text{kg}°\text{C}} \times 400°\text{C}$$

$$= 4560 \, \text{J}$$

Water has a very high specific heat capacity. This means that it absorbs a lot of heat energy when it warms up. Conversely, water gives out a lot of heat when it cools down. Several ways in which the high specific heat capacity of water is useful to us are described below:

- We are made mostly of water. This means that if we suddenly exercise, and our muscles produce a lot of heat energy, then we do not warm up too quickly.

- Water is very important for keeping our houses warm. A house central-heating system pumps hot water around the house. The hot water can release a lot of energy as it flows through radiators. If water had a low specific heat capacity, then radiators a long way from the boiler would never be warm.

- Water is capable of absorbing a lot of heat energy, so it is useful for cooling car engines (Figure 3).

- In Britain we live on an island. The fact that we are surrounded by water has a great effect on our climate. Water is so difficult to warm up and cool down that our weather is neither extremely hot nor extremely cold. Land can be warmed up or cooled down much more quickly than sea, so the temperaure in the middle of a continent can change rapidly. You can see from the map on page 111 that Canada and Russia have worse winters than we do, although they are no further north.

Substance	Specific heat capacity (J/kg°C)
water	4200
alcohol	2400
ice	2100
concrete	800
glass	630
steel	450
copper	380
lead	130

Table 1

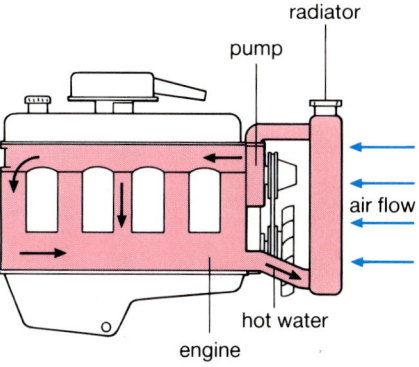

Figure 3
A car cooling system

Questions

1 This question refers to Figure 2 which shows how the temperature of a 2 kg steel block changes as heat is supplied to it.
(a) Make a copy of the graph.
(b) Mark on the graph how the temperature would have changed if:
(i) the mass was 4 kg, (ii) the mass was 1 kg, (iii) the insulating material was removed from the metal. Label these graphs (i), (ii), (iii).
2 A night storage heater contains 60 kg of concrete. How much heat is required to warm the concrete from 10°C to 40°C?
3 Chen and Sally are sitting beside a lake. Chen is throwing stones into the lake. Here is an extract from their conversation.

Chen: Just think of all that kinetic energy the stone has. It is all converted into heat energy when it makes a splash in the lake. If I keep this up the lake will get warmer.
Sally: Don't be stupid, it would take all day!
Evaluate their conversation.
4 Julie has an outdoor swimming pool which is 20 m long, 10 m wide and 2 m deep.
(a) When the pool is full, what is the volume of water in it?
(b) The density of water is 1000kg/m³: what mass of water is there in the pool?
(c) It is necessary to warm the water from 17°C to 22°C. How much heat energy does that require?

(d) During a hot summer spell the sun shines on the pool for an average of 8 hours a day. While the sun shines, the pool absorbs energy at a rate of 60 kW. How much energy is absorbed each day?
(e) Roughly how many days does it take the pool to warm up to 22°C? (Ignore any heat losses at night.)
(f) Julie decided to warm the pool with electrical heaters. How much would it cost her to warm up her pool by 5°C? The electricity board charges 2p for 1 MJ of energy.

5 Latent Heat

During its manufacture, steel is toughened by rapid cooling. Water, evaporating from the steel's surface, takes heat away.

In the last unit you learnt that when heat energy is supplied to a substance it warms up. However, this is not always the case. Heat energy must be supplied to melt ice or to boil water. You have seen pans of water boiling on a cooker. The cooker gives heat to the water but the boiling water never gets any hotter than 100°C. The energy that is supplied is used to evaporate molecules. The energy required to turn 1 kg of a liquid at its boiling point into 1 kg of vapour is known as the **specific latent heat of vaporisation**. It is measured in joules per kilogram (J/kg).

Figure 1 shows a simple experiment that allows you to measure how much heat energy is used to boil water. A beaker of water is boiling on top of an electric balance. The heater is supplying heat energy at a rate of 1000 W. In a time of 250 s the reading on the balance dropped from 1000 g to 900 g. From this we can see that 100 g (or 0.1 kg) boiled away in that time.

How much energy is needed to turn 1 kg of boiling water at 100°C into steam at 100°C (Figure 2)?

$$\text{Heat supplied} = 1000\frac{\text{J}}{\text{s}} \times 250\,\text{s}$$
$$= 250\,000\,\text{J}$$

This boiled away 0.1 kg. So ten times as much energy, 2 500 000 J, will boil away 1 kg of water (Figure 2).

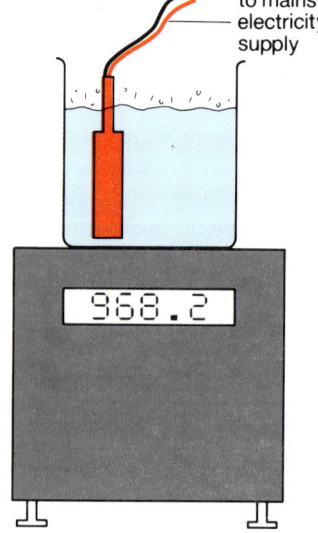

to mains electricity supply

968.2

Figure 1
An experiment to measure how much heat energy is used to boil water

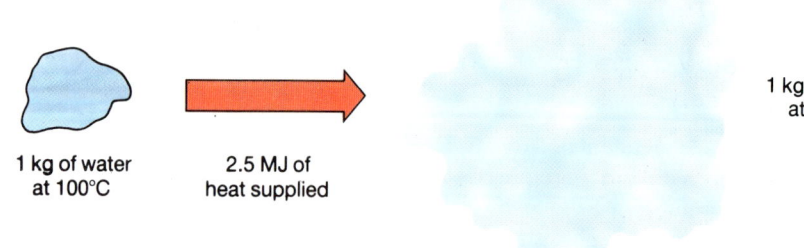

1 kg of water at 100°C

2.5 MJ of heat supplied

1 kg of steam at 100°C

Figure 2
The specific heat of vapourisation of water is 2.5 MJ/kg

Melting and casting

Casting is a very important process which is used in many different industries. For example, pipes, valves, guns and pistons are all manufactured by casting metal. Figure 3 shows you how this process works. Molten metal is poured into a mould which has been made exactly to the shape of a pipe. When the metal has solidified and cooled again the mould can be removed. Usually the moulds are made from sand which has been made rigid with a resin.

If you were in business making metal castings you would find you had a large electricity bill. You would be using a very large amount of energy to heat up and melt your metal. Figure 4 shows a graph of how the temperature of a piece of aluminium changes as heat is supplied to it in a furnace. Over the region *AB* the heat energy is used to warm up the aluminium. At *B* the aluminium begins to melt. By *C* the aluminium has melted completely and the temperature begins to rise again. You can see that it takes almost as long to melt the aluminium as it does to warm it up. So the energy required to melt the aluminium is large. This energy is called the **latent** (or hidden) **heat of fusion**. The *specific latent heat of fusion* of aluminium is the heat energy required to melt 1 kg of aluminium.

Ice also needs a lot of energy to melt it, which is why if you want a cold drink you will put a couple of ice cubes into your glass. Also an ice pack is a very effective way of removing heat from body tissues. If you have a bad bruise, try an ice pack to help reduce the swelling.

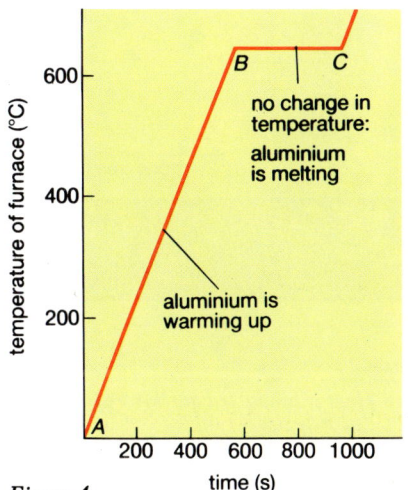

Figure 4

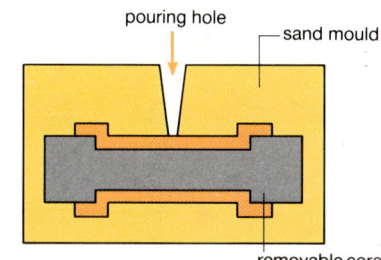

Figure 3
Metal casting
(a) The finished pipe

finished pipe

(b) How it is cast

Questions

1 When is it possible to supply heat to a substance without a temperature change?

2 This question refers to the aluminium that was heated in the furnace. How its temperature changed is shown in Figure 4. Some additional data is shown in the table below.

- Mass of aluminium heated = 100 kg
- Time of heating = 1000 s
- Power of heaters in the furnace = 100 kW

(a) At what temperature did the aluminium melt?
(b) How long did it take to warm the aluminium up to its melting point?
(c) How much heat was supplied to the aluminium in that time?

(d) Use the equation:
Heat supplied = $m \times s \times (T_2 - T_1)$
to calculate the specific heat capacity of aluminium.
(e) How much heat was supplied to melt the aluminium?
(f) How much heat energy is required to melt 1 kg of aluminium?
(g) Figure 4 shows how the temperature changes in an imaginary ideal furnace, which loses no heat to its surroundings. Real furnaces do lose heat. Make a copy of Figure 4 and add to it a second graph to show how the temperature would change in a real furnace.

3 Icebergs come from the Arctic or Antarctic ice caps; they are made from fresh water. It has been suggested that icebergs could be towed from the Arctic to North Africa to supply water to countries that suffer from drought.

Comment on this proposal.

4 A sample of molten wax is put in a boiling tube and allowed to cool. A graph of the temperature of the wax against time is shown below.
(a) What is the melting point of the wax?
(b) Explain why the temperature remains constant for about 15 minutes.

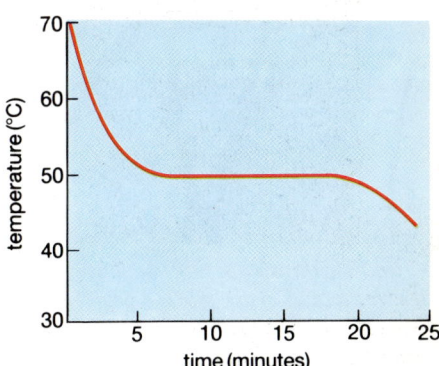

6 Conduction

When you walk around your house in bare feet you will notice that your feet feel warm as long as you stay on a carpet. But if you go into a kitchen which has tiles on a concrete floor your feet will soon feel cold. You notice this effect because the tiles are good **conductors** of heat. A good conductor of heat will carry heat away from your body quickly; this makes you feel cold.

Table 1 shows you materials that conduct heat well and those that are poor conductors or **insulators**. All metals are good thermal conductors and things like plastic, wood and air are insulators. When heat is transferred by conduction, hot atoms pass some of their kinetic energy to colder neighbouring atoms. Metals contain electrons that are free to move. When one end of a metal rod is heated, energy can be carried away from the hot end of the rod by fast moving electrons (Figure 1(a)). In a thermal insulator there are no electrons which are free to move, and heat is transferred more slowly by hot atoms (Figure 1(b)) bumping into colder atoms.

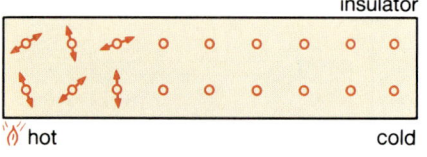

Figure 1
(a) Heat is conducted quickly by fast-moving electrons

(b) Heat is conducted slowly by atoms bumping into each other

Material	Conductivity (W/mK)	
copper	385	good conductors
iron	72	
concrete	5	
glass	1	
brick	0.6	
water	0.6	
fat	0.046	
wool	0.04	
air	0.025	poor conductors

*Table 1. This table showing **conductivities** allows you to compare how well two different materials will conduct heat. A concrete floor will take heat away from your feet far faster than a woollen carpet*

Keeping warm

If you look at Table 1 you can see that fat, wool and air are all poor conductors of heat. It is no surprise then, that these three substances are important for keeping warm-blooded animals (including us) at the right temperature. If you have ever been swimming in the North Sea you will know how rapidly you get cold. Your skin is in contact with cold water and you are losing heat by conduction. In really cold Arctic waters you could not survive more than a few minutes. Seals, however, spend all of their lives in cold water. They are protected from losing heat by conduction by a very thick layer of fat (blubber) which surrounds all of their body.

Birds have feathers and other animals have fur or hair. Fur and feathers are poor conductors of heat, but the way they reduce heat loss is by trapping a thick insulating layer of air.

We have few hairs to keep us warm. If we stand naked, moving air currents soon take heat away from us. However, wearing clothes traps air, so keeping us warm.

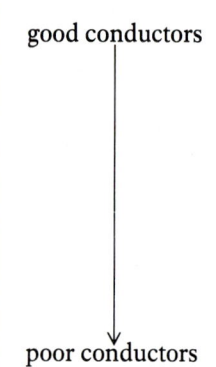

This song thrush has fluffed out her feathers to trap a layer of air. Air is a poor conductor of heat and so she manages to keep warm even in cold weather

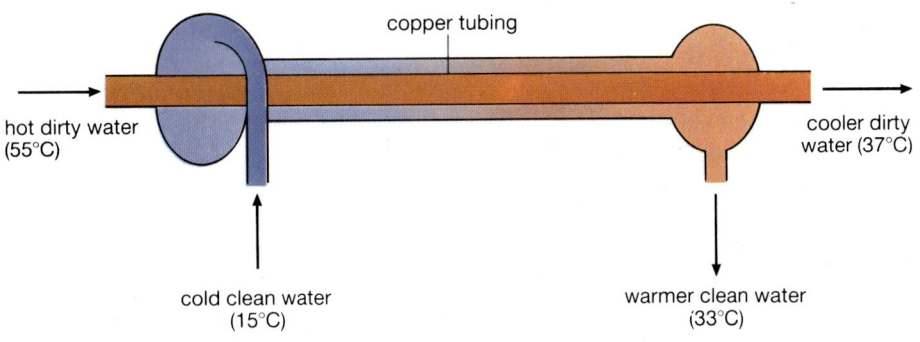

Figure 2
A heat exchanger for a laundry

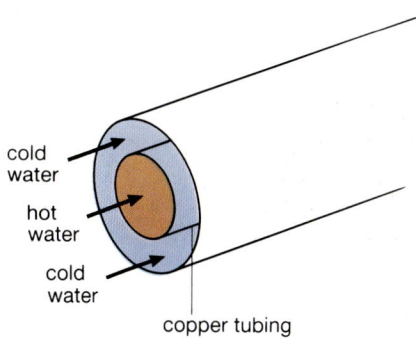

Figure 3
Heat exchanger wall made of metal

Heat exchange

Figure 2 shows a double pipe heat exchanger that is being used in a laundry. Hot dirty water from the washing goes through the central pipe. Clean cold water flows round the outside of the pipe. The idea is to warm up the clean water using heat from the dirty water. This helps to reduce the laundry's heating bill.

Figure 3 shows part of the wall of another heat exchanger. The following points make sure that as much heat is removed from the hot liquid as possible:
- The wall of the container should be made from a good conductor (usually metal).
- The wall should be thin, so that heat is conducted rapidly.
- The surface area should be large to conduct more heat.
- The temperature difference between the cooling water and the hot liquid should be large.
- More heat is extracted from the hot liquid if it flows slowly through the heat exchanger.

Questions

1 Here is part of a conversation between Peter, Ramesh and Lorraine, who are talking about keeping warm in winter.

Peter: I think string vests are ridiculous, they are just full of holes. How can they possibly keep you warm?

Ramesh: I heard on the radio that it was better to wear two vests than one. How can they do any good?

Lorraine: It's not the vest that keeps you warm, it's the air underneath them.

Evaluate this conversation.

2 Opposite you can see two hot water tanks full of water at 40°C. The walls of each tank are made from the same material.

(a) One face of the small tank loses heat at a rate of 1000 J/s. In total how much heat does the small tank lose per second?

(b) Tank A has a larger surface area so it loses more heat per second. How much heat per second does this larger tank lose?

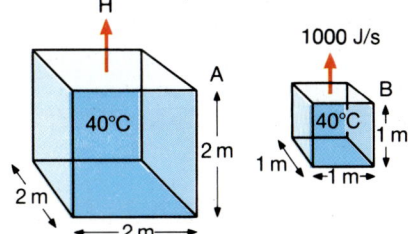

(c) For each tank work out the ratio:

$$\frac{\text{heat lost per second}}{\text{volume of tank}}$$

Which tank will cool down more quickly?

(d) In winter small animals are more likely to freeze to death than larger animals. Explain why.

3 This question refers to the laundry's heat exchanger shown in Figure 2. What effect will the following changes have on the final temperature of the clean water that is being warmed up?

(a) Iron tubing is used instead of copper.

(b) Thinner copper tubing is used for the wall of the heat exchanger.

(c) The length of the heat exchanger is increased.

(d) The hot dirty water is made to run faster.

4 Why do power stations need heat exchangers?

7 Convection

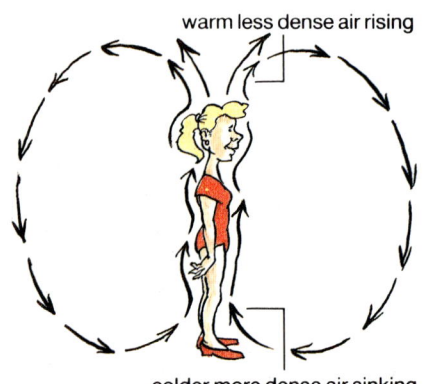

Figure 1
Convection currents near a person in a room

In July 1987 Richard Branson made an adventurous crossing of the Atlantic travelling in a hot air balloon. After a 3000 mile journey from America he landed in the Irish Sea. To make a hot air balloon fly you need a burner which heats up the air inside the balloon. When air is heated it expands and its density becomes less. When hot air is surrounded by colder, more dense air it rises. This is the same as when a cork floats to the surface of water.

Figure 1 shows how air circulates if you stand, with most of your clothes off, in the middle of a room. Your body heats up air next to it, which then rises. Colder air flows down near the walls to replace the warmer air. The currents of air that flow are called **convection currents**.

Heat can be transferred by convection in *liquids* and *gases*, but not in solids. Although most liquids and gases do not transfer heat very well by conduction, they do transfer heat quickly by convection. Convection currents allow large quantities of a hot liquid or gas to move and give heat to a colder part. If we wore no clothes our bodies would lose lots of heat by convection. By wearing clothes we trap a layer of air which acts as an insulator.

Exactly the same idea is used to reduce heat losses from your house. Your loft is full of air and convection currents there can cause a large amount of heat loss. Lofts can be insulated with felt or glass fibre. To stop convection in the cavity between inner and outer walls, it is possible to pump in polystyrene foam. The foam has lots of air trapped in it and so is a poor conductor of heat. Double glazing also traps air, in a layer between two pieces of glass (Figure 2).

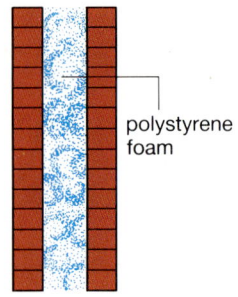

Figure 2
(a) *Cavity wall insulation*

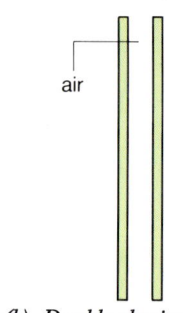

(b) *Double glazing*

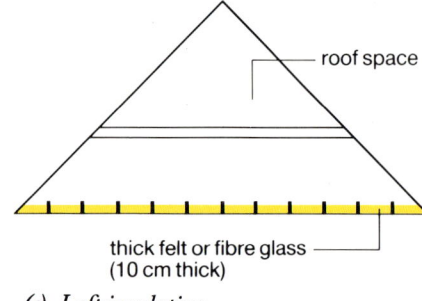

(c) *Loft insulation*

Convection currents also play an important part in our weather systems. Winds are convection currents on a large scale. You have probably noticed when you have been sunning yourself on the beach that there is usually a sea breeze (Figure 3). In the daytime the land warms up quickly. This causes a rising air current and a breeze blowing in from the sea. Seabirds are quick to take advantage of these rising currents of air, and can soar upwards without flapping their wings.

Figure 3
The large specific heat capacity of the sea means that it stays cool. The land warms quickly, and warm air rises above the land

Convection in water

When water is heated it too will transfer heat energy by convection. Figure 4 shows a simple way that you can show this in a laboratory. Place some potassium permanganate crystals at the bottom of a flask of water. When heat is supplied from underneath the dissolved potassium permanganate rises showing the path of the currents.

It is possible to circulate water around the heating system of a house using convection currents. Figure 5 shows such a heating system. The boiler must be put at the lowest point in the house. The hot water rises upwards to the roof. The water then feeds the radiators and finally the cool water is returned to the boiler.

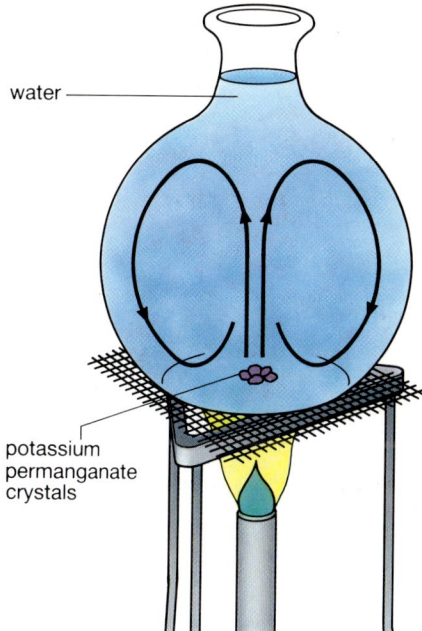

Figure 4
A laboratory experiment to show convection currents in water

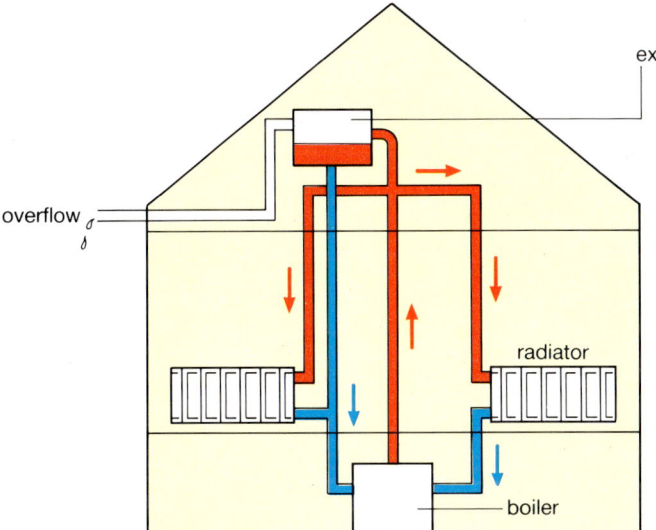

Figure 5
This system of circulating water is suitable for a small house only. In a large house it is more usual to use a pump to help push the water round

Questions

1 Heat cannot be transferred by convection in a solid. Explain why.
2 Figure 3 shows the direction of a sea breeze in daytime. At night, land cools down more quickly than the sea. Draw a diagram to show the direction of the sea breeze at night.
3 About a hundred years ago in Cornish tin mines, fresh air used to be provided for the miners through two ventilation shafts shown below.
(a) To improve the flow of air, a fire was lit at the bottom of one of the shafts. Explain why.
(b) One day, Mr Trevethan, the owner of the mine, had an idea. To make the ventilation even better he lit another fire at the bottom of the second shaft. Comment on this idea.

4 This question is about the cost of insulating a house. At the moment the house has no loft insulation, cavity wall insulation or double glazing.
(a) Use the data provided to decide which method of insulation provides the best value for money.
(b) How many years do you have to wait before the double glazing has paid for itself?

- Yearly heating bill for house: £700
- Cost of double glazing: £3000
  This would reduce the fuel bill by 20% a year
- Cost of loft insulation: £450
  This would reduce the fuel bill by 30% a year
- Cost of cavity wall insulation: £1200
  This reduces the fuel bill by 25%

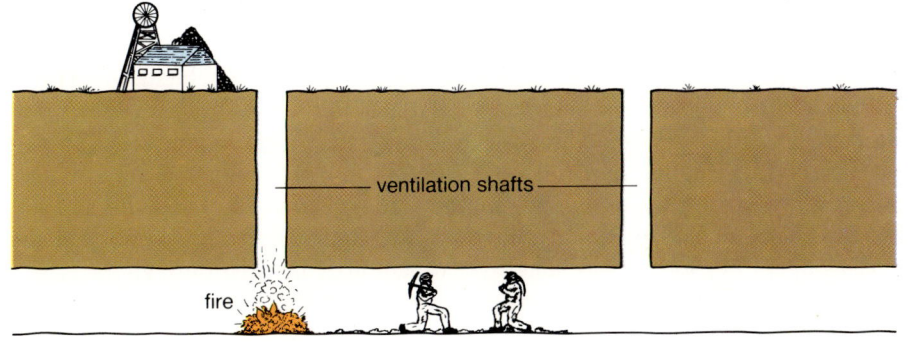

8 Radiation

Radiation is the third way by which heat energy can be transferred from one place to another. Heat energy from the Sun reaches us by radiation. The Sun emits electromagnetic waves which travel through space at high speed. Light is such a wave, but the Sun also emits a lot of **infra-red** waves (or infra-red radiation). Infra-red waves have longer wavelengths than light waves. It is the infra-red radiation that makes you feel warm when you lie down in the sun. Infra-red radiation travels at the same speed as light; as soon as you see the Sun go behind a cloud you feel cooler.

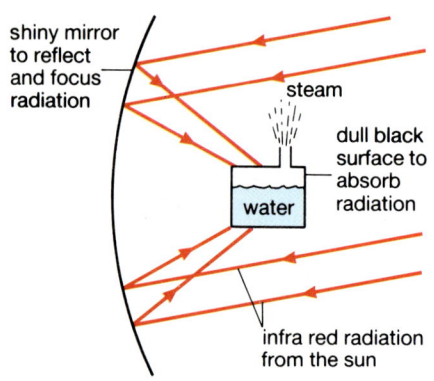

Figure 1
A solar furnace

Good and bad absorbers

Infra-red radiation behaves in the same way as light. It can be reflected and focussed using a mirror. Figure 1 shows the idea behind a solar furnace. The shiny surface of the mirror is a poor absorber of radiation, but a good reflector. Radiation is absorbed well by dull black surfaces. So the boiler at the focus of the solar furnace is of a dull black colour.

Good and bad emitters

Figure 2 shows an experiment to investigate which type of teapot will keep your tea warm for a longer time. One pot has a dull black surface, the other is made out of shiny stainless steel. Radiation that is emitted from the hot teapots can be detected using a thermopile and a sensitive ammeter. When you do the experiment you will find that the black teapot emits more radiation than the shiny surface. So a shiny teapot will keep your tea warmer than a black teapot.
- Black surfaces are good absorbers and good emitters of radiation
- Shiny surfaces are bad absorbers and bad emitters of radiation

Figure 2
A sensitive instrument called a thermopile can detect radiation

The greenhouse effect

In Italy, where the average temperature in summer is about 5°C higher than in Britain, tomatoes grow very well. It is a great help to a tomato grower in Britain if he uses a greenhouse. On a warm day the temperature inside a greenhouse can be 10°C or 15°C higher than outside (Figure 3). Infra-red radiation from the sun passes through the glass of the greenhouse, and is absorbed by the plants and soil inside. The plants radiate energy, but the wavelength of the emitted radiation is much longer. The longer wavelengths of radiation do not pass through the glass and so heat is trapped inside the greenhouse. The temperature rises until the loss of heat through the glass by conduction balances the energy absorbed from the Sun.

This butterfly loses heat very rapidly because it has a very large surface area and a relatively small volume. To warm itself up it opens its wings to absorb sunlight. To cool off, it folds them shut

Some people worry that a similar greenhouse effect could be happening in our atmosphere. As we continue to burn fossil fuels we are filling our atmosphere with carbon dioxide and other chemicals. As these chemicals absorb long wavelength radiation emitted from the Earth's surface, the average temperature of the Earth increases. The planet Venus shows the greenhouse effect in a big way. Its atmosphere is mostly carbon dioxide, and its average surface temperature is about 460°C, hot enough to melt some metals.

Figure 3

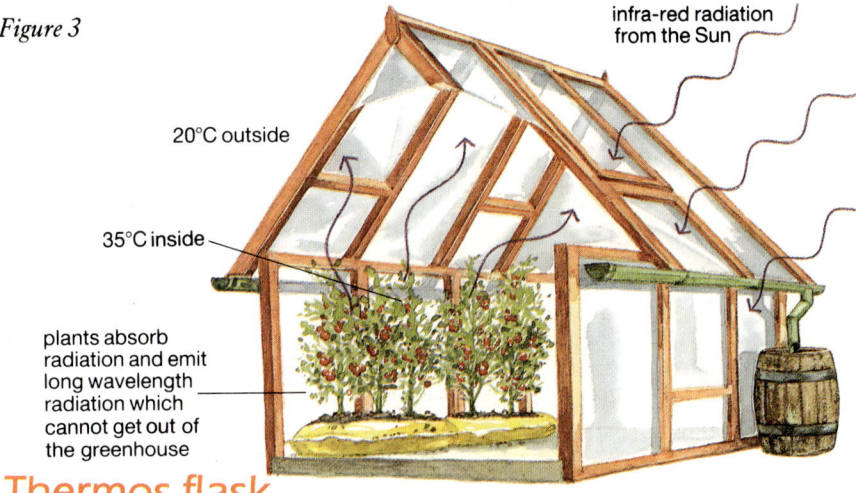

infra-red radiation from the Sun

20°C outside

35°C inside

plants absorb radiation and emit long wavelength radiation which cannot get out of the greenhouse

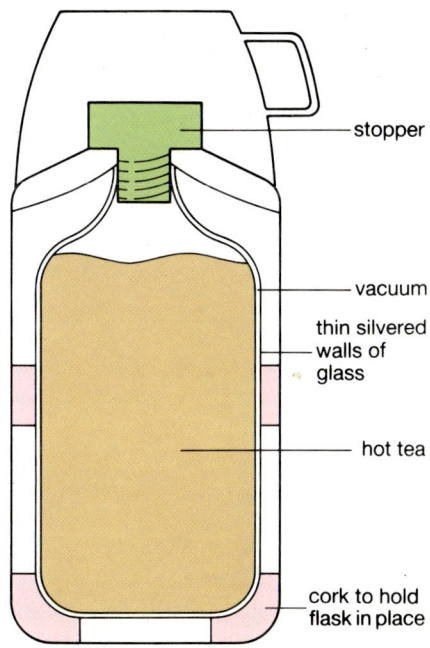

stopper

vacuum

thin silvered walls of glass

hot tea

cork to hold flask in place

Figure 4
A thermos flask

Thermos flask

A thermos flask keeps things warm by reducing heat losses in all possible ways. The flask is made with a double wall of glass and there is a vacuum between the two walls. Conduction and convection cannot take place through a vacuum. The glass walls are thin, so that little heat is conducted through the glass to the top. Heat can be radiated through a vacuum, but the glass walls are silvered like a mirror so that they are poor emitters of radiation. The stopper at the top prevents heat loss by evaporation or convection currents (Figure 4).

Questions

1 Some casserole dishes that are used in ovens are black, but the outside of an electric kettle is shiny. Can you explain why?

2 Tracy and Abdul are talking about greenhouses. Neither of them has studied much physics, so they are not too sure how greenhouses work. Read their conversation and correct any errors in their thinking.

Tracey: I saw that Mr Brown put some aluminium foil up inside his greenhouse. I know that aluminium is a good conductor, so he has probably put it there to conduct more heat into his greenhouse.

Abdul: No. I don't agree. I think he has put the foil there to keep the greenhouse cooler, but I don't understand why it works.

3 This question is about controlling the temperature of a greenhouse. The greenhouse can lose heat by radiation and conduction when its windows and doors are closed. Heat can also be lost by convection when a window is opened. Information about heat losses on a particular day is shown in the graph. On this day the greenhouse is absorbing 10 kW of power from the Sun.
(a) Explain why the graph (opposite) shows that the air temperature outside the greenhouse is 20°C.
(b) At what rate must the greenhouse be losing heat if the temperature inside it is constant?

(c) At 38°C how much heat is lost by:
(i) radiation, (ii) conduction? Explain why 38°C is the steady temperature with the door and windows closed.
(d) The window is now opened. Use the graph to work out the new steady temperature of the greenhouse.

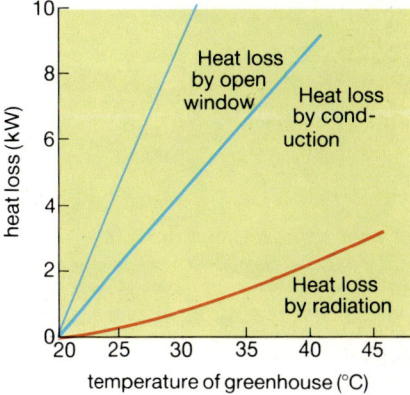

Heat loss by open window

Heat loss by conduction

Heat loss by radiation

temperature of greenhouse (°C)

heat loss (kW)

9 Warming-Up Exercises

- Ground floor area = 100 m²
- Ceiling area = 100 m²
- Area of brick wall = 200 m²
- Area of windows and doors = 40 m²

U-values (W/m²°C)

Uninsulated house	● Uninsulated roof	2
	● Cavity wall	1
	● Floor without carpets	1
	● Windows, single glazed doors	5
Insulated House	● Insulated roof	0.3
	● Cavity filled wall	0.5
	● Floor with carpets	0.3
	● Windows, double glazed doors	2.5

Keeping warm in the house

Suppose you are a heating engineer who wants to know what size of central heating system to install in a house. You can calculate the heat losses from a house using **U-values**. If a material transfers heat well it has a *high* U-value; if the material is a good insulator it has a *low* U-value. You can see some values for U-values above. An engineer can calculate the rate of loss of heat using this equation:

$$\text{Rate of heat loss (Watts)} = U\text{-values} \times \text{area} \times \text{temperature difference}$$
$$= \frac{W}{m^2\,°C} \times m^2 \times °C$$

We can use the data provided to calculate the rate at which heat is lost from a poorly insulated house. The outside temperature is 0°C and the house is to be kept at 15°C.

Heat loss per second, $H = U \times A \times T$

- Ceiling area: $\quad H = 2 \times 100 \times 15 = 3000\,\text{W}$
- Walls: $\quad H = 1 \times 200 \times 15 = 3000\,\text{W}$
- Floor: $\quad H = 1 \times 100 \times 15 = 1500\,\text{W}$
- Windows: $\quad H = 5 \times 40 \times 15 = 3000\,\text{W}$

Total 10 500 W

So you would need to install a heating system capable of producing about 10 kW. If the house owner used a gas central heating system continuously at this rate it would cost about £25 per week.

Keeping cool

If you have ever done any long-distance running, you will know that you get very hot indeed and that your efforts cause you to sweat a lot. You will see from the data provided on the next page that marathon runners can suffer severe problems due to temperature changes of the body or loss of water (dehydration).

Double glazing is needed to reduce heat loss in buildings that have large areas of glass

- First we will use the data (bottom right) to calculate how much water one runner, Sarah, loses in a three hour marathon when the temperature is 18°C. At this temperature she loses heat due to evaporation at a rate of 600 W.

  Over 3 hours the heat lost is:

  $$\text{Heat lost} = 600\frac{\text{J}}{\text{s}} \times (3 \times 3600)\text{s}$$

  $$= 6\,500\,000\,\text{J}$$

  $$= 6.5\,\text{MJ}$$

  But 2.5 MJ of heat evaporates 1 kg of water,

  so she loses $\frac{6.5}{2.5} = 2.6$ kg of water in the race.

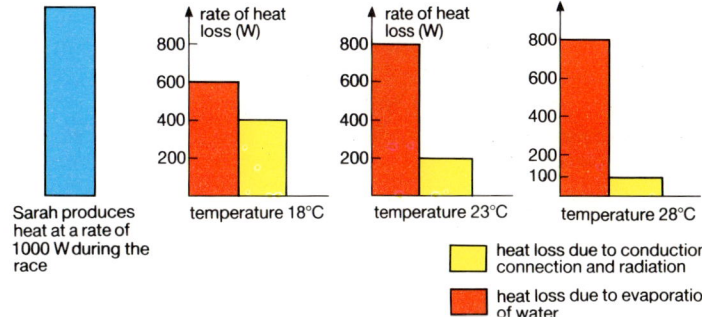

Figure 1

- When marathon runners finish a race they are wrapped up in bags to stop them cooling too quickly. When Sarah finishes her race and rests, her body will only be *producing* heat at a rate of 100 W, but she is still *losing* it at a rate of 1000 W (600 W due to evaporation, 400 W due to other processes). So she loses heat from her body at a rate of 900 W. If this is allowed to happen even for 10 minutes, she could become very cold. In 10 minutes:

  $$\text{heat lost} = 900\,\text{J} \times (10 \times 60)\text{s} = 540\,000\,\text{J}$$

  We can calculate her loss in temperature using:

  $$\text{Heat lost} = \text{mass} \times \text{specific heat capacity} \times \text{temperature drop}$$

  so temperature drop $= \dfrac{H}{ms} = \dfrac{540\,000\,\text{J}}{60\,\text{kg} \times 4200\,\text{J/kg°C}}$

  $$= 2.1°\text{C}$$

Questions

House heating

1 Use the data provided to work out the rate of heat loss from the same house, if it is insulated well.

2 Some people think that it is more important to insulate the walls than to put in double glazing. This seems sensible because the windows are small. Comment on this idea.

3 How much would it cost to heat a well-insulated house per week, if the heating runs for only six hours per day? Assume that the temperature difference between inside and outside is still 15°C.

Marathon running

4 Explain why as the air temperature rises, a marathon runner loses more heat by evaporation of sweat, and less by conduction, convection and radiation.

5 If Sarah loses as much as 3 kg of water, she will suffer severe effects from dehydration.
(a) Explain why she could be in danger if the temperature is more than 23°C.
(b) What action can a runner take to avoid dehydration?

6 (a) What will happen to Sarah's body temperature if she runs when the air temperature is 28°C? Assume that she has no means of cooling other than those suggested in the graphs.
(b) How much extra heat energy is she producing: (i) per second, (ii) per hour?
(c) Calculate by how much her body temperature will have increased after one hour.
(d) Explain why Sarah is unlikely to finish a marathon if the temperature is as high as 28°C.

Keeping warm at the end of a marathon. The shiny blanket is a poor radiator of heat

Marathon running data

- Mass of runner = 60 kg
- Specific heat capacity of water = 4200 J/kg°C
- Specific latent heat of vaporisation of water = 2.5 MJ/kg
- At rest an athlete produces heat energy at a rate of 100 W
- Average body temperature = 37°C
- The body can only work normally between 33°C and 41°C

SECTION F: *STUDY QUESTIONS*

1 A test tube containing water is heated just below the water surface. The water at the top begins to boil but it takes some time before the water at the bottom of the tube becomes hot.

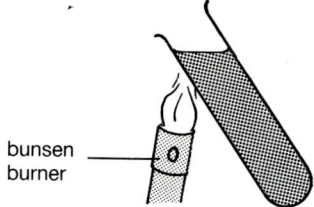

bunsen burner

What does this experiment show?
A Water expands when heated.
B Water can be made to boil at below 100°C.
C Water is a bad conductor of heat.
D Water absorbs heat radiation.
E Convection currents occur in water.

MEG

2 Why does a person coming out of the sea on to the beach quickly feel cold?
A Water evaporates from the skin.
B Water is a good conductor of heat.
C Convection occurs in the air.
D Air is a bad conductor of heat.
E Heat is lost by radiation from the skin.

MEG

3 A solid is heated at a steady rate until it becomes a gas. The diagram shows the graph of temperature against time.

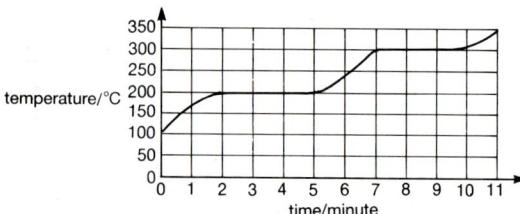

How long does it take (in minutes) before the substance begins to boil?
A 2   B 5   C 6   D 7   E 10

NEA

4 Long hot-water pipes sometimes have bends in them as shown below.

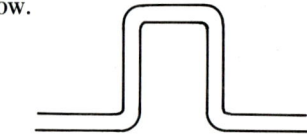

What are the bends for?
A To reduce the speed of the water
B To give the pipe more rigidity
C To allow for pressure changes
D To allow for expansion of the pipe
E To allow for contraction of the water

MEG

5 In which one of the following examples is convection likely to be the main method of heat transfer?
A Food being cooked in a microwave oven
B Energy from the Sun reaching the Earth
C A room being heated by a radiator
D Heating a wire with a soldering iron
E Food being cooked under an electric grill

LEAG

6 Of the following, the best example of heat transfer by conduction is
A from the Sun to the Earth.
B from the boiler to the hot water storage tank in the domestic hot water system.
C through glass into a greenhouse.
D from a fire to a person in a room.
E from a hotplate to the contents of a saucepan.

NEA

7 Solar heating panels use the Sun's radiation to warm water. They are blackened because dark surfaces
A are good reflectors of heat.
B are good radiators of heat.
C are good absorbers of heat.
D are good conductors of heat.
E are good at transferring heat by convection.

MEG

8 In an experiment in which equal masses of lead and a lead–tin alloy called solder were allowed to cool, the following cooling curves were obtained.

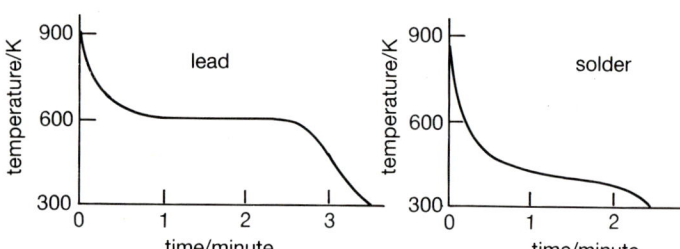

(a) For the lead:
(i) what was the physical state of the lead after thirty seconds,
(ii) what was happening to the lead at 600 K,
(iii) what is 600 K in Celsius units,
(iv) why did the temperature of the lead remain constant at 600 K for some time?
(b) State **two** important differences between the lead and the solder which can be deduced from these cooling curves.
(c) Solder is often used by plumbers to repair burst pipes by filling the fracture with solder. Why is solder better than lead for this purpose?

NEA

9 Two solid objects X and Y have the same mass but the specific heat capacity of X is twice that of Y. If each object is supplied with 20 J of heat.
A the temperature rise of X will be half that of Y.
B the temperature rise of X will be twice that of Y.
C each experiences the same temperature rise but this cannot be calculated unless the mass is known.
D the temperature of each object rises by 10 K.
E the temperature of each object rises by 20 K.

NEA

10

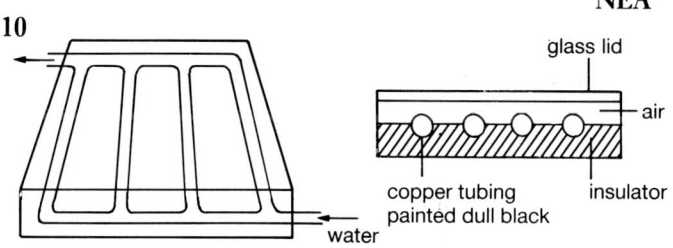

(a) A simple solar panel consists of thin copper tubing painted dull black through which water flows slowly. The tubing is half embedded in insulating material and above it is a layer of air trapped by a glass lid.
(i) Why is the tubing made of thin copper and painted black?
(ii) What material would be appropriate as the insulator and what is its purpose?
(iii) How does the glass lid improve the efficiency of the panel?
(iv) What is the best position and inclination of such a panel?
(b) (i) If a system of such panels can heat 200 kg of water daily from 10→35°C, calculate the energy taken in by the water.
(Specific heat capacity of water = 4200 J/kgK).
(ii) The maximum daily energy arriving at the Earth's surface is 30 MJ per m². Calculate the area of absorbing surface required if 20% of the energy striking the surface is absorbed.
(iii) What limits the usefulness of such panels in Britain?

WJEC

11

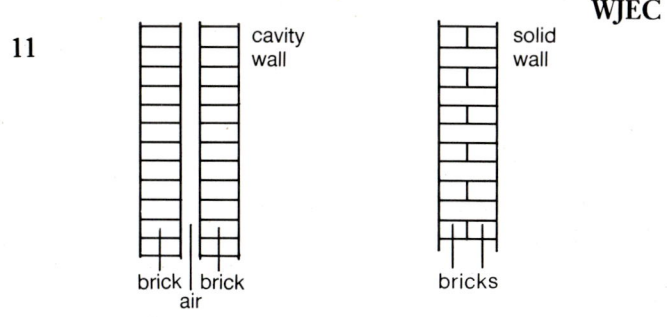

(a) Why do air-filled cavity walls keep a house warmer in winter than solid brick walls?
(b) Why does filling the cavity with plastic foam keep the house even warmer?
(c) Explain how a hot water radiator heats a room.

MEG

12 Diagram (i) illustrates an instrument used to measure the time that the Sun shines during a day. The blackened glass bulb contains mercury and is supported inside an evacuated glass case. Diagram (ii) shows how the connecting wires are arranged inside tube A.

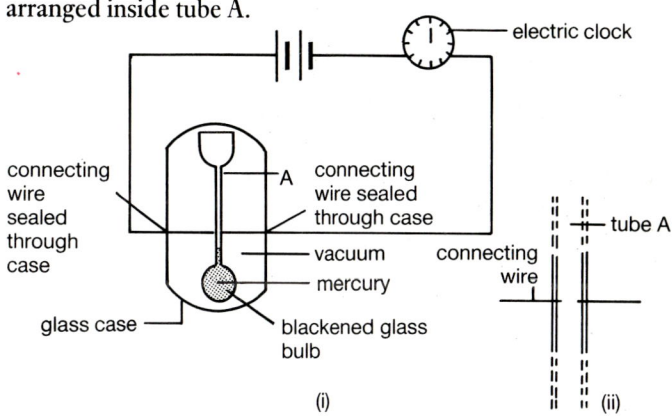

(a) How does energy from the Sun reach the mercury? Give a reason for your answer.
(b) Explain why the clock starts when the Sun shines.
(c) Why is tube A of small cross-sectional area?
(d) Explain why blackening the bulb ensures that the mercury level falls rapidly when the Sun ceases to shine.

MEG

13 This diagram shows a circuit in which an immersion heater is used to heat a metal block. The heater supplies energy at the rate of 50W.

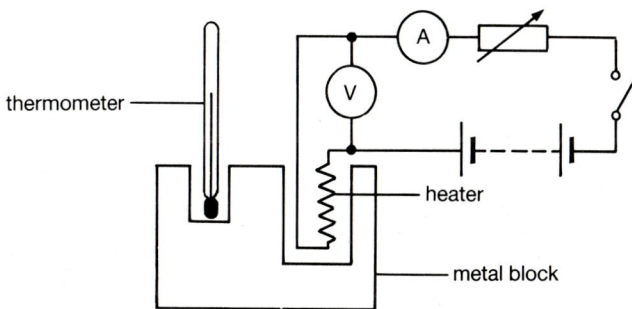

(a) How many joules of energy are supplied each second?
(b) The heater is switched on for 300 seconds. During this time the temperature rises from 20°C to 50°C.
(i) Calculate the energy supplied.
(ii) If the mass of the block is 2 kg, calculate the energy required to raise the temperature of 1 kg of the material by 1°C.
(iii) What name is given to the quantity you have calculated in (b)(ii)?
(c) When the experiment is performed as shown above, the measured rise in temperature is smaller than expected.
(i) Why is the measured rise smaller?
(ii) How would you change the apparatus to make the measured rise in temperature closer to the expected value?

NEA

14 Mr Adams is a neurosurgeon. Recently he had to operate on Polly. She had a giant aneurysm. An aneurysm is like a balloon that has blown out from the side of an artery in the brain. Aneurysms occur when the wall of an artery is weak. They are very dangerous since they could burst and bleed into the brain.

Polly's aneurysm was so large that it was necessary to stop the blood flowing into the brain. At normal body temperatures the brain dies in a few minutes without a blood supply. But at 15°C the brain can last for nearly an hour without a blood supply. To cool her down Polly's blood was passed through a heart/lung machine. The graphs shows how her body temperature changed during the day. The operation was successful and Polly has now recovered completely.

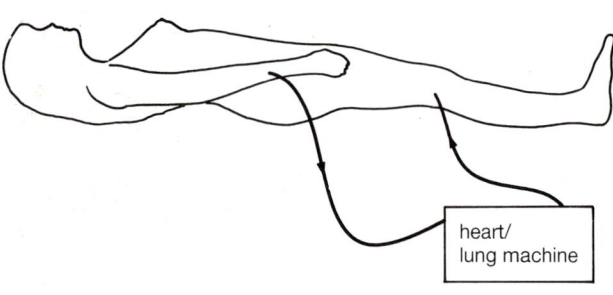

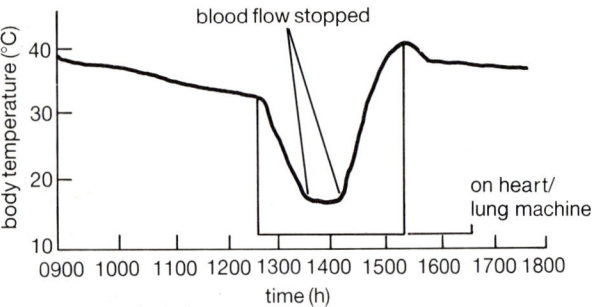

(a) Use the graph to work out how long it took Mr. Adams to remove the aneurysm.
(b) Between 1420 h and 1520 h Polly was warmed up again. By how much did her temperature rise in that time?
(c) Polly's mass is 60 kg. How much heat was needed to warm her up again? Assume that 4000 J of heat energy warms up 1 kg of Polly by 1°C.
(d) What was the power of the heater in the heart/lung machine?

15 A metal block of mass 2.5 kg requires 25 000 J of heat to change its temperature from 15°C to 25°C. The specific heat capacity of the metal in J/kg °C, is
A 400   B 1000   C 5000   D 250 000
E 2 500 000

NI

16 Meteors are small pieces of matter made mostly of iron. Like the Earth, meteors are in orbit around the Sun. Meteors travel very quickly and can cause a lot of damage when they hit something. The craters on the Moon were made by meteors. Fortunately few meteors hit the surface of the Earth. This is because our atmosphere slows them down in a very short time. Some data about a meteor are shown below.

> • Mass of meteor = 0.01 kg
> • Speed of meteor entering atmosphere = 30 000 m/s
> • Specific heat capacity of iron = 500 J/kg°C

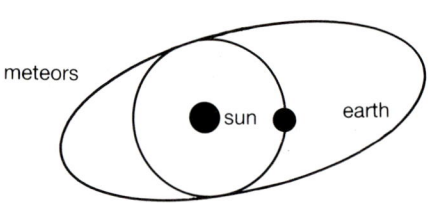

(a) Calculate the kinetic energy of the meteor as it enters the atmosphere. Use the formula:
$$\text{Kinetic energy} = \tfrac{1}{2} \times \text{mass} \times (\text{speed})^2$$
(b) Explain why the meteor slows down in the atmosphere.

Andrew and Kate are having a discussion about meteors. Andrew says: "the kinetic energy of the meteor is turned into heat energy. I worked out that the final temperature of the meteor is about 900 000°C, using the equation:
Energy transformed = mass × specific heat capacity × temperature change."
(c) Use the data to show how Andrew got this answer. What assumption did he make?

Kate says: "I read in a book that a meteor only reaches a temperature of about 20 000°C in the atmosphere. Your answer must be wrong. This is because you have forgotten that the meteor is losing heat."
(d) Discuss what Kate says. Can you think of ways that the meteor will lose heat?
(e) Explain why very few meteors reach the surface of the Earth.

17 The time taken to cook an egg is
• proportional to the mass of the yolk
• proportional to the distance between the yolk and the shell
• inversely proportional to the surface area of the shell.
Elaine decides to cook an ostrich egg for breakfast. Ostrich eggs are 3 times as long as hens' eggs. Elaine likes her hens' eggs boiled for 5 minutes. How long should she boil the ostrich egg for?

SECTION G
Waves and Sound

Everyone knows about water waves, but did you know that your vision and hearing also depend on waves?

1 Introducing Waves

Aerials transmit radiowaves. These convey energy and information . . .

Waves do two important things; they carry energy and information. You have seen ocean waves crashing into a sea wall at high tide. Those waves certainly carry energy.

When you watch television you are taking advantage of radiowaves. These waves carry energy and information from the transmitting station to your house. Light and sound waves carry energy and information from the television set to your eyes and ears.

Waves on slinkies

One of the best ways for you to learn about waves is to see them moving along on a stretched 'slinky' or spring. Figure 1 shows a slinky lying on the floor. When you move your hand from side to side some humps move away from you along the slinky. Although the wave moves along the slinky, the movement of the slinky itself is from side to side. If you tie a piece of string to the slinky, you will see that it moves in exactly the same way as your hand did to produce the waves. This sort of wave is called a **transverse wave**. The particles carrying the wave in the slinky move at right angles to the direction of wave motion. Water ripples on the surface of a pond and light waves are examples of transverse waves.

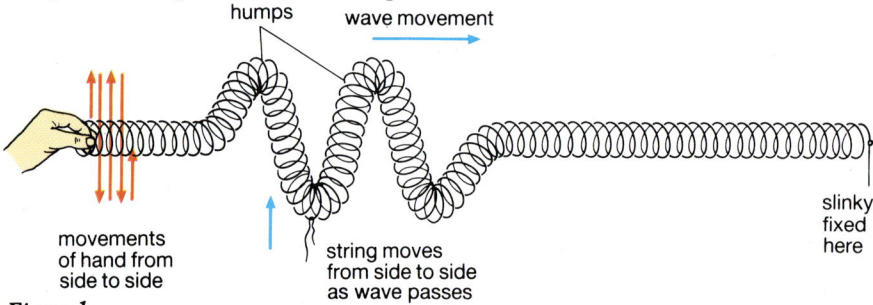

Figure 1
A transverse wave moving down a slinky

You produce a different kind of wave when you move your hand backwards and forwards along the slinky (Figure 2). Your hand compresses and then expands the slinky. The wave is made up of compressions and expansions which move along the slinky. This time a piece of string tied to the slinky moves backwards and forwards along it. Again, this is how your hand moved to produce the waves. This sort of wave is called a **longitudinal wave**. The particles carrying the wave in the slinky move backwards and forwards along the direction of wave motion. Sound waves are longitudinal.

. . . which your television receives

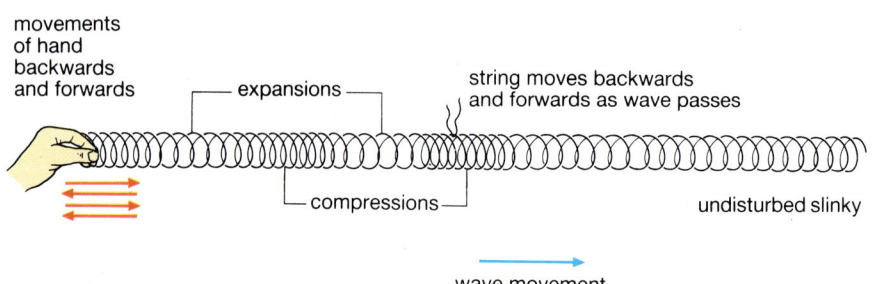

Figure 2
A longitudinal wave moving down a slinky

Sea waves can carry a lot of kinetic energy, particularly during a storm. This is released when the waves break against a sea wall, and the wall can be badly damaged

Describing waves

Figure 3 is a graph showing the displacement of a slinky along its length at one moment. The arrows on the graph show the direction of the motion of the slinky; a larger arrow means a larger speed.

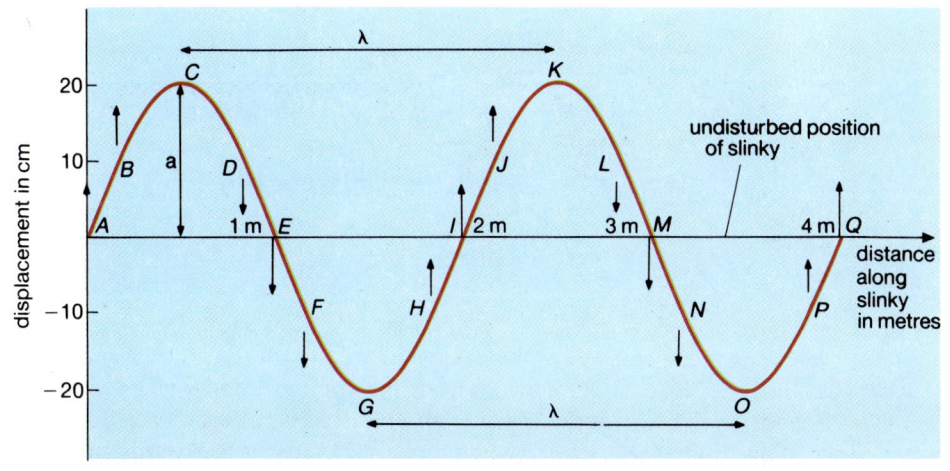

Figure 3
The displacement of a slinky along its length. The arrows on the graph show the direction of motion of the slinky

- **Phase**. Points B and J are moving in phase. They are moving in the same direction, with the same speed. They also have the same displacement away from the undisturbed position of the slinky. F has been displaced in the opposite direction to B and J, and is moving in the opposite direction. F is out of phase with B and J. However, F moves in phase with N.
- The **wavelength** of a wave motion is the shortest distance between two points that are moving in phase. You can think of a wavelength as the distance between two humps. We use the greek letter λ (*lambda*) for the wavelength.
- The **amplitude** of a wave is the greatest displacement of the wave away from its undisturbed position. You can think of the amplitude as the height of a hump.
- The **frequency**, f, of the wave is the number of complete waves produced per second. There are two complete waves in Figure 3. The unit of frequency is waves per second or *hertz* (Hz). 1 kHz means 1000 Hz.
- The **time period** of a wave, T, is the time taken to produce one complete wave.

Explosions produce shock waves which can be used to knock down buildings, as you can see in the photograph above. Sometimes the shock wave travels to other buildings, where it can smash windows

Wave velocities

The **velocity** of a wave, v, is the distance travelled by a wave in one second. The velocity of waves down a particular slinky is the same for all wavelengths. Figure 4(a) shows waves moving on a slinky with frequency 3 Hz and wavelength 0.4 m. In one second three waves have been produced, so the distance travelled by the first wave is $3 \times 0.4 = 1.2$ m. The wave velocity is 1.2 m/s. For any wave (see Figure 4(b)) we can calculate the wave velocity using the formula:

$$\text{Velocity} = \text{frequency} \times \text{wavelength}$$
$$V = f \times \lambda$$

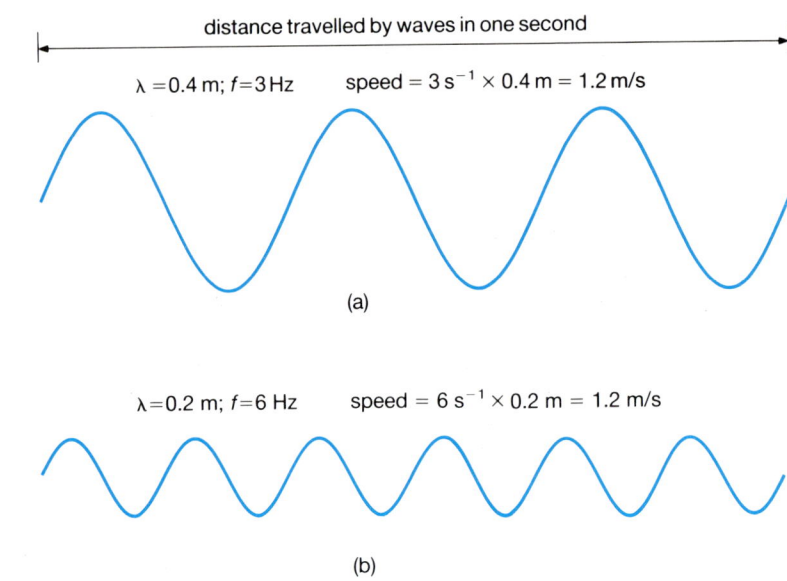

Figure 4
The speed of waves along a slinky is the same for all wavelengths

Questions

1 Explain the difference between longitudinal and transverse waves.
2 This question refers to the graph in Figure 3, on the previous page.
(a) What is the wavelength of the wave?
(b) What is the amplitude of the wave motion?
(c) The frequency of the wave motion is 2 Hz. What is the time period of the wave?
(d) Calculate the speed of the wave.
(e) Give a point moving in phase with:
(i) I, (ii) B, (iii) M.

(f) Make a sketch of the wave motion in Figure 3. Use the arrows, showing the direction of movement of the particles in the slinky, to draw in the position of the slinky a short time later.
(g) In which direction is the wave moving?
3 Make a sketch of a longitudinal wave of a slinky, and mark in a distance to show one wavelength.
4 A radio station produces waves of frequency 200 kHz and wavelength 1500 m.

(a) What is the speed of radio waves?
(b) Another station produces waves with a frequency 600 kHz. What is their wavelength?
5 Why do the wave pulses shown below contain some information?

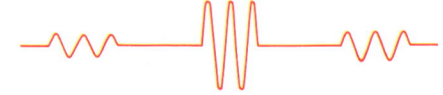

2 Water Waves

You can learn more about the nature of waves by studying ripples that move over the surface of water. It is easy to show that these ripples are transverse waves. If you look at a floating cork you will see it bob up and down as ripples travel along the surface of the water (Figure 1).

You can produce water waves in a ripple tank, by lowering a dipper into the water (Figure 2). A motor vibrates the dipper up and down to produce waves continuously. A beam of wood produces straight waves, and a small sphere produces circular waves. If you shine a light from above the tank you will see bright and dark patches on the screen below. These patches show the positions of the crests and troughs of the waves.

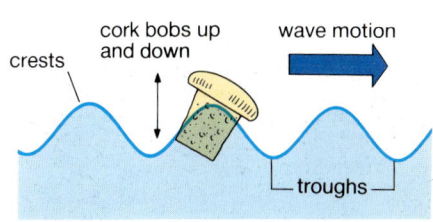

Figure 1
Ripples on water are transverse waves

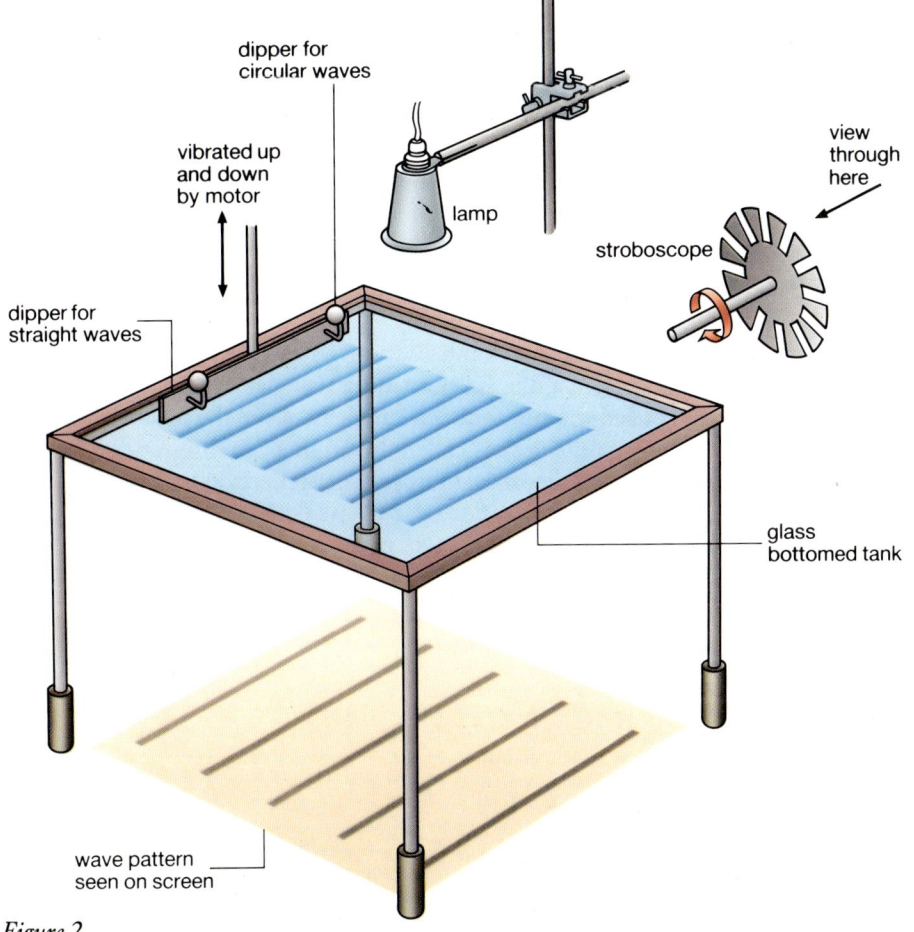

Figure 2
A ripple tank

Using a stroboscope

Water waves move quite quickly and it can be difficult to see them. However, if you look through a rotating stroboscope you can make the waves appear stationary. The stroboscope is a disc with 12 slits in it. If you rotate the disc twice a second you will see the ripple tank 24 times a second. The waves will appear stationary when the dipper produces waves with a frequency of 24 Hz. Each time there is a slit in front of your eye one wave has moved forwards to the position of the next wave.

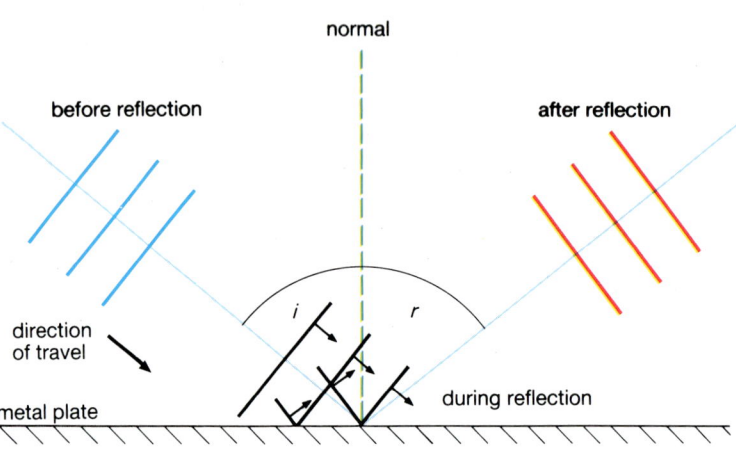

Figure 3
Reflection of waves off a plane surface; angle of incidence, i = angle of reflection, r.

Reflection

In Figure 3 you can see some waves approaching a straight metal barrier in a ripple tank. On the diagram, a line is drawn at right angles to the surface of the barrier. This line is called the **normal**. The angle between the normal and the direction of travel before reflection is called the **angle of incidence**, i. The angle between the normal and the direction of travel after reflection is called the **angle of reflection**, r. When waves are reflected, i always equals r. Figures 4 and 5 show some further examples of water waves being reflected.

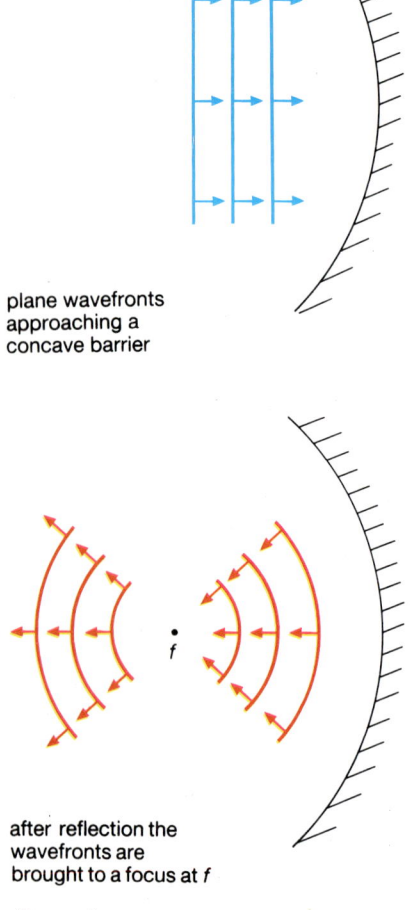

plane wavefronts approaching a concave barrier

after reflection the wavefronts are brought to a focus at f

Figure 4
Reflection of waves by a concave barrier

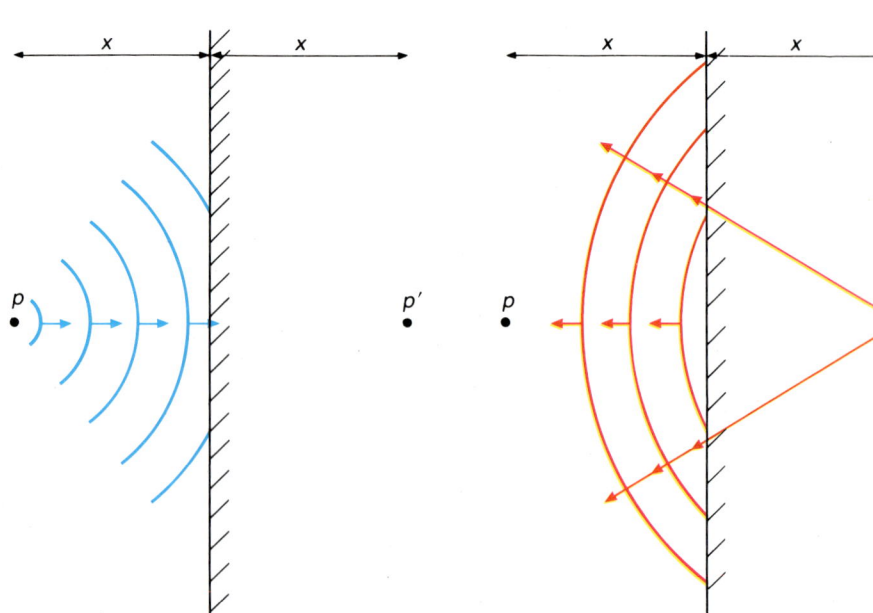

Figure 5
(a) Circular wavefronts spread out from a point

(b) After reflection, the waves appear to have come from a point behind the barrier

Refraction

In Figure 6 you can see some waves going from a region of deep water to shallow water. A region of shallow water in a tank can be made with a glass plate. In shallow water, waves travel more slowly than they do in deep water. There must be the same number of waves passing through both the deep and shallow regions. This means the frequency of the waves is the same. The speed of the waves is given by the equation $V = f \times \lambda$. So when the wave slows down in shallow water the wavelength must be less. In Figure 7 you can see some waves approaching a region of shallow water at an angle. They slow down and change direction. This is called **refraction**.

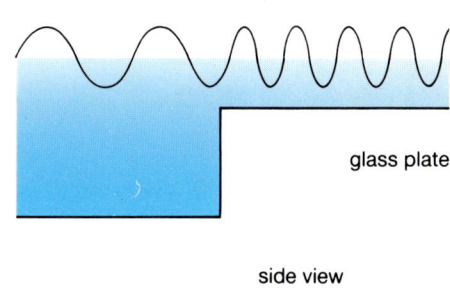

side view

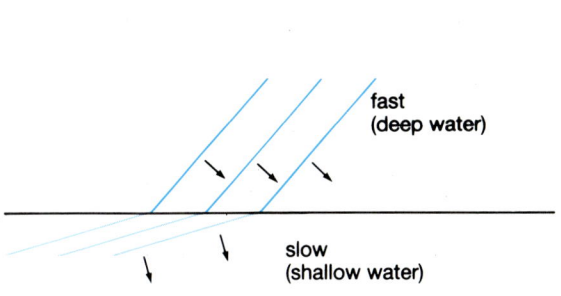

Figure 7
Water waves are refracted when they enter shallow water

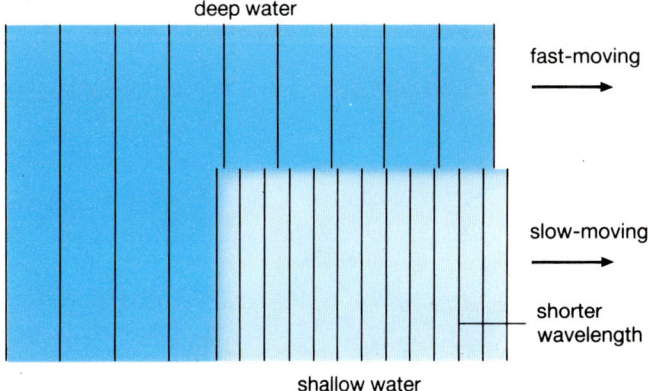

Figure 6
Waves going from deep water to shallow water

Questions

1 Draw careful diagrams to show how the waves are reflected in the following cases.

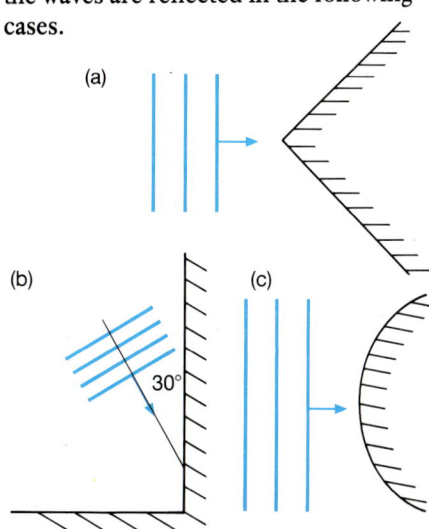

(a)

(b)

(c)

2 At the top of the next column is a bird's eye view of a ripple tank. The tank is tilted so that there is deep water one end, and shallow water at the other.

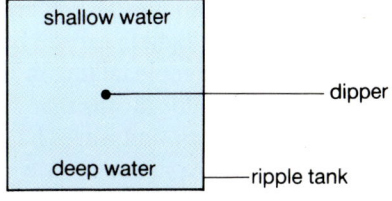

Draw a diagram to show the shape of the ripples produced by the dipper in the middle.

3 Draw careful diagrams to show how the waves are refracted in the following cases.

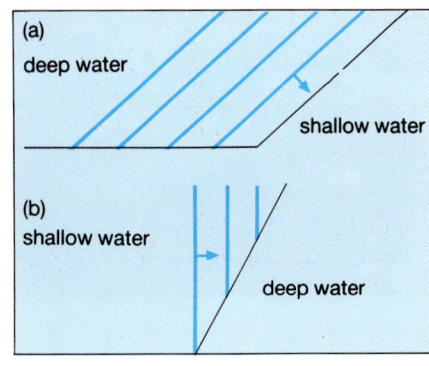

4 A propellor with two blades rotates twenty times a second. You look at it through a hand stroboscope with 10 slits.

(a) Explain why the propellor looks stationary if you rotate your stroboscope twice a second.

(b) What will you see if you rotate your stroboscope at these frequencies:
(i) 1 Hz (ii) 4 Hz (iii) 2.1 Hz?

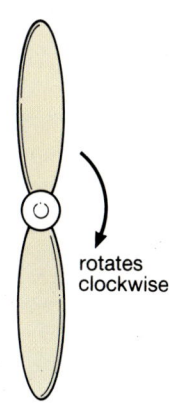

rotates clockwise

137

3 Diffraction and Interference

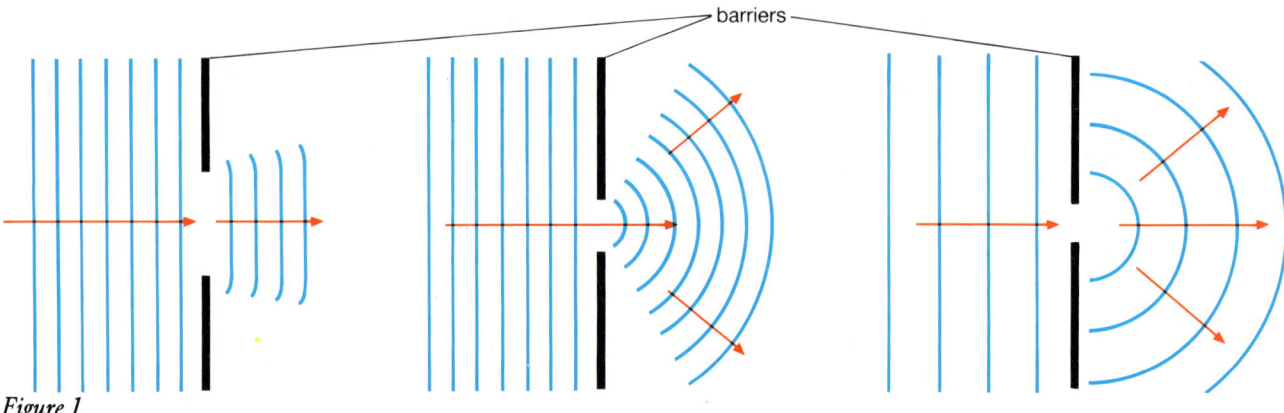

barriers

Figure 1
(a) *Small wavelength, large gap*   (b) *Small wavelength, smaller gap*   (c) *Large wavelength, small gap*

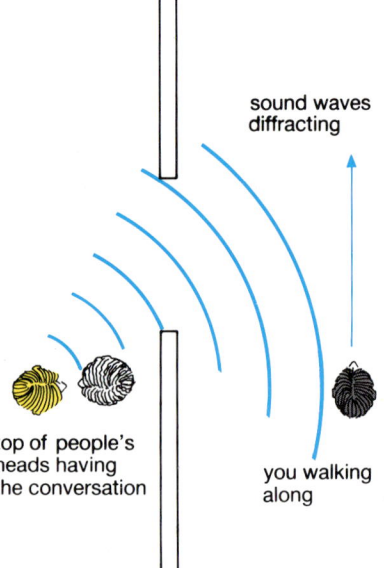

sound waves diffracting

top of people's heads having the conversation

you walking along

Figure 2
A 'bird's eye view' of a conversation overheard through a doorway

Diffraction

When waves pass through a small hole, they spread out. This is called **diffraction**. You can see the diffraction of water waves in a ripple tank. Figure 1 shows you what happens when waves go through a series of gaps in barriers.

In Figure 1(a) the gap is large in comparison with the wavelength of the waves. The waves only spread out a little. In Figure 1(b) the gap is smaller. The waves spread out more. In Figure 1(c) the wavelength is larger than the gap and the waves now spread out completely.

Sound also diffracts. This tells us that sound is carried by waves. Figure 2 shows the sort of position you might find yourself in. You are walking down a corridor and you can hear two people talking though an open door, but you cannot see them. The sound waves do not travel in a straight line. They spread out and change direction as they pass through the doorway. That is why you can hear them talking even though you are a long way down the corridor.

We also think that light is carried by waves. So why can't we see round corners? You cannot see the people talking in Figure 2, because light waves have very small wavelengths. This means that when light waves go through a doorway they hardly diffract at all. Sound waves have much longer wavelengths, so they diffract through the doorway.

In this photograph you can see water waves diffracting as they pass through a harbour entrance. Water waves have a large wavelength, and so they spread out a lot when they diffract

Interference

Some interesting patterns can be produced in a ripple tank when two small dippers make waves together. The two dippers produce a series of crests and troughs. These spread out and overlap as shown in Figure 3.

There are some places where the waves from the two dippers arrive *in phase*. This means that two crests or two troughs arrive at the same time. The two crests move the water upwards to make a larger wave. So in some places, the water moves up and down *more* than it does with only one dipper working. We call this **constructive interference**.

At other places in the ripple tank, waves from the two dippers are arriving *out of phase*. This means that when a crest arrives from one dipper a trough arrives from the other. This time the effect of the two waves is to cancel each other out. The water does not move at all; it is as if the dippers have been switched off. We call this **destructive interference** (Figure 4).

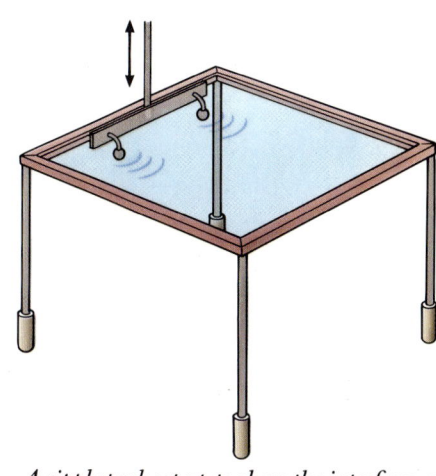

A ripple tank set up to show the interference of waves

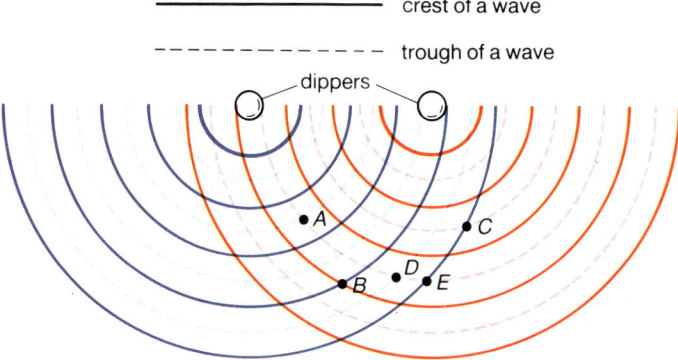

Figure 3

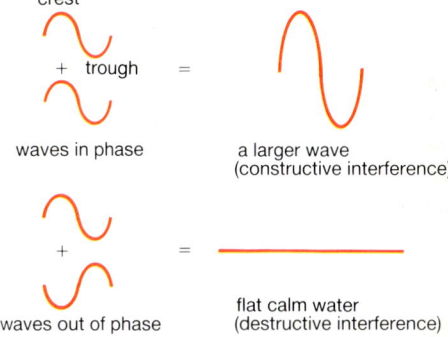

Figure 4

Questions

1 Astronomers use radio telescopes to detect radio waves from distant galaxies. A lot of radio waves have wavelengths a few metres long. Radio telescope dishes are about 100 m across. This means that the radio waves are only diffracted by a small amount. Explain why.

2 In an experiment in a ripple tank, straight water waves are produced with a frequency of 20 Hz. The waves travel at a speed of 40 cm/s through a gap in a barrier of width 1 cm.
(a) Calculate the wavelength of the waves.
(b) Draw a diagram to show how the waves would spread through the gap. Explain how you decided what to draw.
(c) Draw another diagram to show what happens when the gap is made 4 cm wide.

3 In Figure 3 which of the points A, B, C, D, E will show (i) constructive interference (ii) destructive interference?

4 The two diagrams below show slinkies with moving humps on them. Draw diagrams to show the shape of the slinkies (i) when the humps meet (ii) after the humps have passed through each other.

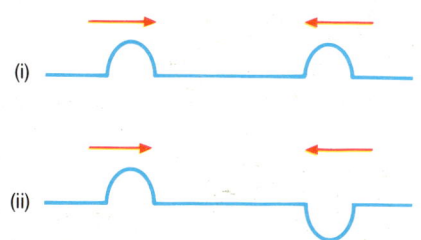

5 Use the information below to sketch a graph of the displacement of the point A against time. Your time axis should cover the next two seconds after the instant shown in the diagram.

- The wavelength of both sets of waves is 0.5 m.
- The amplitude of the waves is 0.2 m.

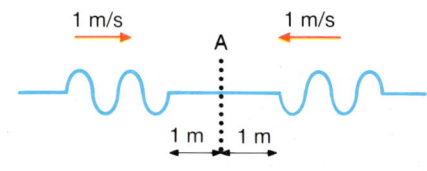

4 Sound Waves

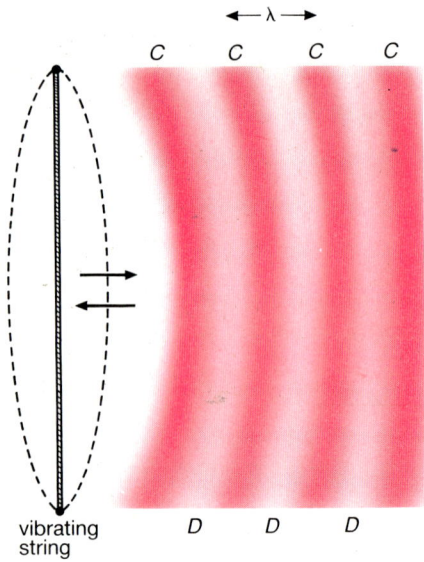

C = compression
D = decompression

Figure 1
A vibrating guitar string

Speeds of sound (m/s)	Material
330	air
1500	water
5000	steel

Table 1

The bell in the jar will make a noise as long as the jar is full of air. If we pump the air out, we cannot hear the noise any more, but we can still see the bell working

Making and hearing sounds

When you pluck a guitar string the instrument makes a noise. If you put your finger on the string you can feel the string vibrating. Sounds are made when something vibrates. The vibrations of a guitar string pass on kinetic energy to the air. This makes the air vibrate.

Sound is a longitudinal wave. Molecules in air move backwards and forwards along the direction the sound travels in. In Figure 1 you can see that when the guitar string moves to the right it compresses the air on the right hand side of it. When the string moves to the left the air on the right expands. The string produces a series of **compressions** and **decompressions**. In a compression the air pressure is greater than normal atmospheric pressure. In a decompression the air pressure is less than normal atmospheric pressure.

Compressions and decompressions travel through air in the same way that compressions will move along a slinky. Your ear detects the changes in pressure caused by sound waves. When a compression reaches the ear it pushes the ear drum inwards. When a decompression arrives, the ear drum moves out again. The movements of the ear drum are transmitted through the ear by bones. Then nerves transmit electrical pulses to the brain.

Speed of sound

The speed of sound depends on which material it travels through. Sound waves are transmitted by molecules knocking into each other. In air sound travels at about 330 m/s. In solids and liquids, where molecules are packed more tightly together, sound travels faster (Table 1). In a vacuum there are no molecules at all. Sound cannot travel though a vacuum, although light can.

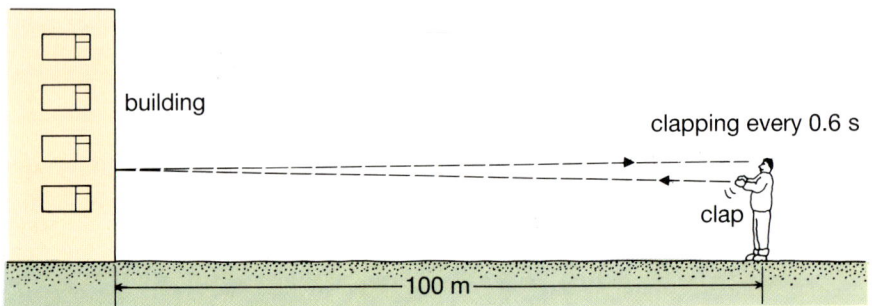

Figure 2
Measuring the speed of sound through air

Figure 2 shows a simple way for you to measure the speed of sound through air. Stand about 100 m away from a building and clap your hands. Sound waves will be reflected back to you from the building. When you hear the echo, clap again so that you clap in time with the echo. A friend, watching you clapping your hands, times 10 claps in 6 seconds. Now, you know that the sound took 0.6 s to travel 200 m (to the wall and back again).
So the speed of sound is given by:

$$V = \frac{d}{t}$$

$$= \frac{200\,\text{m}}{0.6\,\text{s}} = 330\,\text{m/s}$$

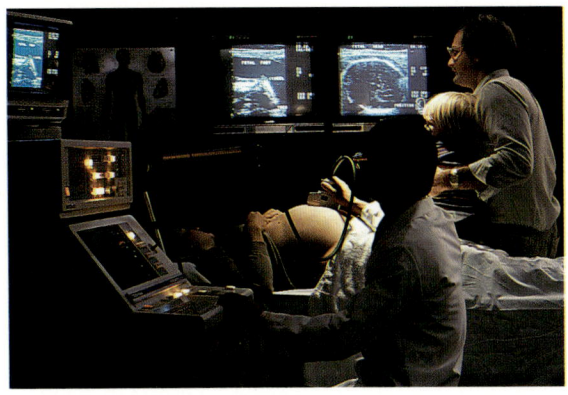

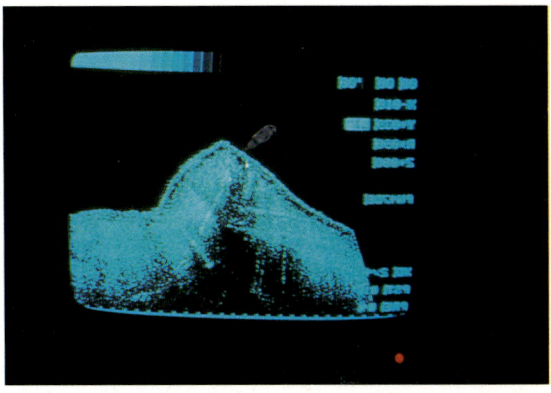

Ultrasonic depth finding

We use the name **ultrasound** to describe very high frequency sound waves. These waves have such a high frequency that we cannot hear them. Because ultrasound has a high frequency, the waves have a short wavelength. This means that it is possible to produce a narrow beam of ultrasound without it spreading out due to diffraction effects.

Ships use beams of ultrasound for a variety of purposes. Fishing boats look for fish, destroyers hunt for submarines or explorers chart the depth of the oceans. Figure 3 demonstrates the idea. A beam of ultrasonic waves is sent out from the bottom of a ship. The waves are reflected from the sea bed back to the ship. The longer the delay between the transmitted and reflected pulses, the deeper the sea is (Figure 4).

In the picture on the left, ultrasound emitted by the probe placed on the mother's stomach is reflected by the fetus. A computer builds up a picture from those reflected waves, which, unlike X-rays, are perfectly safe. In this photograph the whole family views its new member, a healthy baby boy. The photograph on the right shows the result of an ultrasound scan to check for breast tumours

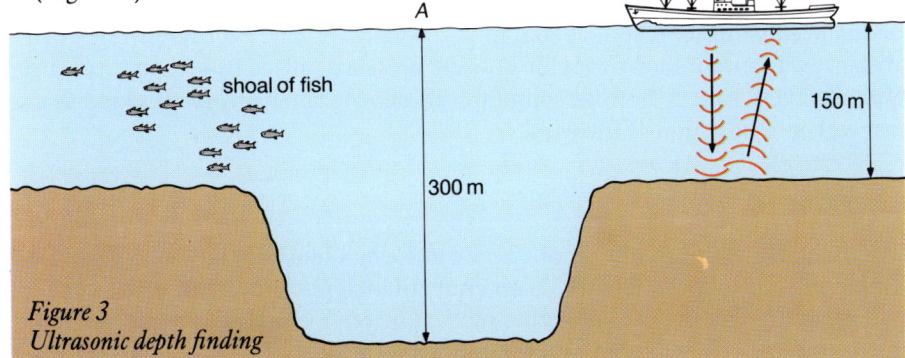

Figure 3
Ultrasonic depth finding

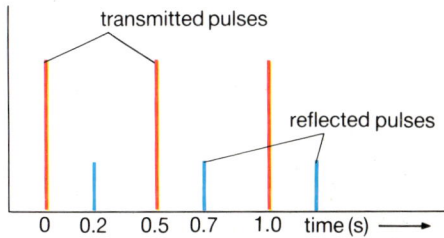

Figure 4
Ultrasound pulses for a depth of 150 m

Questions

In these questions you will need to refer to the table of speeds given on page 140.

1 The frequency of sound that comes out of your mouth is on average about 250 Hz.
(a) What is the wavelength of that sound?
(b) Explain why sound is diffracted when it comes out of your mouth.

2 The ship in Figure 3 sends out short pulses of ultrasound of frequency 50 kHz (50 000 Hz) every 0.5 s.
(a) The length of each pulse is 0.01 s. How many waves are emitted in that time?
(b) Use the information in Figure 4 to show that ultrasound travels through water at 1500 m/s.
(c) Sketch a graph to show the time interval between transmitted and reflected pulses when the ship reaches point A.
(d) What difficulty is there in

measuring the depth of the sea, when it is 500 m deep?
(e) What is the wavelength of the ultrasound waves?
(f) Why is it important to have a *narrow* beam of ultrasound waves? Explain why it is possible to produce a narrow beam with ultrasound but not ordinary sound waves.

5 Loudness, Quality and Pitch

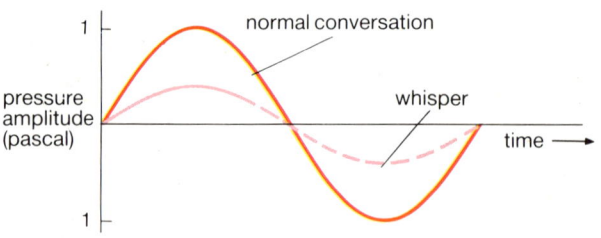

Figure 1
Sound waves caused by the human voice

Loudness

Your ears are very good detectors of energy. A disco produces sound energy at a rate of about 40 W. You find this very noisy. If you were to light a disco with one 40 W light bulb, everyone would complain that it was too dark.

Your ears detect sounds over a wide range of frequencies. You can hear frequencies as low as 20 Hz, and as high as 20 000 Hz. Yours ears are most sensitive to frequencies of about 2000 Hz. So a note at a frequency of 2000 Hz sounds louder than a note of frequency 10 000 Hz that carries the same energy. A loud noise makes your ear drums move a long way, while a very quiet noise has only a small effect on your eardrums. The loudness of a noise depends on the pressure caused by the sound wave. During normal conversation your voice will cause the air pressure to change by about 1 pascal ($1\,\text{N/m}^2$). Your voice produces a pressure wave of amplitude 1 pascal (Figure 1). This is a small change in comparison with atmospheric pressure (100 000 pascals).

Loud noises can be very unpleasant and can damage your hearing. There are laws in force which limit the noise that industrial machinery, cars and aeroplanes are allowed to make. A sonic boom from Concorde can be heard over very great distances. Figure 2 shows roughly how the pressure caused by Concorde changes with distance. Exactly how far sound travels depends on many factors such as the strength and direction of the wind.

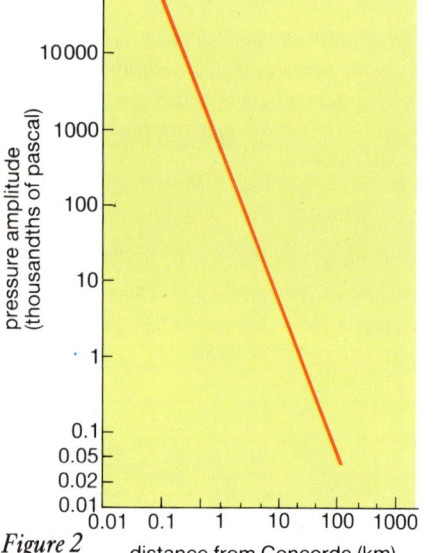

Figure 2

Pitch

We use the term **pitch** to describe how a noise or a musical note sounds to us. Bass notes are of low pitch, treble notes are of high pitch. Men have low-pitched voices, women have voices of higher pitch. The pitch of a note is related directly to its frequency. The higher-pitched notes are the notes with higher frequency.

Figure 3 shows you how you can use a *microphone* and an *oscilloscope* to show that higher-pitched notes have higher frequencies.

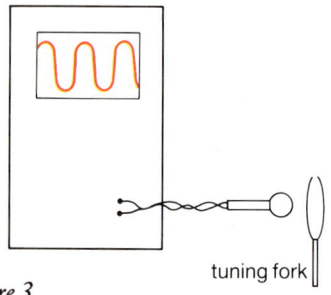

Figure 3
(a) Experiment to investigate frequencies

(b) A note from a tuning fork of low pitch

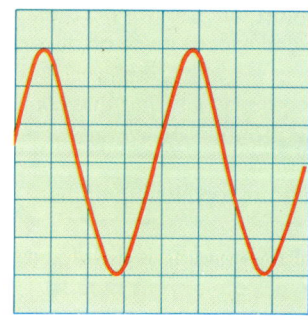

(c) A note from a tuning fork of higher pitch

If you played the same note on these instruments, they would have the same frequency (or pitch). You could still tell which instrument had played the note because each has its own special waveform

Quality

On a piano the note that is called middle C has a frequency of 256 Hz. This note could be played on a piano or a violin, or you could sing the note. Somebody listening to the three different sounds would recognise straight away whether you had sung the note or played it on the piano or violin. The **quality** of the three notes is different. The quality of a note depends on the shape of its waveform. Although two notes may have the same frequency and amplitude, if their waveforms are of a different shape you will detect a different sound (Figure 4).

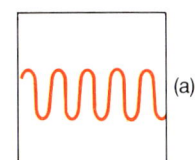 (a)
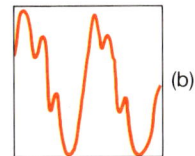 (b)
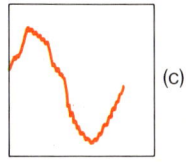 (c)
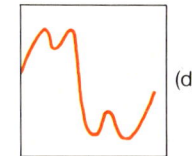 (d)

Figure 4
Different waveforms produced by four different musical instruments. The notes have different qualities.

Questions

1 Use the data below, and Figure 2, to answer these questions.
(a) At approximately what distance from Concorde will the sound level be enough to interrupt a conversation?
(b) Work out roughly how far you need to be from Concorde so that you cannot hear it.

Noise level	Pressure amplitude in thousanths of a pascal
Painfully loud	100 000
Normal conversation	1000
Quiet countryside	1
Too quiet to be heard	0.02

(c) Explain why it is necessary to have laws controlling the noise levels near airports.

2 In Figure 3 the dot on the oscilloscope screen took 0.001 s to cross the screen. What is the frequency of each of the two notes?

3 Look at Figure 4. In each case the same oscilloscope settings were used.
(a) Which note has the highest frequency?
(b) Which note is the loudest?
(c) Which note is the softest?
(d) Which two notes have the same frequency?

4 An electronic synthesizer produces the two pure notes *A* and *B* as shown (right). It produces a third note *C* by adding the two waveforms together.
(a) Copy the two waveforms carefully onto some graph paper.

(b) Add the two waveforms together to produce the waveform of *C*.
(c) Does *C* sound louder than *A* or *B*?
(d) How does the frequency of *C* compare with: (i) *A*, (ii) *B*?
(e) Does *C* have the same quality as *A* or *B*?

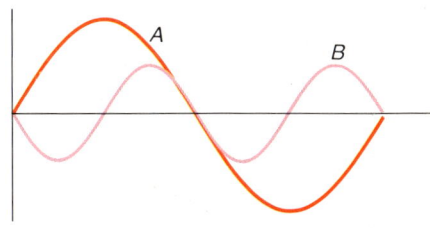

6 Earthquakes

Devastation caused by seismic waves from the T'ang Shan earthquake in 1976

Around the world there are about 800 000 earthquakes every year. You may be rather alarmed to read that. Fortunately most of these earthquakes are too small for us to notice. At the other end of the scale some earthquakes release an enormous amount of energy. One of the largest recorded earthquakes occurred in T'ang-Shan, China in 1976. This earthquake released as much energy as 10 000 nuclear explosions. The effect on the surrounding countryside was devastating and about 700 000 people are thought to have died.

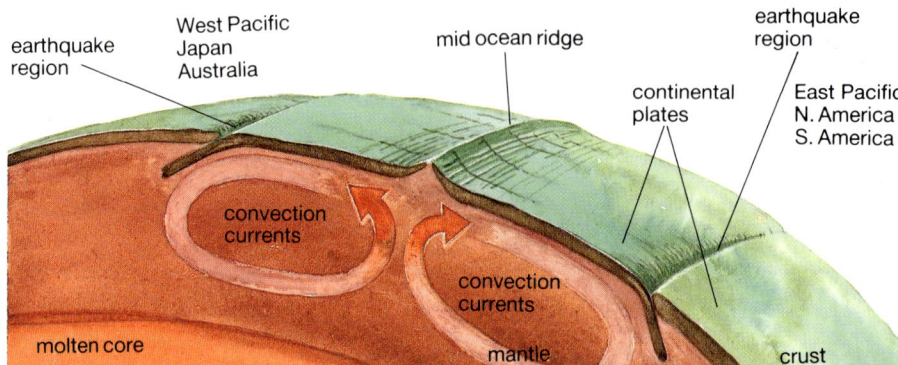

Figure 1
Continental plate movements

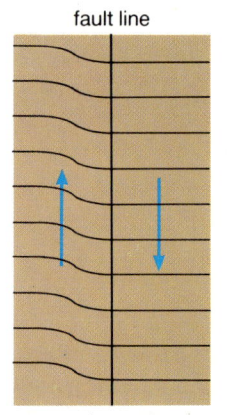

Figure 2
(a) The movement of continents causes the rocks to deform on either side of the fault

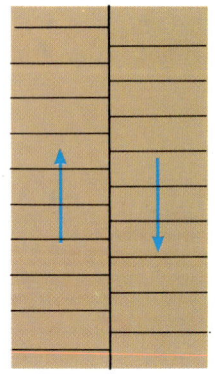

(b) An earthquake occurs, causing the rocks to slip past each other

Why do earthquakes happen?

The inside of the Earth is very hot. At the centre of the Earth is a molten (liquid) **core** made mostly of iron. Heat from the core warms up the solid layer above it called the **mantle**. This heating in the mantle sets up convection currents and hotter, less dense, rocks start moving upwards towards the surface of the Earth. You might be surprised to read that convection occurs in solid rock. The convection currents are very slow-moving; it takes thousands of years for rocks to move any noticeable distance.

Geologists think that convection currents in the mantle cause large parts of the Earth's outer surface (**crust**) to move as well. Figure 1 shows how two 'continental plates' collide on either side of the Pacific Ocean. As these large continental plates move past each other large pressures build up in rocks. Figure 2 shows two layers of rock beginning to slide past each other.

For a while these layers of rock are held in position by large frictional forces. Eventually the frictional forces can hold the rocks in place no longer and the rocks slip past each other. This slipping gives rise to earthquakes. The slipping of rocks occurs at lines called **faults**. One of the best known faults is the San Andreas fault in California. In 1906 this fault moved giving rise to a large earthquake which caused serious damage to San Francisco.

Seismic waves

When earthquakes happen they produce shock (seismic) waves which travel through the Earth. There are three sorts of wave (Figure 3). **Surface waves** roll around the top layers of the Earth like ocean waves. These waves cause the damage to buildings. The waves that go through the Earth are **compressional** (p-) waves and **shear** (s-) waves; p-waves are longitudinal waves which travel through the mantle and the core; s-waves are transverse waves which only travel through the solid mantle.

These seismic waves can be detected using a *seismometer*. A large mass is suspended freely from a beam. When an earth tremor moves the ground the mass stays at rest. A pen can record the relative movement between the Earth and the mass. A chart that shows how the earth movement varies with time is called a *seismograph*. Figure 4 shows how three different seisometers record the seismic waves from one earthquake. The s-, p- and surface waves all move at different speeds. There is a larger interval in time between receiving the s- and p-waves, the further you are away from the earthquake.

The scientific study of earthquakes is very important. It is now possible to predict where earthquakes are likely to happen. In earthquake regions, engineers must build stronger buildings to reduce the chance of serious damage in an earth tremor. However, there is nothing anybody can do to protect against a really big earthquake.

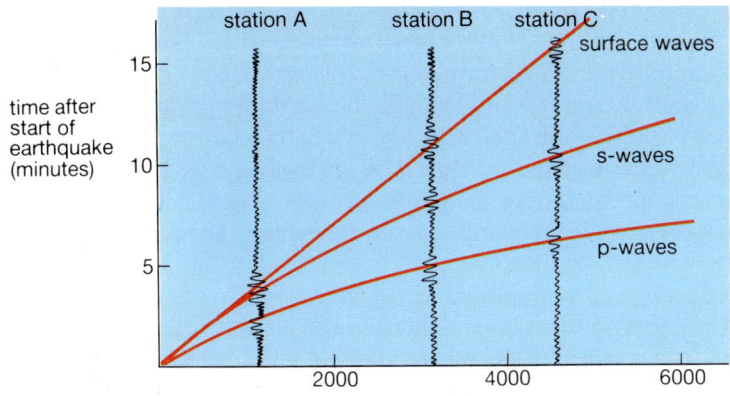

Figure 4
Seismic waves arriving at different times at seismograph stations

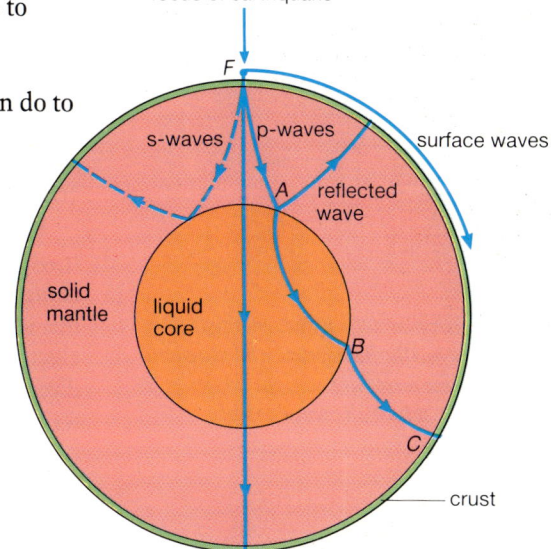

Figure 3
The paths of some seismic waves through the Earth. S-waves cannot travel through the Earth's liquid core

Questions

1 (a) Explain carefully how heat from the centre of the Earth can cause the surface of the Earth to move.
(b) Why do earthquakes occur in some parts of the world but not others?
2 For this question you will need to use the data in Figure 4.
(a) Calculate the speed of surface waves in km/s. Are these faster or slower than p-waves?
(b) How far away from the centre of the earthquake are stations A and B?
(c) The map (right) shows the positions of stations A and B on the west coast of America. Use the map to suggest in which city the earthquake happened.
(d) You are at a distance of 2000 km from the earthquake. What time delay is there between receiving p- and s-waves?

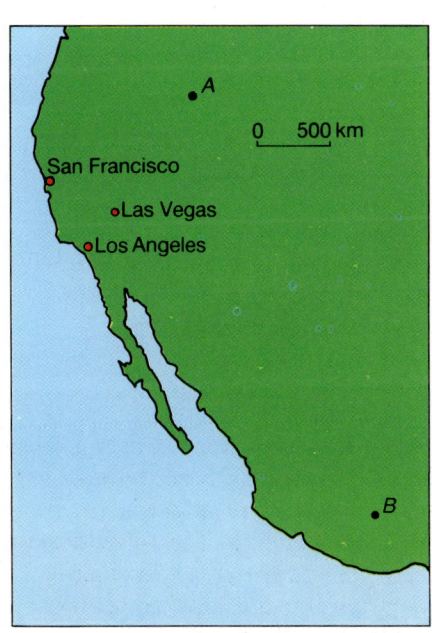

(e) You detect surface waves ten minutes after receiving p-waves. How far from the centre of the earthquake are you?
(f) Why do you think that the Americans know when the Russians test a nuclear bomb underground?
3 This question is about the path of seismic waves through the Earth shown in Figure 3.
(a) When p-waves enter the core at point A they are refracted. How can you tell from the diagram that the waves travel more slowly in the core than they do in the mantle?
(b) Draw a diagram to show the refraction of the p-waves at A.
(c) As p-waves travel further into the core their speed increases. How can you tell that from the path they follow, AB?

7 Electromagnetic Waves

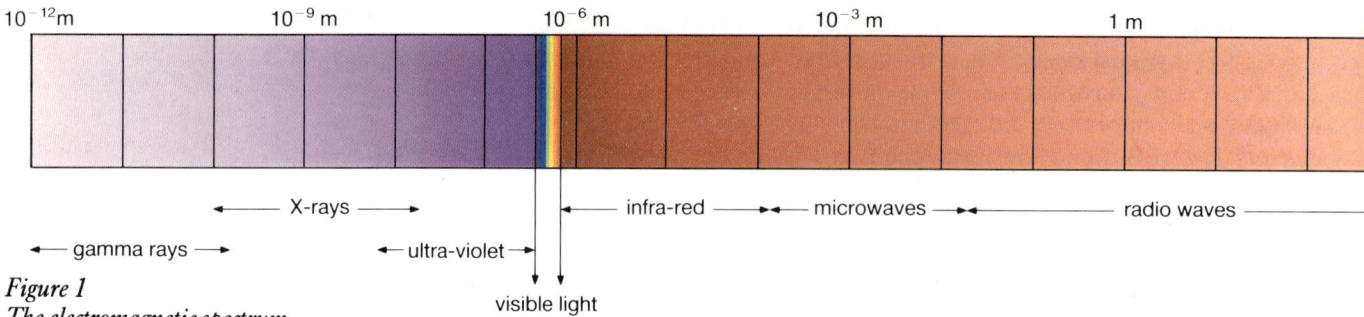

X-rays

gamma rays

ultra-violet

infra-red

microwaves

radio waves

visible light

Figure 1
The electromagnetic spectrum

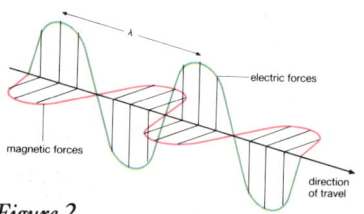

electric forces

magnetic forces

direction of travel

Figure 2
In an electromagnetic wave, energy is carried by oscillating electric and magnetic forces. These forces are at right angles to the direction in which the wave travels

Sunbathers use ultraviolet rays from the sun to get a tan. Staying in the sunshine (or under a sun lamp) a long time can be dangerous, however, as the ultraviolet rays cause skin cancer

You will already have heard of radio waves and light waves. These are two examples of **electromagnetic waves**. There are many sorts of electromagnetic wave, which produce very different kinds of effect. Figure 1 shows you the full *electromagnetic spectrum*. The range of wavelengths in the spectrum stretches from 10^{-12} m for gamma rays to about 2 km for radio waves.

All the waves you have met so far travel through some material. Sound waves travel through air, seismic waves travel through the Earth, water ripples travel along the surface of water. Electromagnetic waves can travel through a vacuum; this is how energy reaches us from the Sun. The energy is carried by changing electric and magnetic forces. These changing forces are at right angles to the direction in which the wave is travelling. So electromagnetic waves are transverse waves (Figure 2).

Electromagnetic waves show the usual wave properties. They can be reflected and refracted. They show diffraction and interference effects. In a vacuum all electromagnetic waves travel at the same speed of 3×10^8 m/s. However, electromagnetic waves do not all travel at the same speed when they travel in a material. For example, different colours of light travel at different speeds in glass.

This radio dish transmits international telephone calls using 0.1 metre wavelength radiowaves

- **Radio waves**. Radio 4 broadcasts on longwave use a wavelength of 1500 m; Capital Radio uses a wavelength of 194 m. These radio waves are produced by high frequency oscillations of electrons in aerials. Radio waves with wavelengths of a few hundred metres are used in local and national radio (see Figure 3).

  Radio waves with wavelengths of a few centimetres are used to transmit television signals and international phone calls. If you make a phone call to America your radio signals are sent out into space by large aerial dishes, like the one you can see in the photograph opposite. These signals are received by a satellite in orbit around the Earth. Then the signals are relayed to another aerial dish in America. It is important to use short wavelengths for international communications, so that a narrow beam of waves can be directed towards the satellite. Long wavelengths would be diffracted, so not much energy would reach the satellite.

- **Infra-red waves** have wavelengths between about 10^{-4} m and 10^{-6} m. Anything that is warm will lose energy by giving out infra-red radiation. You lose some heat energy by radiation. You can certainly feel the infra-red radiation given out, or *emitted*, by an electric fire. As things get hotter they also emit electromagnetic waves of even shorter wavelength. The Sun sends out light and ultra-violet waves.

  Infra-red photography can be used to measure the temperature of objects. The hotter something is, the more infra-red radiation it gives out.

- **X-rays** have wavelengths of about 10^{-10} m. These rays can cause damage to body tissues, so your exposure to them should be limited. X-rays are now widely used in medicine. X-rays of short wavelengths will pass through body tissues but will be absorbed by bone. Such rays can be used to take a photograph to see if a patient has broken a bone.

  Slightly longer wavelength X-rays are used in body scanning. These X-rays are absorbed by body tissues and doctors can build up a picture of the inside of a patient's body. This allows doctors to investigate whether a patient has cancer.

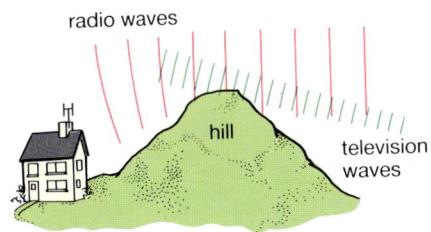

Figure 3
Long and medium wavelength radio waves will diffract around hills and houses. However, waves used for TV signals are of short wavelengths. These will not bend around hills so well; this house will have poor TV reception

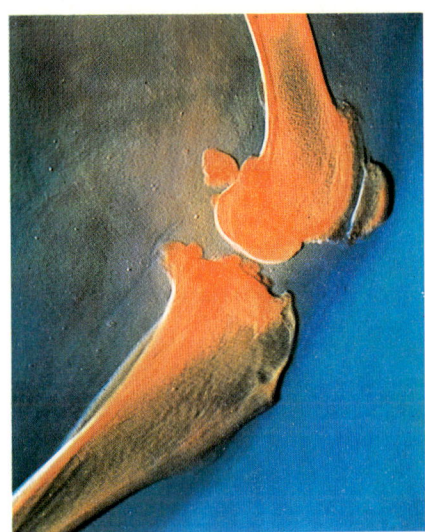

This false colour X-ray photograph shows a severely arthritic knee joint. Techniques like this are valuable for diagnosing injury or illness

Questions

1 In the table below you can see some data showing typical values of wavelengths and frequencies for different types of radio wave.

Type of radio wave	Wavelength (m)	Frequency (MHz)
Long	1500	
Medium	300	
Short	10	
VHF		100
UHF		3000

(VHF = very high frequencies, UHF = ultra high frequencies).

(a) Copy the table. Then use the equation $V = f \times \lambda$ to fill in the missing values.
(b) Which wave would you use for (i) a local radio station (ii) television broadcasts to the USA. Explain your choice.

2 The diagram (right) shows a side view of a radio dish. Explain why the receiver is placed some distance away from the dish. (A diagram may make your answer clearer.)

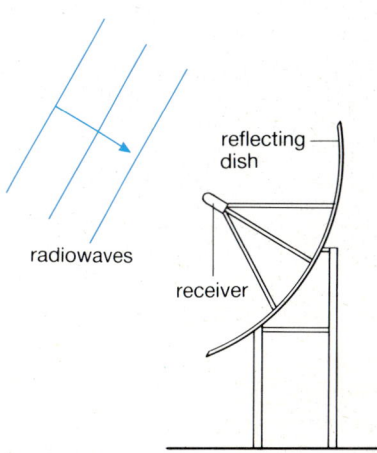

8 Light as a Wave

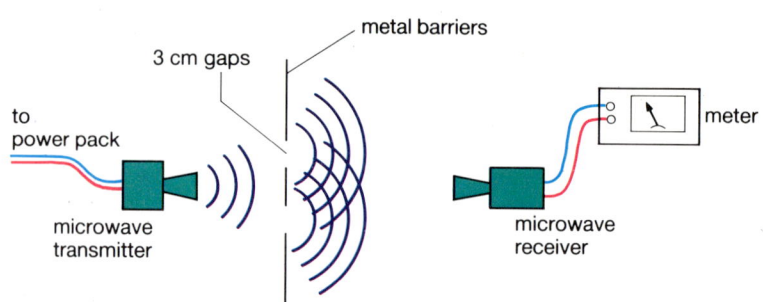

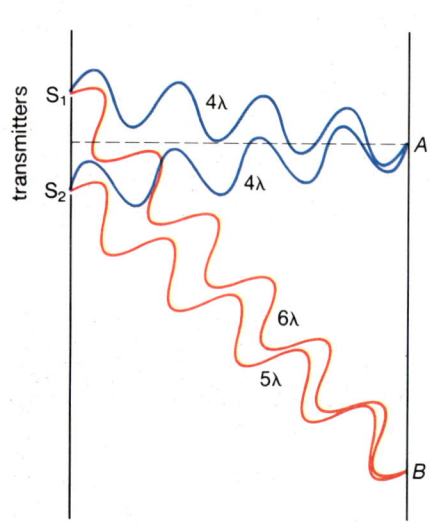

Figure 2
Constructive interference occurs at both points, A and B

The brilliant colours of the peacock are due to interference effects. Light is reflected from fine ribbings on the feathers. At certain angles constructive interference occurs for one colour of light

Figure 1
An experiment to detect the interference of microwaves

Microwave interference

Electromagnetic waves with a wavelength of 3 cm (**microwaves**) can be used to demonstrate interference effects in a laboratory. Figure 1 shows the sort of arrangement you can use. Microwaves from a transmitter are directed towards two small gaps in a metal barrier. The gaps should be about 3 cm wide, then the microwaves diffract out through these gaps. On the other side of the barrier the microwaves will overlap. This allows them to interfere in the same way that water waves do in a ripple tank. There will be places where the microwave receiver detects *constructive interference*. In other places the receiver detects little energy due to *destructive interference*.

Figure 2 shows you how constructive interference can occur in more than one place. Waves travel the same distance to A. So they arrive in phase there, and A is a point where you detect constructive interference. You also detect constructive interference at B. This time the waves from S_1 travel further than those from S_2 by one extra wavelength. The *path difference* between the two sets of waves is one wavelength.

You can use this idea to work out the wavelength of microwaves. For example you might measure with a ruler that $S_1B = 18$ cm and $S_2B = 15$ cm.

> Path difference $S_1B - S_2B = 1$ wavelength
>
> So 1 wavelength $= 18$ cm $- 15$ cm
>
> $= 3$ cm

In general, constructive interference occurs if path difference $= n\lambda$

destructive interference occurs if path difference $n\lambda + \dfrac{\lambda}{2}$

(n is a whole number, 0, 1, 2, 3 etc.)

Interference of light

Figure 3 shows you how you can study the interference of light. It is the same sort of idea you used to study microwave interference. To make interference happen you must produce two beams of light that overlap. Light has a very small wavelength. So you need to use two very small slits, as shown in the diagram. The slits must be very narrow (about 0.1 mm wide) so that light is diffracted by them. The slits must also be placed close together so that the beams of light from each slit overlap.

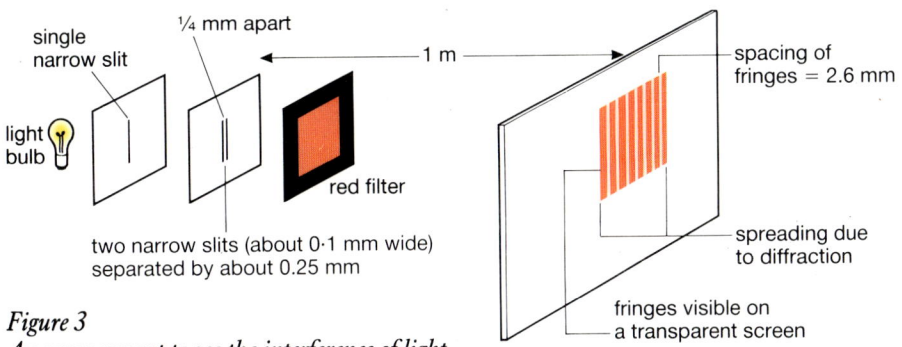

Figure 3
An arrangement to see the interference of light

Figure 4

In a slightly darkened room it is quite easy to see **interference fringes**. With a red filter in place, you will see alternate red and dark lines. The red lines correspond to places of constructive interference; here the red light waves arrive in phase. At the dark places the light waves arrive out of phase; this is destructive interference.

The light wavelengths are too small for you to measure with a ruler as you did with microwaves. However, there is a formula which will help you.

$$\text{wavelength} = \frac{\text{fringe spacing} \times \text{slit separation}}{\text{distance from slits to screen}}$$

Using the information in Figure 3, the wavelength for red light is:

$$= \frac{(2.6 \times 10^{-3}\,\text{m}) \times (2.5 \times 10^{-4}\,\text{m})}{1\,\text{m}}$$

$$= 6.5 \times 10^{-7}\,\text{m}$$

Look at Figure 4 which shows fringes using red, green and blue light. From the formula above you can see that the wavelength is proportional to the fringe spacing. So red light has the longest wavelength and blue light the shortest.

When white light is used, you see a whole series of colours. This is because white light is made up of all colours. Each colour has a different wavelength. At some angles red light interferes constructively, while at other angles blue light does, and so on.

Questions

1 This question refers to Figure 4. The red, green and blue fringes were obtained using exactly the same apparatus but different filters. The wavelength of the red light is 6.5×10^{-7} m. Use the diagrams to work out the wavelength of green and blue light.

2 This question is about the interference fringes shown in Figure 3.
(a) The screen is moved further away from the slits. What difference does that make to: (i) the spacing of the fringes, (ii) the brightness of the fringes?
(b) The slits are moved further apart. What difference does that make to the spacing of the fringes?
(c) The slits are made slightly narrower. What difference does that make to: (i) the spacing of the fringes, (ii) the brightness of the fringes, (iii) the number of fringes that you can see?

3 Opposite, you can see a thin layer of oil floating on a puddle of water. You can see a ray of light that is partly reflected from the surface of the oil, and then partly reflected from the water surface. Explain why if you look at oil on water you can see patches of colour.

4 All waves show interference effects. Design an experiment to show that sound waves interfere. Explain how you are going to observe this interference.

SECTION G: *STUDY QUESTIONS*

1 (a) The drawing shows a ship 800 m from a cliff. A gun is fired on the ship. After 5 seconds the people at the front of the ship hear the sound of the gun again.

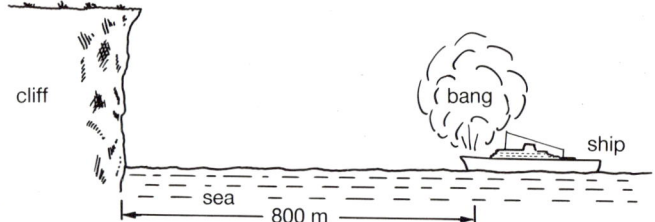

1

(i) What is the name of this effect?
(ii) What happens to the sound at the cliff?
(iii) How far does the sound travel in 5 seconds?

(iv) $\text{Speed} = \dfrac{\text{distance travelled}}{\text{time taken}}$

Use this equation to calculate the speed of sound.

(b) The diagram below shows three people standing around a house.

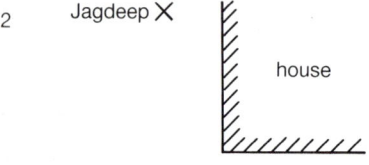

2

(i) Who could see Johnny?
(ii) Who could hear Johnny?
(iii) To answer (i) you have assumed that light travels in straight lines. What did you assume about sound when you answered (ii)?

(c) The diagrams below show experiments which could be done in a ripple tank.

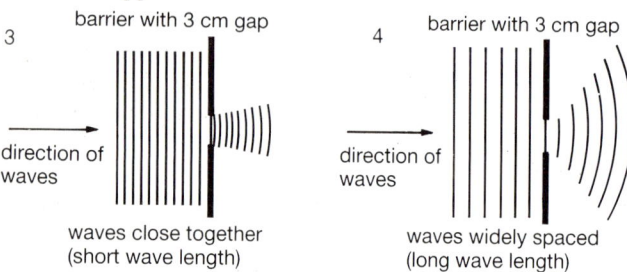

3   4

You will see that the waves spread out after going through the gap in the barrier. This effect is called diffraction.
In which experiment did waves spread out less?

(d) Diagram 5 shows a ship searching for a submarine. It sends our narrow beams of sound such as *AB* and *AC*. Ordinary sound is not satisfactory. Ultrasound has to be used.

5

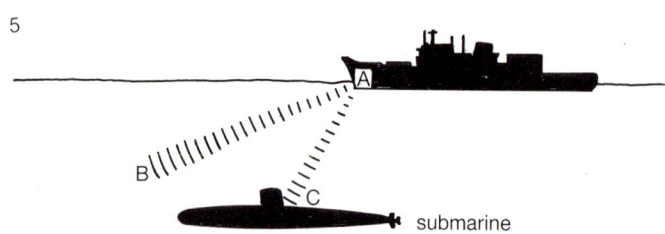

(i) What is ultrasound?
(ii) Why is ultrasound necessary?

(e) The diagram below shows an oscilloscope on the ship.

6

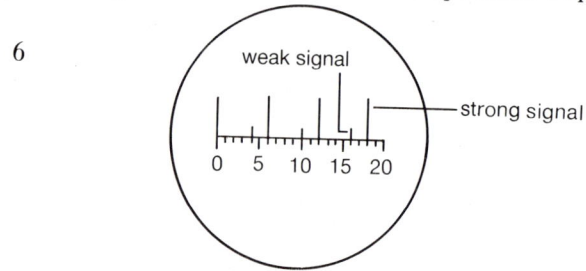

The numbers on the scale (diagram 6) show time in tenths ($\frac{1}{10}$ths) of a second. Pulses of ultrasound are sent out every 6 tenths of a second. These are shown on the oscilloscope as strong signals. Weaker signals are received when ultrasound is directed on path *AC*. (diagram 5)

(i) How much time passes between the strong signal going out and the weak signal coming in?
(ii) Ultrasound travels with a speed of 1500 m/s in water. How far is the submarine from the ship?

(f) Diagram 7 shows ultrasound being used to obtain an image of an unborn baby.

7

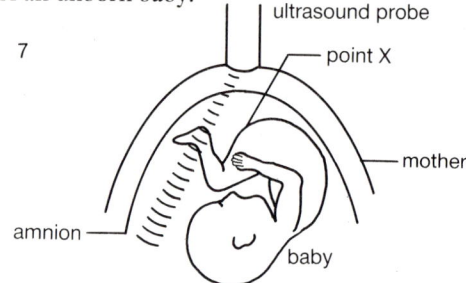

(i) Why is ultrasound used, rather than X-rays?
(ii) When ultrasound goes from one type of material or tissue to another it may be reflected. Explain why ultrasound is reflected at *X*.

(g) This graph shows how the length of a typical baby increases in the womb.

8

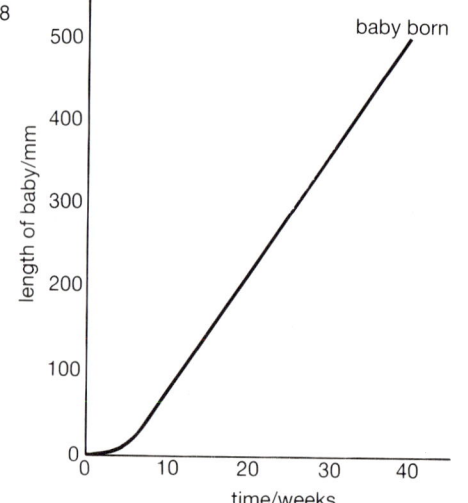

(i) If the ultrasound method shows that the size of a baby is 200 mm how old is it likely to be?

(ii) Calendar:

January	31 days	July	31 days
February	28 days	August	31 days
March	31 days	September	30 days
April	30 days	October	31 days
May	31 days	November	30 days
June	30 days	December	31 days

If the baby is 200 mm long on January 1st, when is it likely to be born? Show how you work out the answer.

LEAG

2 The diagram shows plane water waves of frequency 3 Hz passing across a region where the water is less deep.

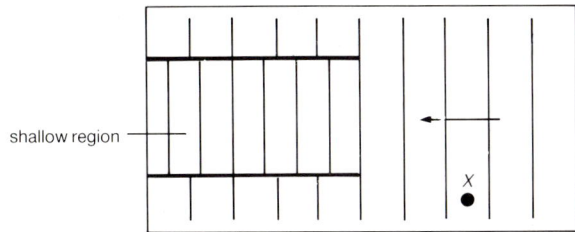

(a) How many waves are being produced each second?

(b) A tiny float is placed at the point marked X. Describe its motion.

(c) What has happened to the wavelength of the waves as they enter the shallow region?

(d) State and explain what you can conclude about the speed of the waves as they enter the shallow region.

(e) Name the device you would use to "freeze" the wave motion.

NEA

3 This question is about microwaves. The diagram below shows cross-sections of the oven of a microwave cooker. In diagram 2 the rotating reflector is shown after a rotation of about 180° from the position shown in Diagram 1.

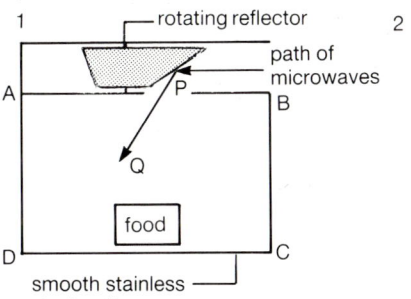
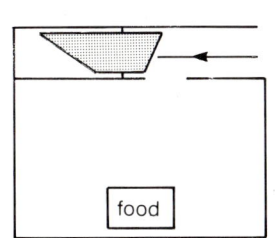

Electromagnetic waves of frequency about 2500 MHz (i.e. 2.5×10^9 Hz) enter the oven as shown. Water molecules within the food absorb the waves and the food becomes hot. Microwaves are reflected by metal but pass through glass, china, dry paper and cardboard with little change.

(a) The microwaves are reflected by the wall of the oven. Why?

(b) The microwaves of path PQ (diagram 1) would be reflected by the walls many times. Copy the diagram of the lower compartment ABCD and draw in the path of the microwave to include the first **three** reflections.

(c) Why does this cooker use a rotating reflector?

(d) What happens to the amplitude of the microwaves as they pass through a bit of food? Why?

(e) A microwave oven uses less energy to cook food than an ordinary gas or electric oven. Why?

(f) Microwaves travel more slowly in food than in air. What effect will this have on the wavelength?

SEG

4 (a) A sound is played into a microphone which is connected to an oscilloscope. The display on the oscilloscope is shown below.

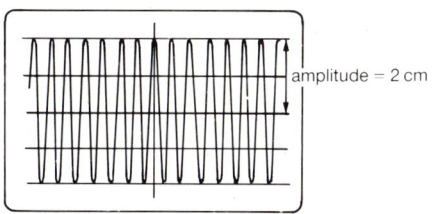

A new sound is played into the microphone and a new display is seen on the oscilloscope as follows:

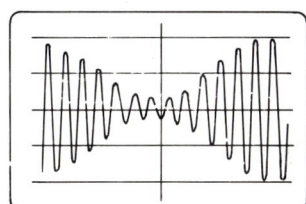

(i) What is the maximum amplitude of the new display?

(ii) The new display shows that the frequency of the sound is constant. How can you say, by looking at the new display, that the frequency is constant?

(iii) Describe the new sound.

(b) The table shows the frequencies and wavelengths of some radio stations.

Station	Frequencies (Hz)	Wavelength (m)
BBC Radio 1	1 091 000	275
BBC Radio 4	200 000	1 500
BBC Radio Birmingham	1 458 000	206
BBC Radio Nottingham	1 584 000	189

(i) Which station has the highest frequency?

(ii) How does the wavelength of this statio compare with the wavelength of the other stations?

(iii) Why should a radio station always transmit at the same frequency?

MEG

STUDY QUESTIONS

5 The speed of deep ocean waves depends on their wavelength. This is shown in the table below.

Speed (m/s)	Wavelength (m)
2.8	5
4.0	10
5.6	20
8.0	40
10.6	70
12.6	100

(a) Plot a graph of speed of the waves (*y*-axis) against their wavelength (*x*-axis).
(b) How fast do waves with a wavelength of 80 m travel?
(c) In a tropical storm near Tonga ocean waves are produced. The wavelength of these waves varies between 10 m and 100 m.
What range of wavelengths will reach Fiji 24 hours after the storm started? (Fiji is 800 km from Tonga.)

6 In the diagram below S₁ and S₂ are two loudspeakers. The speakers are both supplied by the same voltage source. The speakers produce a note of frequency 165 Hz.

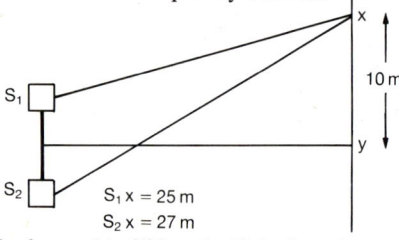

(a) The speed of sound is 330 m/s. Calculate the wavelength of the sound waves.
(b) Edward walks from *Y* to *X*. At *Y* the sound seems loud, and at *X* it seems loud. But in between *X* and *Y* the loudness decreases. Explain Edward's observations clearly.
(c) The connections to one of the speakers are reversed. Explain what Edward hears now as he walks from *X* to *Y*.
(d) The frequency of the note is changed to 660 Hz. Describe what Edward hears as he walks between *X* and *Y*.

7 An earthquake produces seismic waves which travel around the surface of the Earth at a speed of about 6 km/s.
The graph below shows how the ground moves near to the centre of the earthquake as the waves pass.

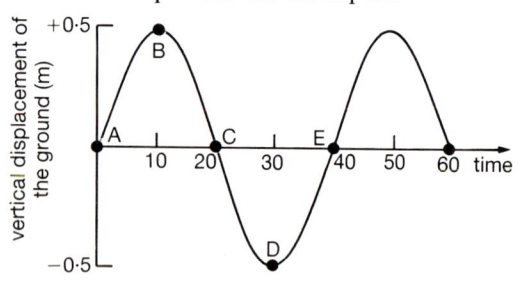

(a) What is the time period of the waves?
(b) What is the frequency of the waves?
(c) Calculate the wavelength of the seismic waves.
(d) Explain why the ground is moving most rapidly at times *A*, *C* and *E*.
(e) When is the ground accelerating at its greatest rate?
(f) Use the graph to estimate the vertical speed of the ground at the time marked *C*.
(g) Make a sketched copy of the graph. Add to it a second graph, to show the ground displacement caused by a second seismic wave of the same amplitude but twice the frequency.
(h) Discuss whether high frequency or low frequency seismic waves will cause more damage to buildings.
(i) The diagram shows seismic waves passing a house. The waves produce ground displacements that have a vertical component *YY'* and a horizontal component *XX'*. Which component is more likely to make the house fall down? Explain your answer.

8 Ivan lives in Akkani in the Soviet Union; he is a radar operator. His radar works like this. A small pulse of radar waves is sent out from his set. If the waves hit a plane they are reflected. A CRO shows the pulse of waves leaving (A) and the waves returning (B) (see diagram below). One day Ivan detects an FB-111A bomber flying over Nome in Alaska. The diagram shows the trace on his CRO.

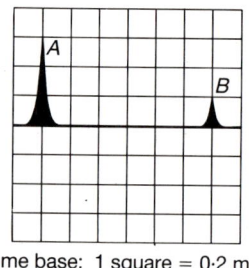

time base: 1 square = 0·2 ms
1 ms = 0·001 s

(a) Use the diagram to work out the time taken for the radar waves to go out to the plane and back.
(b) The waves travel 300 km in 1 ms. How far is it to Nome from Akkani?
(c) One minute later Ivan notices that the lines A and B are only 5 squares apart. How long will it be before the FB-111A reaches Akkani?
(d) What is the speed of the FB-111A, in km/h?

SECTION H

Light and Optics

The Sun emits more energy as light than it does in any other wavelengths. Animal eyes are specially adapted to take advantage of this

1 Rays and Shadows

An eclipse of the Sun. In this photograph you can see the luminous atmosphere of hot gases that surround the Sun, called the corona. Normally the Sun is so bright that the corona is invisible

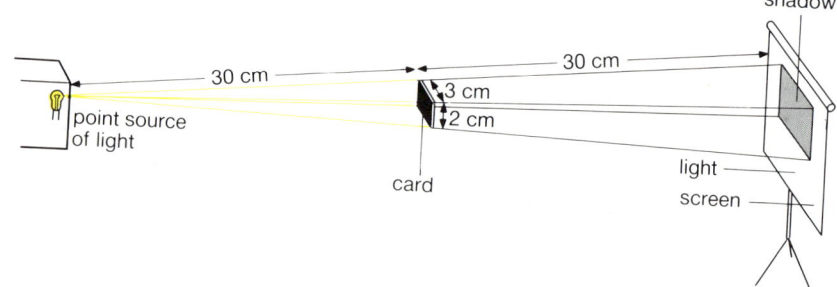

Figure 1

Shadows and eclipses

Shadows are formed when something blocks the path of light. In Figure 1 you can see a piece of card held in front of a small source of light. Some light misses the card and travels on, in a straight line, to the screen. A sharp shadow is formed on the screen behind the card.

Not all shadows are so sharp. If the source of light is large then the shadows have two parts. In the middle of the shadow there is a dark part called the **umbra**. Around the edges of the shadow there is a lighter region called the **penumbra**. A good example of shadow formation is provided by eclipses of the Sun and Moon.

An eclipse of the Sun occurs when the Moon passes between the Earth and the Sun. The Moon is a lot smaller than the Sun but it is closer to us. It is just possible for the Moon to cover the Sun completely. When this happens there is a **total eclipse** of the Sun. During a total eclipse of the Sun the sky goes black and it is possible to see stars. It is only possible to see a total eclipse of the Sun if you are in the umbra of the shadow (Figure 2). If you are inside the penumbra of the

A partial lunar eclipse. The upper left portion of the Moon has entered the umbra part of the shadow cast by the Earth

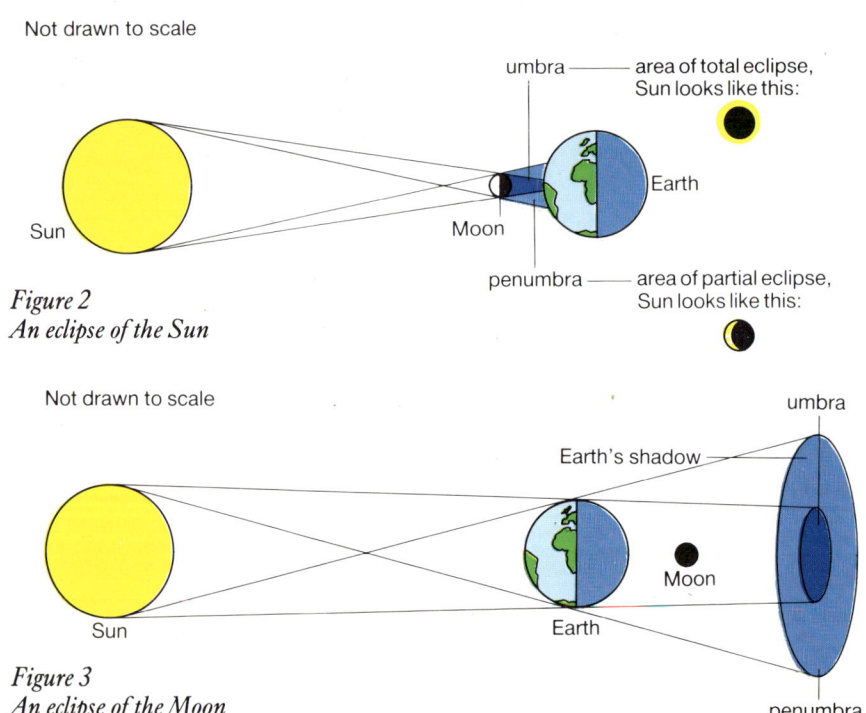

Figure 2
An eclipse of the Sun

Figure 3
An eclipse of the Moon

shadow you will only see a **partial eclipse** of the Sun. Only part of the Sun is covered during a partial eclipse. An eclipse of the Moon happens when the Moon passes behind the Earth and into the Earth's shadow (Figure 3).

The pinhole camera

You can make a simple pinhole camera out of a cardboard shoe box. A small hole is put in one end. The other end of the box should be removed and a piece of tissue paper put in its place. If you now take the box outside and point it at some trees, you will see an image of them on the tissue paper. If you want to take a photograph of the trees, you must use a light-proof box. The photograph is made by allowing the light to fall on to a piece of photographic paper instead of the tissue paper (Figure 4).

The light from the tree travels through the pinhole in a straight line. This causes the image to be upside down. Provided the hole is small the image of the tree will be sharply defined. If the hole is too big the tree will look blurred.

The size of the image lets you work out the height of the tree, which is 30 m from the pinhole.

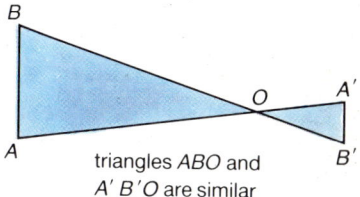

triangles ABO and $A'B'O$ are similar

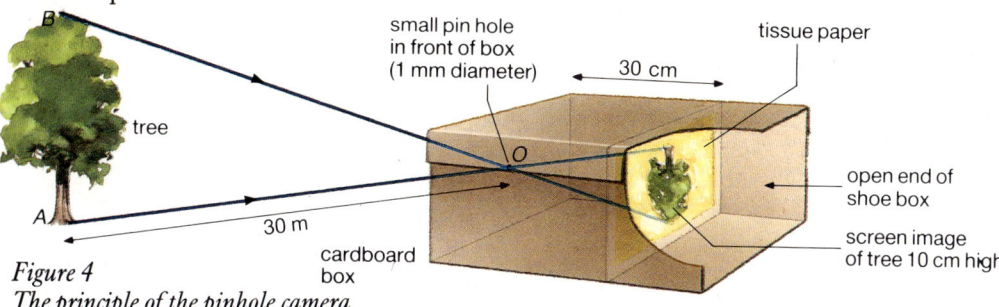

Figure 4
The principle of the pinhole camera

So $\dfrac{AB}{AO} = \dfrac{A'B'}{A'O}$

$AB = \dfrac{A'B'}{A'O} \times AO$

$= \dfrac{10\,\text{cm}}{30\,\text{cm}} \times 30\,\text{m}$

$= 10\,\text{m}$

Questions

1 Work out the area of the shadow in Figure 1.
2 The diagram below shows how an annular eclipse of the Sun can happen. During an annular eclipse the Moon is further away from the Earth than in a total eclipse. Sketch how the Sun would appear when viewed from: (i) X and, (ii) Y.

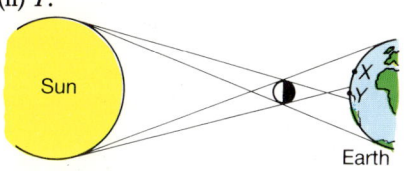

3 Venus moves in an orbit closer to the Sun than the Earth's orbit. In the diagram you can see Venus in three positions marked V_1, V_2, V_3. On the right, X, Y and Z show how Venus looked when seen on three occasions through a telescope. Match X, Y and Z to the positions of Venus.

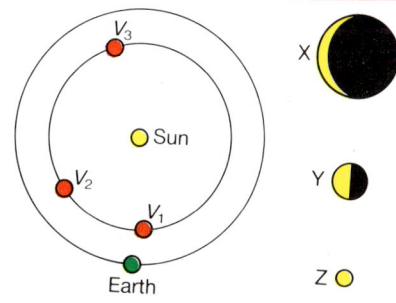

4(a) Explain why making the hole larger in a pinhole camera makes the image more blurred. Illustrate your answer with a diagram.
A student used a pinhole camera to form an image of the Sun. She investigated how the size of the Sun's image depended on the size of the pinhole. The table shows her results.

(b) Plot a graph of image size (y-axis) against hole diameter (x-axis).
(c) The student has made an incorrect measurement of the diameter of the hole. Which measurement is wrong and what should it have been?
(d) Use the graph to predict what the diameter of the Sun's image would be for a very small hole.
(e) Use your answer to part (d) and the extra data provided to calculate the Sun's diameter.

- Distance of Earth to Sun: 150 million km
- Length of pinhole camera: 500 mm

Diameter of Sun's image (mm)	6.5	8.0	9.5	11.0	12.5
Diameter of hole (mm)	2.0	3.5	4.0	6.5	8.0

2 Reflection of Light

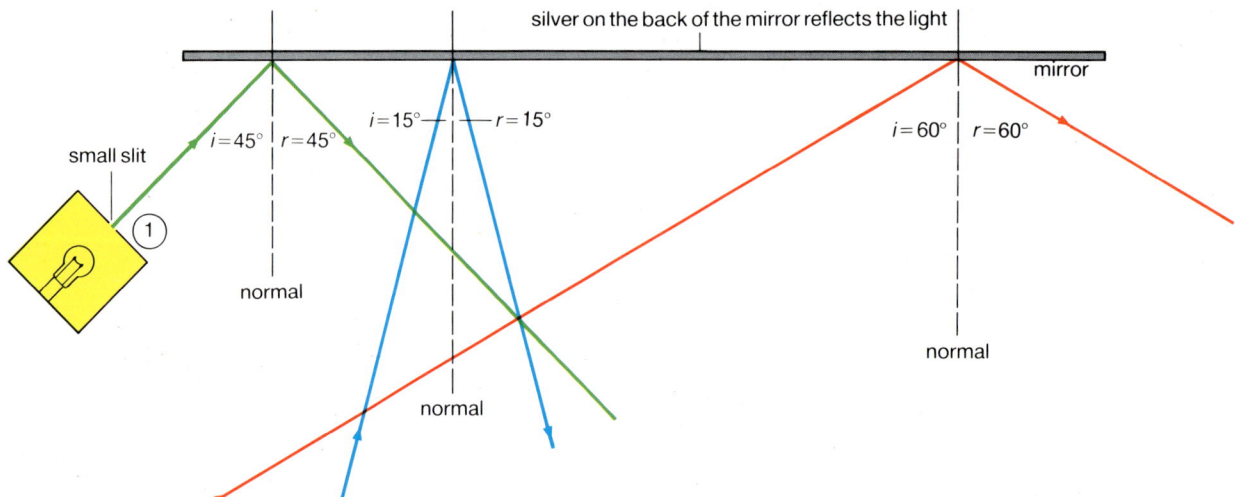

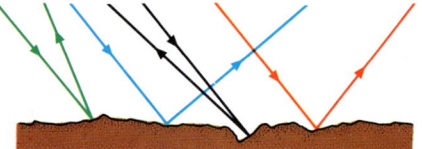

Figure 1
Reflection of light rays from a mirror

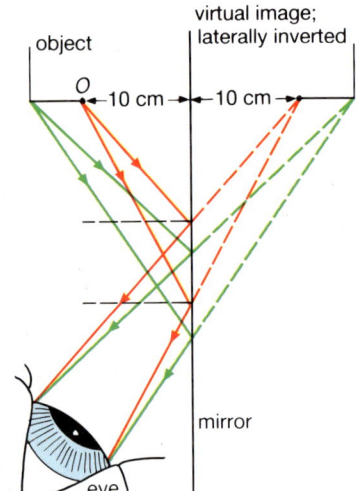

Figure 2
Light is reflected from a rough surface in all directions.

You may use a mirror every day for shaving or putting on make-up (or both!). Mirrors work because they reflect light. In Figure 1 you can see an arrangement for investigating how light is reflected from a mirror. A ray box is used to produce a thin beam of light. Inside the ray box is a light bulb; light is allowed to escape from the box through a thin slit.

Before the light ray strikes the mirror it is called the **incident ray**. The *angle of incidence, i*, is defined as the angle between the incident ray and the normal. The **normal** is a line that passes at right angles through the mirror surface. After the ray has been reflected it is called the **reflected ray**. The angle between the normal and this ray is alled the *angle of reflection, r*.

There are two important points about reflection of light rays that can be summarised as follows:
- The angle of incidence always equals the angle of reflection; $i = r$.
- The incident ray, the reflected ray and the normal always lie in the same plane. All surfaces can reflect light. Shiny smooth surfaces produce clear images. Figure 2 shows that light is reflected in all directions from a rough surface so that there is no clear image.

An image in a plane mirror

We can use the rules about reflection to find the **image** of an **object** in a plane (flat) mirror. In Figure 3 the object is a letter L. Rays from the L travel in straight lines to the mirror where they are reflected ($i = r$). When the rays enter your eye they appear to have come from behind the mirror. This sort of image is called a **virtual image**. Your brain thinks that there is an image behind the mirror, but the L is not really there. You cannot put a virtual image onto a screen. An image that can be put onto a screen (like the one in a pinhole camera) is called a **real image**.

Figure 3
Seeing an image in a mirror

This is an example of some mirror writing

An example of some mirror writing

You can also see that the image appears to be the same distance behind the mirror as the object is in front of it. The image also appears to be back-to-front. You will have seen this effect when you look into a mirror. When you lift your right hand, your image lifts its left hand. The image is said to be **laterally inverted**:

● An image in a plane mirror is virtual, laterally inverted and the same size as the object.

Figure 4 shows how it is possible to produce the illusion of a ghost in a play. The technique is known as 'Pepper's Ghost'. A large sheet of glass is placed diagonally across the stage. The audience can see a wall through the glass on the darkened stage. An actor is hidden from the audience in the wings. He is brightly illuminated so that his image is reflected by the glass for the audience to see. When the actor walks off the stage, his image (the ghost) appears to leave by walking through the wall.

The reflection of light can be used to produce amusing results, like this shop window trick

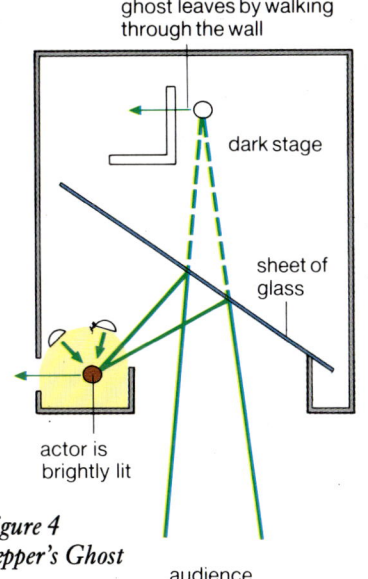

Figure 4
Pepper's Ghost

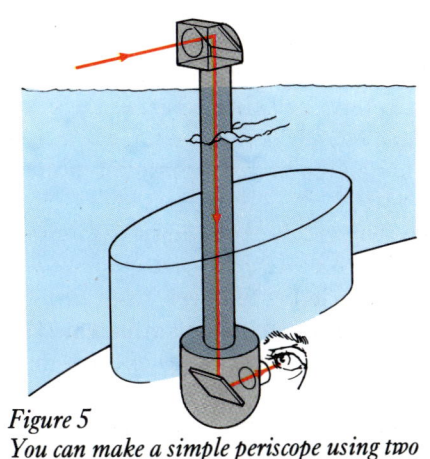

Figure 5
You can make a simple periscope using two mirrors

Questions

1 (a) Describe the image that you will see when you look through the periscope as shown in Figure 5.
(b) At what angle must the mirrors be fitted into the periscope?
2 John runs towards a mirror at 5 m/s. At what speed does his image approach him?
3 (a) How many 10p coins will the eye see reflected in these mirrors?
(b) Draw diagrams to show how each of the images are formed.
4 Show how the word

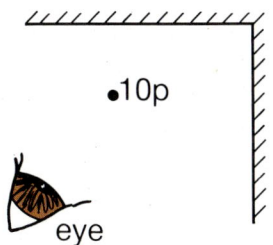

CALCULATOR would look when reflected in a mirror.
5 The diagram on the right shows a split image range finder that fits into a camera.

(a) Explain why you see two images.
(b) How do you adjust the range finder to focus the camera?

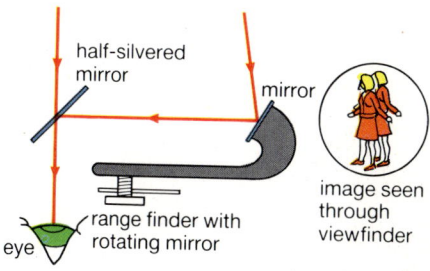

3 The Refraction of Light

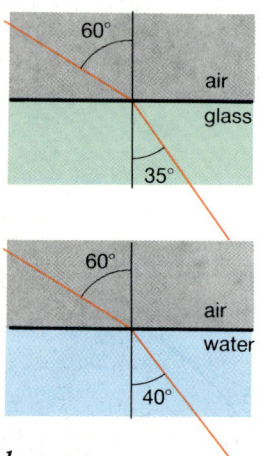

Figure 1
Light travels more slowly in glass than it does in water. So a light ray bends more when it goes into glass

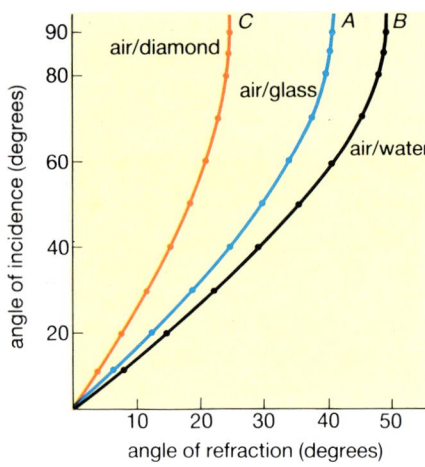

Figure 3
These graphs show how r depends on i when light travels from air into (a) glass (b) water and (c) diamond

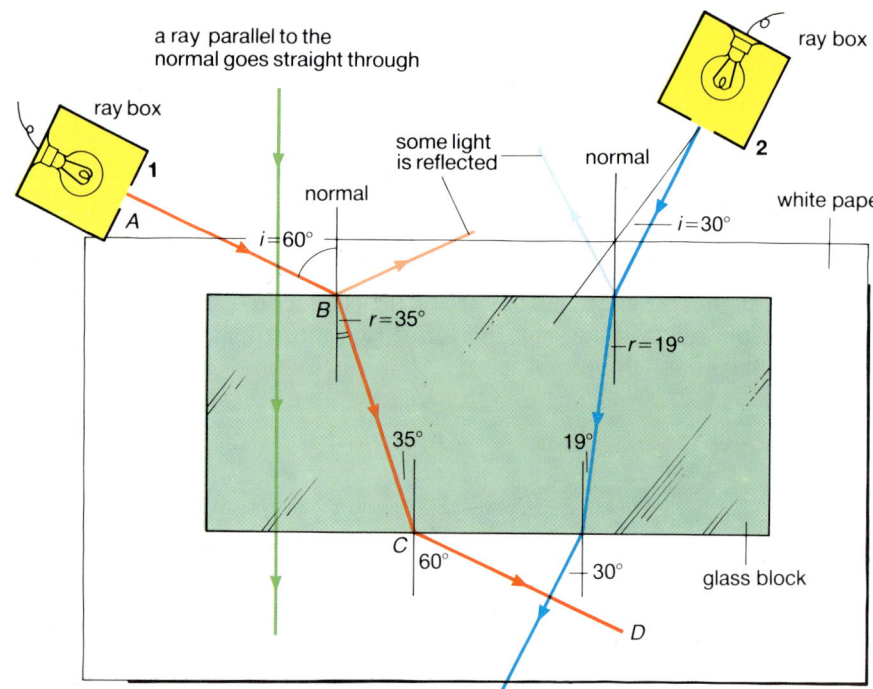

Figure 2
Refraction of light rays by a glass prism

When a light ray travels from air into a clear material such as glass or water, you can see the ray change direction. This is called **refraction**. Refraction happens because light travels faster in air than in other substances (Figure 1).

The amount by which a light ray bends when it goes from air into another material depends on two things:
- What the material is.
- The angle of incidence.

Figure 2 shows you an experiment to study how light rays bend when they go into a block of glass. The angle i between the normal and the incident ray, AB, is the angle of incidence. The angle r between the normal and refracted ray, BC, is the angle of refraction. The experiment shows these points:
- The light ray is bent towards the normal when it goes into the glass. The angle of incidence is greater than the angle of refraction.
- When the light ray leaves the block of glass it is bent away from the normal.
- If the block has parallel sides, light comes out at the same angle as it goes in.

Real and apparent depth

When a light ray leaves water and goes out into air it is refracted. This effect makes a pond look more shallow than it really is. In Figure 4 Susan is leaning over a pond to look at a fish. Light rays from the fish travel up to the surface of the water. At the water surface these rays are bent away from the normal. When these rays enter the eye, Susan imagines that these rays come from I not O. What Susan sees is a *virtual image* of the fish. This image is closer to the surface than the fish itself.

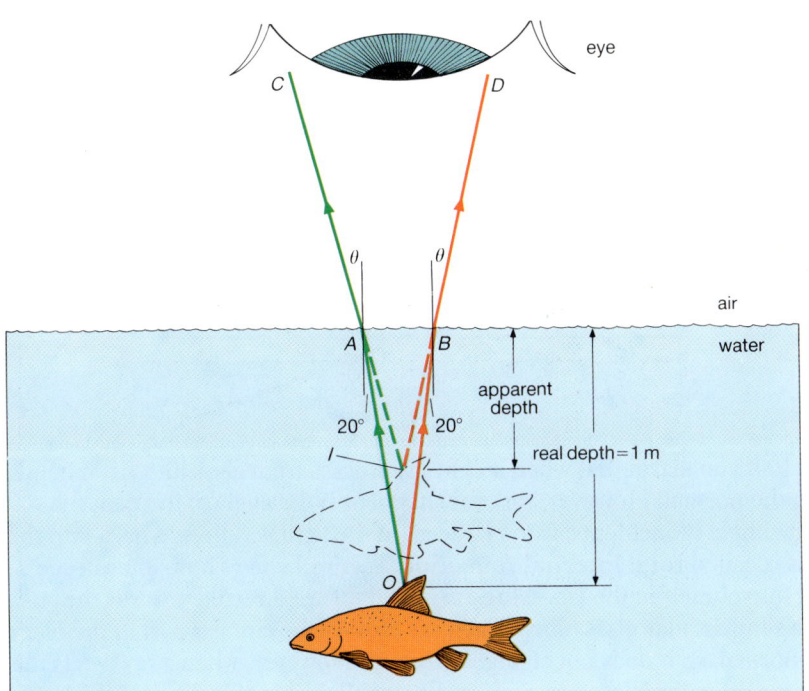

Figure 4
Real and apparent depth

Can you draw a ray diagram to explain why James' fishing net seems to be bent?

Refraction by prisms

You learnt some simple rules about refraction from the experiment shown in Figure 2. You can apply these rules to predict what will happen to light rays going into any shape of block. In Figure 5 you can see a light ray passing through a triangular glass **prism**. Notice that as the ray goes into the prism it is bent towards the normal; as it leaves the prism it is bent away from the normal.

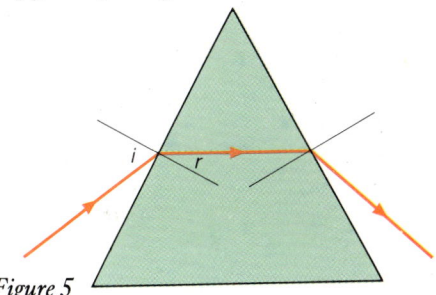

Figure 5
Light is refracted by a triangular prism

Questions

1 When a light ray goes into glass it bends towards the normal; when it comes out it bends away from the normal. Use this rule to sketch the path of the rays through the blocks in these cases.

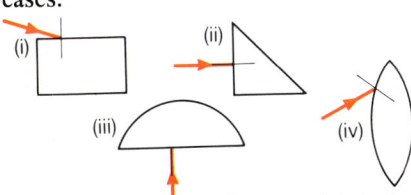

2 (a) Make a copy of Figure 4. Mark in only the rays *OB* and *OA*. Use a protractor to measure the angle 20° as shown.

(b) Use the information in Figure 3 to work out what the angle θ should be.
(c) Now draw in the rays *BD* and *AC*.
(d) Use your scale diagram to calculate the apparent depth of the fish.

3 Draw diagrams to explain why a swimming pool of a constant depth looks shallower at the far end.

4 When a light ray goes from air into a clear material you see the ray bend. How much the ray bends is determined by the **refractive index** of the material.
(a) Look at the table of data. How is the refractive index of a material related to the speed of light in it?
(b) A light ray strikes three materials with angle of incidence of 60°. These materials are: (i) glass, (ii) water, (iii) diamond. Use Figure 3 to calculate the angle of refraction in each case.
(c) Which bends light more, glass or perspex? Perspex refractive index = 1.4.

Material	Speed of Light ($\times 10^8$ m/s)	Refractive Index
Air	3.0	1
Glass	2.0	3/2
Water	2.25	4/3
Diamond	1.25	2.4

4 Total Internal Reflection

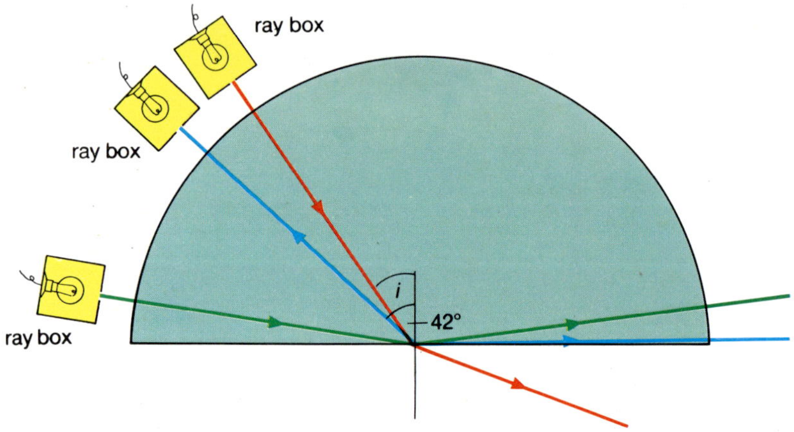

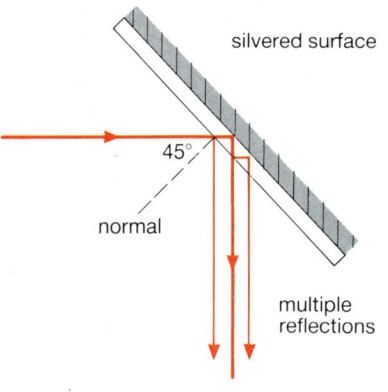

Figure 1
Refraction and reflection in a glass block

Figure 2
(a) Reflection from a mirror

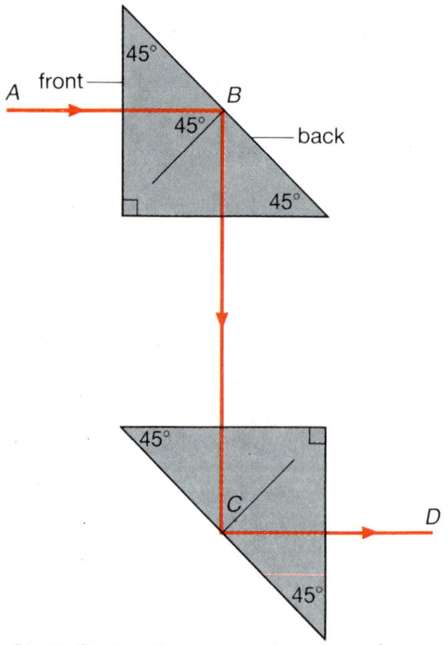

In the last unit you learnt that when a light ray crosses from glass into air, it bends away from the normal. However, this only happens if the angle of incidence is small. If the angle of incidence is too large all of the light is reflected back into the glass. This is called **total internal reflection**. Figure 1 shows how you can see this effect for yourself in the lab. Three rays of light are directed towards the centre of a semicircular glass block. Each ray crosses the circular part of the block along the normal, so it does not change direction. However, when a ray meets the plane surface there is a direction change. For small angles of incidence, the ray is refracted. Some light is also reflected back into the block. At an angle of 42° the ray is refracted along the surface of the block. This is called the **critical angle**. If the angle of incidence is greater than this critical angle then all of the light is reflected back into the glass. The critical angle varies from material to material. While it is 42° for glass, for water it is about 49°.

Total internal reflection in prisms

An ordinary mirror has one main disadvantage: the silver reflecting surface is at the back of the mirror. So light has to pass through glass before it is reflected by the mirror surface. This can cause several weaker reflections to be seen in the mirror, because some light is reflected back off from the glass/air surface (Figure 2(a)).

These multiple reflections can be a nuisance, for example in a periscope. We can avoid the extra reflections by using prisms. In Figure 2(b) the light ray *AB* meets the back of the glass prism at an angle of incidence of 45°. This angle is *greater* than the critical angle for glass, so the light is totally reflected. There is only one reflection because there is one surface. Total internal reflection by prisms is also put to use inside binoculars and cameras.

Refraction and cars

Refraction and reflection are put to use in your car. It is important that your rear lights are clearly visible to the car behind you. At the same time they must not dazzle the driver of a following car. Figure 3 shows how this is achieved. The cover of the rear light is made with a series of points. Any light that is travelling directly backwards is refracted to the side. A similar shape of plastic is used in the reflectors on the back of the cars and bicycles. This time the light from the headlights of a car passes straight through a plane plastic surface (Figure 4). Then total internal reflection occurs at the inside surfaces of the pointed plastic.

(b) Reflection from two prisms to make a periscope

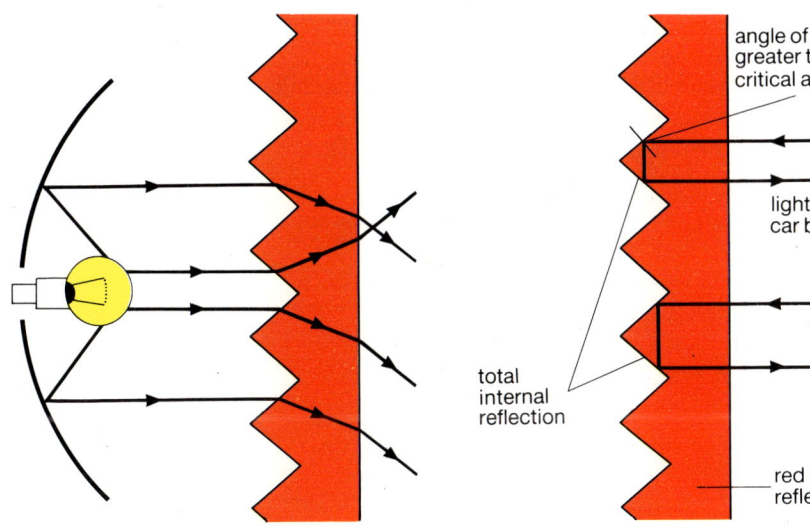

Figure 3
A car rear-light cover

Figure 4
A reflector for a car or a bicycle

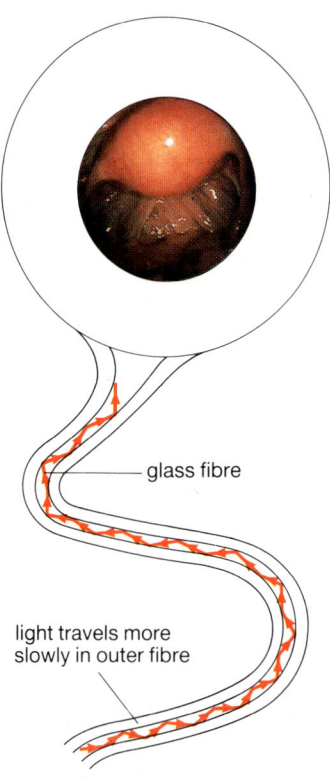

Figure 5
Internal reflection traps light inside glass fibres. This diagram shows an endoscope image of the uterus

Optical fibres

Glass fibres are now used for carrying beams of light (Figure 5). The fibres usually consist of two parts. The inner part (core) carries the light beam. The outer part provides protection for the inner fibre. It is important that light travels more slowly in the outer part. Then the light inside the core is trapped due to total internal reflection.

Surgeons use a device called an *endoscope* to examine the inside of patients' bodies. This is made of two hollow light tubes. One carries light down inside the patient, and the other tube allows the surgeon to see what is there. Optical fibres are also in use by British Telecom. A small glass fibre, only about 0.01 mm in diameter, is capable of carrying hundreds of telephone calls at the same time. These fibres will soon replace the old copper cables in telephone systems.

Questions

1 Turn back to the last unit. Explain how the graphs in Figure 3 can help you work out critical angles. What are the critical angles of glass, water and diamond?

2 Explain, with the help of a diagram, what the fish in the diagram below will see as he looks upwards.

3 Below, you can see a ray of light entering a five-sided prism. Copy the diagram and draw the path of the ray through the prism.

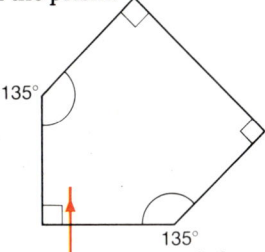

4 The diagram below shows sunlight passing through a prismatic window, that is used to light an underground public convenience.

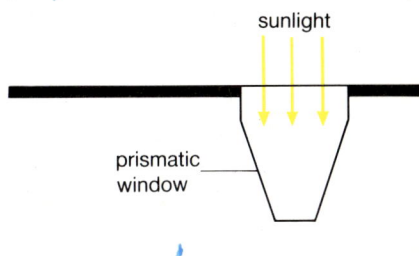

(a) Copy the diagram and show the path of the rays through the window.
(b) Explain why the window is shaped this way.

5 Converging Lenses

Every day of your life you use a converging lens; there is one in each of your eyes. There are also converging lenses in many optical instruments such as telescopes, cameras and slide projectors.

Converging lenses are usually made out of glass and they have two spherical surfaces. When a light ray enters the glass it is refracted towards the normal, and then away from the normal when it leaves (Figure 1).

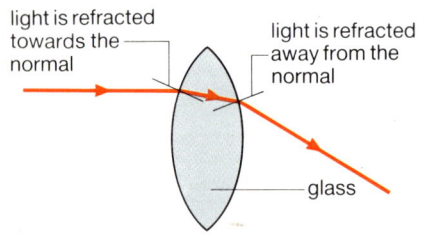

Figure 1
Refraction by curved lens surfaces

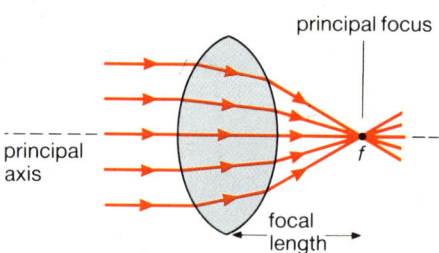

Figure 2
(a) A fat lens is a strong lens; it has a short focal length. Its curved surface refracts the light through a large angle

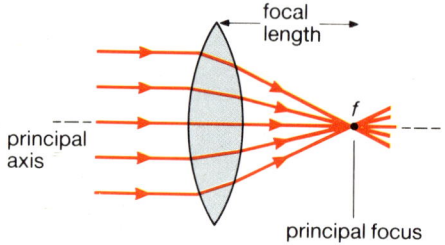

(b) A thin lens is a weak lens; it has a longer focal length than the strong lens

In Figure 2 you can see what happens to a lot of rays which are parallel to the **principal axis**. Each ray is refracted by a different amount, depending on where it meets the lens. After these rays have passed through the lens they converge and meet at a point. This point is called the **principal focus** of the lens. The **focal length** of the lens is the distance between the lens and the principal focus. Each lens has two principal focuses. If the rays were to come from the right in Figure 2(a), they would come to a focus on the left of the lens.

Figure 3 shows how a lens can focus rays that are not parallel.

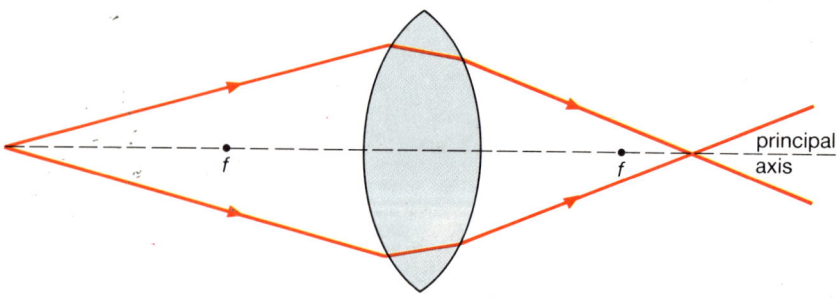

Figure 3
A lens can also focus rays that are not parallel; this time the rays meet behind the principal focus

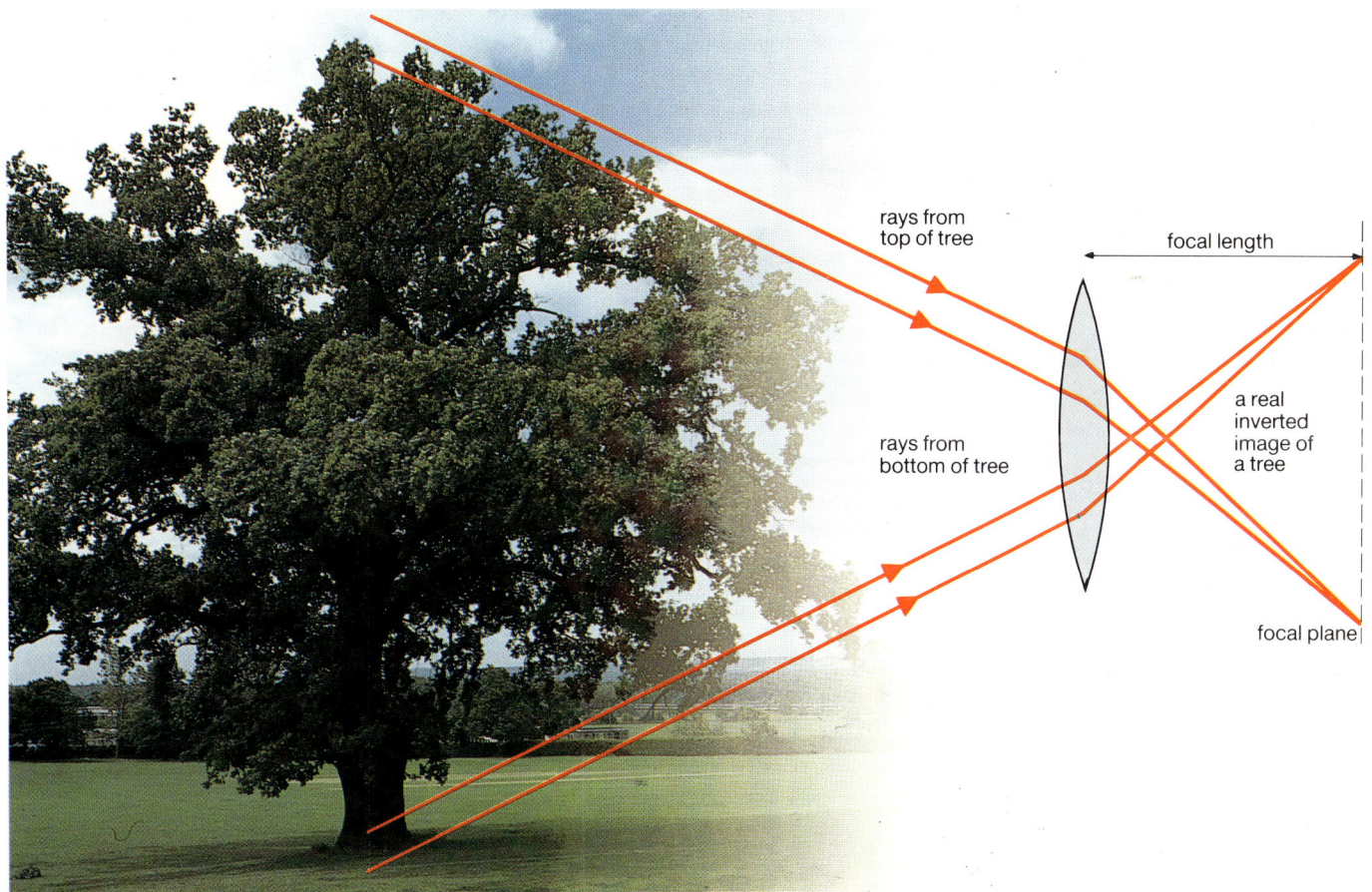

rays from
top of tree

focal length

rays from
bottom of tree

a real
inverted
image of
a tree

focal plane

Figure 4
*This lens forms a real, inverted image of a
distant tree. The distance between the image
and the lens is called the focal length. Note
that in this diagram the refraction of light
is treated as if it happens in the middle of
the lens. To simplify matters this is used
from now on*

Forming a real image with a lens

Figure 4 shows you how a lens can be used to make an image of a distant object
such as a tree. It is a real image, since it can be put on to a screen. The diagram
shows how the rays from the top and bottom of the tree cross over and so an
inverted (upside down) image is formed.

The tree is very far away. The rays coming from the top of the tree all make
nearly the same angle with the lens. This means that the rays are all nearly
parallel. This provides you with an important rule.

Rays from a distant object entering a lens are parallel. These parallel rays come
to a focus, at a distance equal to the focal length, behind the lens. By measuring
the distance between the lens and the screen, you can measure the focal length of
a lens.

Finding the image

If you know the focal length of a lens and the position of an object, you can work
out where the lens will form an image of that object. You can construct a scale
drawing using any two of the three rays shown in Figure 5(a) overleaf:

- A ray parallel to the principal axis is refracted through the principal focus.
- A ray through the centre of the lens, *C*, does not change its direction.
- A ray through the principal focus on the first side of the lens is refracted
  parallel to the principal axis.

This is the method for making a scale drawing like Figure 5(a):
(1) Draw the principal axis and a thin line to show the lens.
(2) Mark the position of the principal focuses. In this example the focal length is 10 cm.
(3) Mark the position of the object. In this case, it is 20 cm away from the lens.
(4) Draw the three construction rays from the top of the object. The top of the image is where these rays meet. The image can now be drawn in; the bottom of the image lies on the principal axis.

In Figures 5(b) and (c) you can see two other examples of ray diagrams. In all cases the images are real and inverted, but the sizes of the images vary. When the object is a long distance from the lens the image is small and close to the lens. When the object is just outside the focal length of the lens, the image is magnified and a large distance from the lens.

Questions

1 The diagram below shows light rays from a small object O, passing through a lens and forming an image at I.

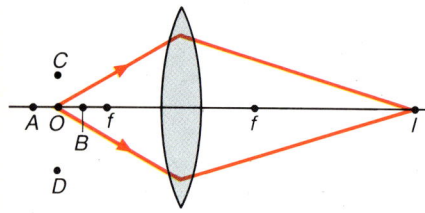

(a) Copy the diagram. Add to the diagram the position of the image, for each of the object positions A, B, C, D. Mark your images A', B', C', D', to correspond to each object position.
(b) What would happen to the image if a piece of card covered the bottom half of the lens?

2 In Figure 5 you can see that the image is magnified if the object is close to the lens. We define the magnification as:

$$M = \frac{\text{height of image}}{\text{height of object}}$$

(a) Work out the magnification for each of the images in Figure 5.
(b) Take careful measurements to prove that this formula is also true:

$$\frac{\text{image height}}{\text{object height}} = \frac{\text{Distance: lens} - \text{image}}{\text{Distance: lens} - \text{object}}$$

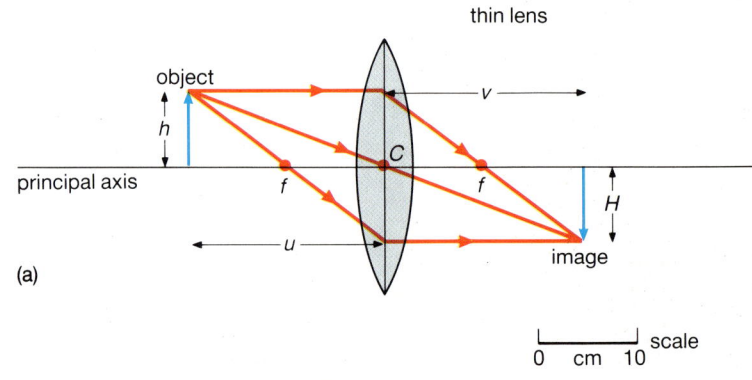

(a)

scale
0   cm   10

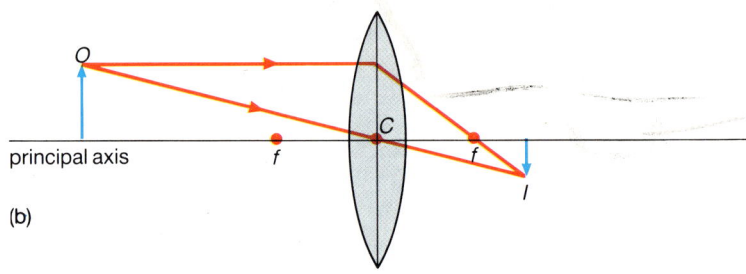

(b)

you can use just two rays to find the image

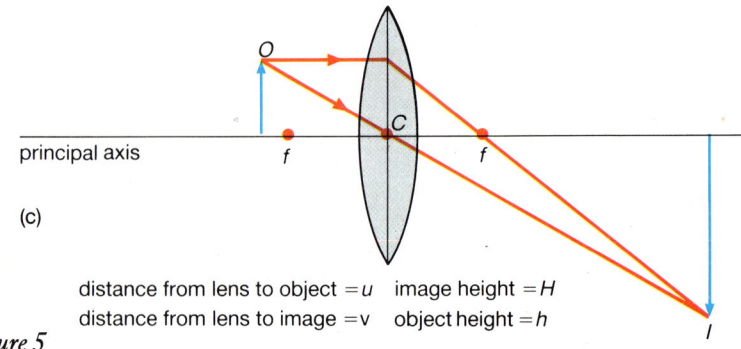

(c)

distance from lens to object $= u$ image height $= H$
distance from lens to image $= v$ object height $= h$

Figure 5

6 Optical Instruments

The camera

A diagram of a simple camera is shown in Figure 1. The purpose of the lens is to project an image of a distant object (a mountain, for example) on to a film.

When you want to take the photograph, pressing a button opens up the shutter to allow light to fall on to the film.

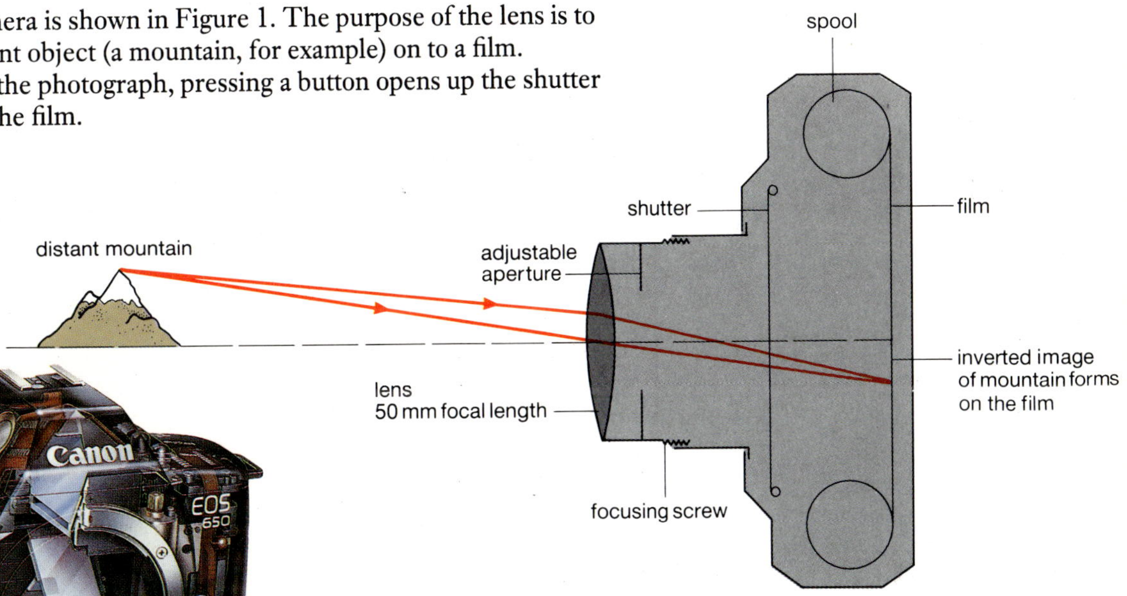

spool

shutter — *film*

distant mountain

adjustable aperture

inverted image of mountain forms on the film

lens 50 mm focal length

focusing screw

Figure 1
A camera

Although modern cameras work on the basic principle of the simple camera in the diagram, they are often much more complex. This camera has features like automatic exposure and focusing, and a motorised film advance

However, before you take a photograph you need to consider these points:
- **Choice of film**. A 'fast' film is more sensitive to light than a 'slow' film. The fast film needs less time to take a photograph than a slow film. The advantage of a faster film is that you can take a photograph when the light is poor, or when something is moving quickly. The disadvantage of a fast film is that it produces poorer quality pictures than a slow film.
- **Focusing**. Your picture must be in focus. To focus, move the lens backwards and forwards with the focusing screw. For distant objects the lens is moved back towards the film. For closer objects you move the lens forwards.
- The **shutter speed** controls the amount of light coming into the camera. On a dark day you might choose a shutter speed of 1/30 s, while on a bright day you can use a faster speed of 1/60 s.
- **f-number**. This refers to the diameter of the aperture. If you set your aperture at f/8, it means that its diameter is 1/8 of the focal length of the lens. The f-number, like the shutter speed, controls the amount of light coming into the camera. A wide gap allows more light in. The f-number also determines the depth of focus of your photograph. A small gap (a small f-number, such as f/22) will give a large depth of focus. This means that things both near and far to the camera will be in focus.

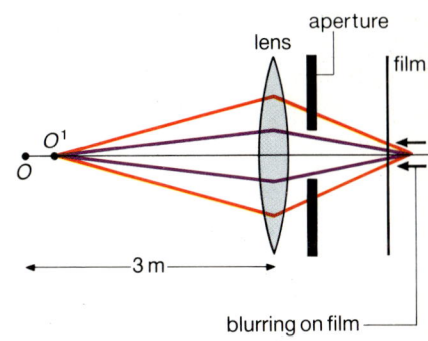

lens *aperture* *film*

blurring on film

Figure 2
The depth of focus is larger if an aperture is used. In this diagram the camera is correctly focused on an object 3 m away. Rays from O^1 will be more out-of-focus if they pass through all of the lens

165

The slide projector

Figure 3 shows how a slide projector works. A brightly illuminated slide Ⓐ is used as an object for the projector lens Ⓑ. This lens projects an image of your slide on to a screen a few metres away. A 500 W light bulb Ⓒ is used to make the slide bright. A concave mirror Ⓓ behind the bulb reflects light forwards. The two condenser lenses Ⓔ then converge the light towards the slide. A heat filter Ⓕ is used to prevent the slide being damaged.

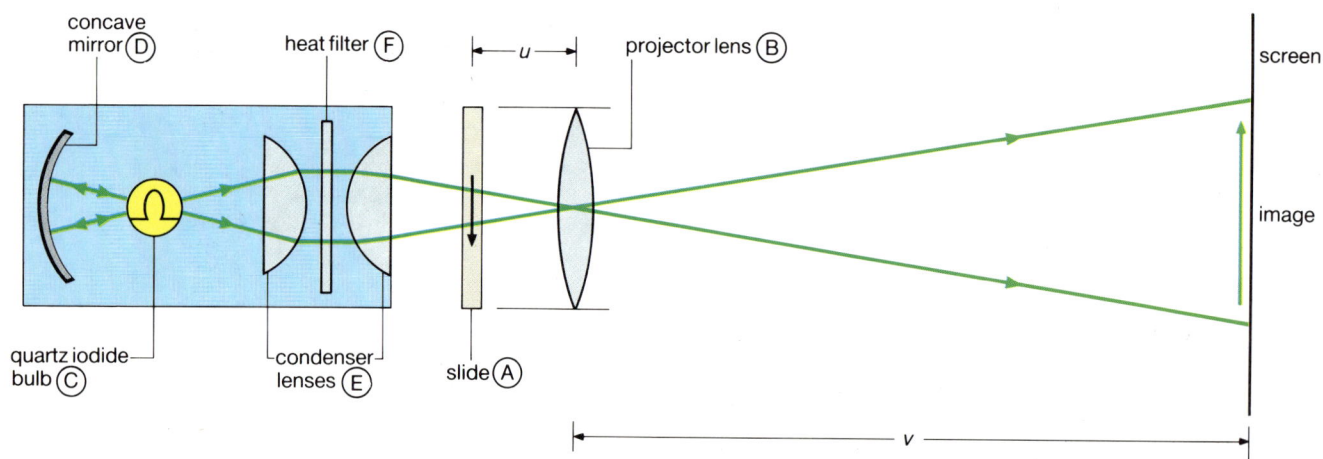

Figure 3
The principle of the slide projector

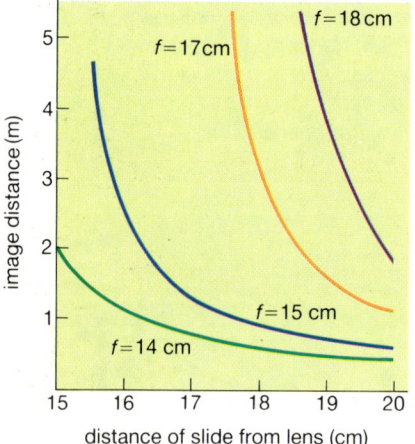

Figure 4
Graphs to show how the distance of the slide from the lens affects the image distance. Graphs for four different lenses are shown

Light bulbs produce infra-red waves as well as light. The heat filter absorbs the infra-red waves, which would heat up the slide. The filter does let light through. The slide projector is cooled by a fan which blows air through it. Vents on top of the projector allow the warm air to escape.

In this slide projector, the distance, u, between the lens and slide can be adjusted between 15 cm and 20 cm. What lens should you choose for the projector? It is important that the slide is just outside the focal length of the projector lens. This makes sure that you see a large image on the screen. So a lens with a focal length of about 15 cm will be the best. Figure 4 gives a series of graphs for different lenses, to show how the distance, u, affects the distance between the image and lens, v.

The Magnifying Glass

Figure 5(a) shows James using a magnifying glass to study a beetle. The magnifying glass is an ordinary converging lens, but the beetle must be placed *inside* the focal length of the lens. The rays from the beetle's head are diverging when they reach the eye. So the rays appear to have come from the point I. James sees an image of the beetle that is virtual, the right way up and magnified.

Figure 5(b) helps to explain why James' beetle appears magnified. You cannot see clearly anything that is closer to your eye than 25 cm. When James looks at the beetle through the magnifying glass, the beetle is closer than 25 cm. However, the image of the beetle is about 25 cm from his eye.

A lens with a short focal length makes a powerful magnifying glass. With a short focal length lens, your eye gets closer to the object that is to be magnified.

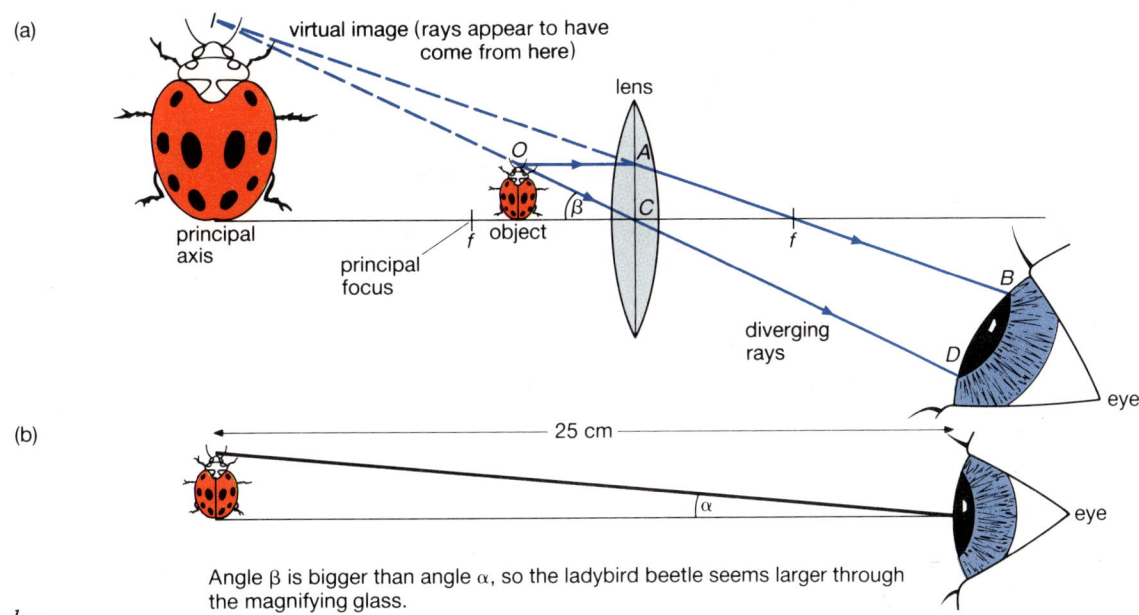

(a)

virtual image (rays appear to have come from here)

lens

principal axis

principal focus

object

diverging rays

eye

(b)

25 cm

eye

Angle β is bigger than angle α, so the ladybird beetle seems larger through the magnifying glass.

Figure 5
A magnifying glass

Questions

1 Look at the three photographs to the right. Explain in each case what factors affected the photographer's choice of:
(i) speed of film, (ii) shutter speed, (iii) f-number.

2 Below you can see some information about possible f-numbers and shutter times for a particular camera.

f-numbers (aperture settings)	shutter times (seconds)
f/22	1/500
f/16	1/250
f/11	1/125
f/8	1/60
f/5.6	1/30
f/4	1/15
f/2.8	1/8
f/2	1/2

(a) What does an aperture setting of f/2 mean?
(b) What does a shutter speed of 1/8 s mean?
(c) On a dull day the camera takes good photographs on settings of f/8 and 1/60 s. When the Sun comes out what two possible changes could you make to the settings to take a photograph?

3 A lens of focal length 15 cm is used in the slide projector in Figure 3.
(a) Use Figure 4 to work out how far the slide is from the lens to project an image: (i) 3 m from the lens, (ii) 5 m from the lens.
(b) A slide measures 35 mm × 23 mm. What is the size of the picture on the screen, when the distance between the projector and the screen is 3 m? This formula may help you solve the problem:

$$\frac{\text{Image height}}{\text{Object height}} = \frac{\text{Image distance } (v)}{\text{Object distance } (u)}$$

(c) The lens in the projector breaks. When you go to buy a new one the shop only has lenses of focal length 14 cm, 16 cm and 19 cm. Which one would you choose? Explain your answer.

Top: f/2; 1/15 second; fast film

Middle: f/4; 1/500 second; fast film

Bottom: f/16; 1/30 second; slow film

7 The Eye

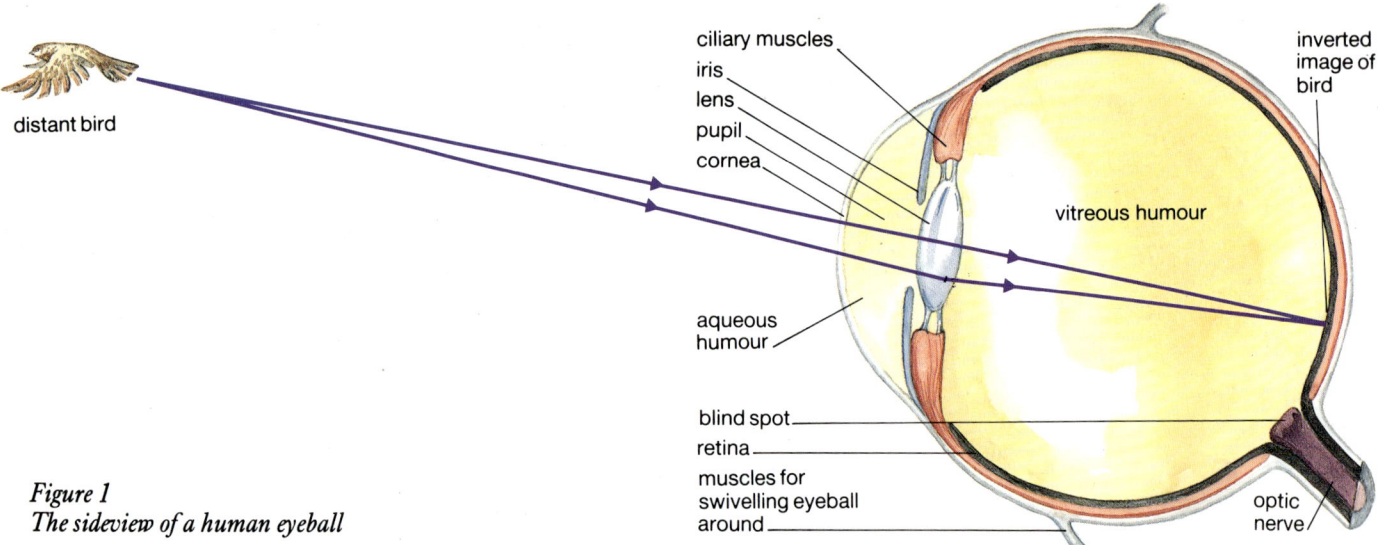

Figure 1
The sideview of a human eyeball

Figure 1 shows you what an eyeball would look like if you could see it from the side. Below are listed the important points about the working of the eye:

- The eyeball is roughly spherical and keeps its shape due to the liquids inside it, the **vitreous humour** and the **aqueous humour**.
- The eye **lens** makes an image of a distant object on the **retina**.
- The retina contains cells which are sensitive to light. Some cells (**cones**) detect different colours, and other cells (**rods**) respond to the brightness of light.
- The **optic nerve** carries signals from the retina to the brain. Although the image on the retina is upside down, the brain sorts this out for us so we can see things the right way up.
- The amount of light that enters the eye is determined by the size of the **pupil**. The **iris** acts like the aperture of a camera. In bright light it closes down to protect the eye. In the dark the iris opens up to allow the eye to gather more light.

Focusing the eye

Your eye is most relaxed when you are looking at distant objects. The eye then focuses parallel rays onto the retina (Figure 2(a)). When you are looking at distant objects, your eyes cannot focus on something that is close to you at the same time. The lens is not strong enough to converge rays coming from nearby on to the retina. To look at something close to you, the eye lens has to change shape. The **ciliary muscles** make the lens fatter (Figure 2(b)). The light is now bent more when it goes through the lens and can be focused on the retina. This focusing process is called accommodation.

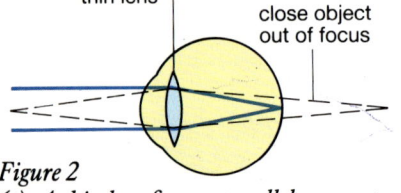

Figure 2
(a) A thin lens focuses parallel rays onto the retina

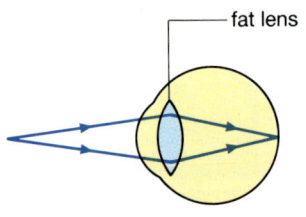

(b) A fat lens is needed to look at objects close to the eye

Wearing spectacles

A normal eye is able to see both near and distant objects clearly. However, it is quite common for people to suffer from either short or long sight:

- Someone who is short-sighted can see things nearby, but cannot focus on distant objects. The problem is that the eye lens is too powerful. Parallel rays from distant objects are focused in front of the retina. This can be corrected by using a **diverging lens**. A diverging lens will spread the rays out, so the eye can now bring them to focus on the retina (Figure 3).

● People who are long-sighted cannot focus on objects that are close to the eye. However, they may be able to see clearly things far away. This time the problem is that the eye lens is too weak. So rays from objects close to converge at a point behind the retina. This sight defect can be corrected by using spectacles with converging lenses (Figure 4).

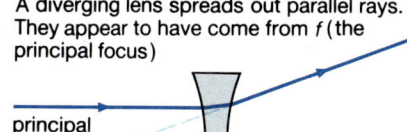

A diverging lens spreads out parallel rays. They appear to have come from *f* (the principal focus)

principal axis

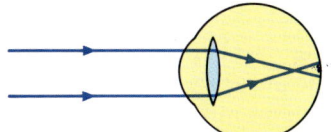

Figure 3
(a) A short-sighted eye cannot see distant objects

(b) A diverging lens corrects short sight

distant objects now look near

(c) A diverging lens

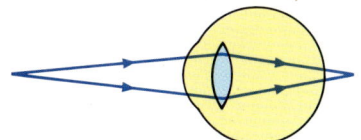

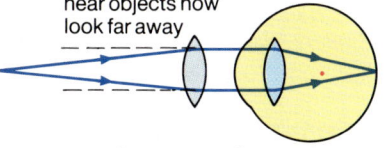

near objects now look far away

Figure 4
(a) A long-sighted eye cannot see objects close to

(b) A converging lens corrects long sight

Binocular vision

An animal that has both eyes at the front of the head is usually a hunter. Those animals that have eyes at the side of the head make good meals for the hunters. Having eyes at the side of the head gives an animal a wide range of vision. That is important if you do not want to get eaten. With two eyes at the front of your head, your brain gets two slightly different views of everything. This is very helpful when it comes to judging distances. If you are hunting your supper like the lioness in the photograph, you need to know how far away it is. Try putting a pencil in each hand and touching the points together; you will find it easy with both eyes open but difficult if you close one eye.

Fast food, Kenyan style!

Questions

1 (a) Suggest three ways in which the eye and the camera are similar.
(b) What is different about the way a camera and an eye focus light?
2 Instead of spectacles some people wear contact lenses to solve their eyesight problems. Contact lenses are curved pieces of plastic that fit directly on to the cornea. Below you can see three shapes of contact lens.

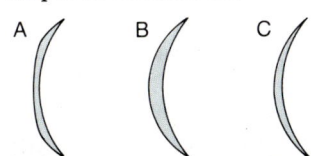

(a) Which side of each lens sticks on the eye? (the left or the right side)

(b) Which lens(es) could be used to correct for: (i) long sight, (ii) short sight?
3 (a) The eye in Figure 4 is long-sighted. How can you tell from the diagram 4(b) that the eye can see distant objects normally?
(b) Peggy wears *bifocal* spectacles. Without her bifocals Peggy can only see things about 4 m away from her eyes clearly. Explain how her bifocals help.

diverging lens for looking at distant objects

converging lens for reading

4 A surgeon in Russia has suggested that it is possible to cure short sight with an operation. His idea is to cut the muscles that control the shape of the eye lens. What are the advantages and disadvantages of this idea?
5 Close your right eye and look at the blue dot. Move your eye closer to the page until the red dot disappears. What causes this blind spot?

8 More Optical Instruments

An astronomical telescope

In Figure 1 you can see how to make an astronomical telescope using two converging lenses. This telescope combines two ideas that you have already met. The large **objective lens** is the one you point at the Moon (or stars). This lens forms an inverted real image of the Moon. This image will be at the principal focus of the objective lens.

The **eye lens** is now used as a magnifying glass to produce an enlarged image of the Moon. The eye lens does not turn the image the right way up. So you see a final image which is virtual, upside down and magnified. It does not matter if the image is upside down to an astronomer. It would be confusing, though, if you used this telescope to look at something on the Earth.

The Isaac Newton reflecting telescope in the Canary Islands

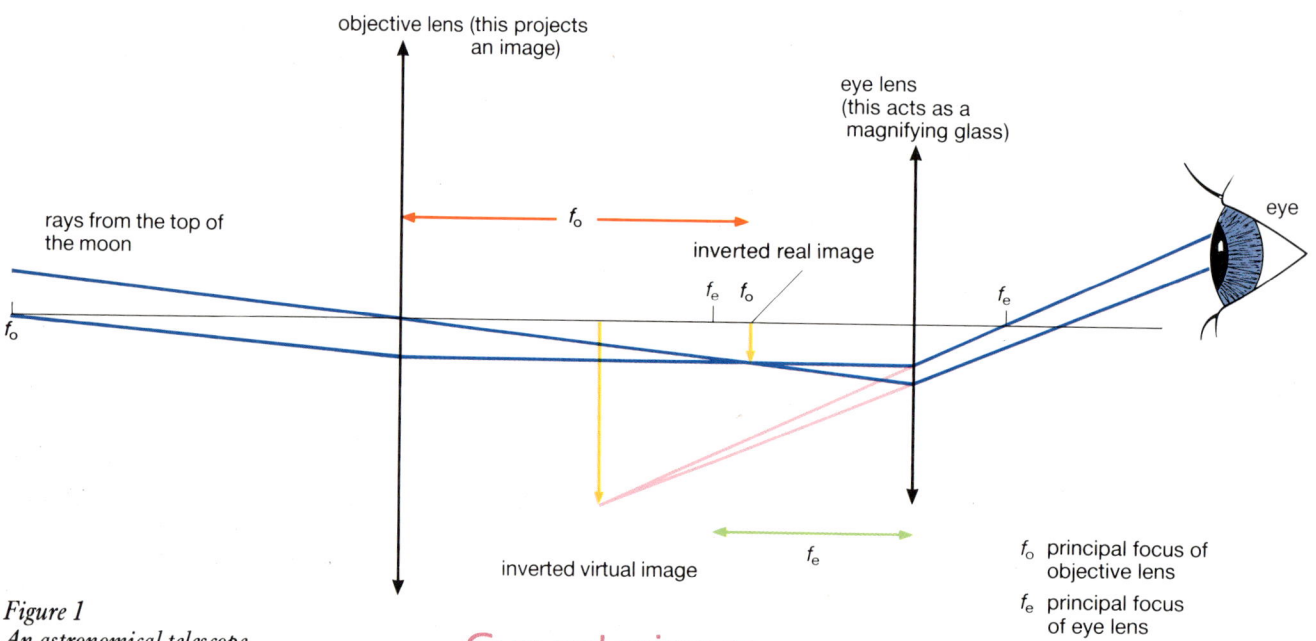

objective lens (this projects an image)

eye lens (this acts as a magnifying glass)

eye

rays from the top of the moon

f_o

inverted real image

f_e f_o

f_e

f_e

inverted virtual image

f_e

f_o principal focus of objective lens

f_e principal focus of eye lens

Figure 1
An astronomical telescope

Curved mirrors

When a mirror with a nearly spherical surface is used, parallel rays are brought to a focus (Figure 2). The point f is called the principal focus of the mirror. Such a mirror can be used in a **reflecting telescope**. The principle is the same as for the astronomical telescope. The **converging mirror** produces a real image of a distant star. An eye lens is used to magnify the image. The main advantage of reflecting telescopes is that they can be very large. The world's largest reflecting telescope has a mirror of diameter 5 m.

Concave mirrors are also used widely as reflectors. By putting a light bulb near the principal focus f, a parallel beam of light is produced. Such mirrors are found in torches, car headlamps, search lights and electric fires. They are also used as shaving or make-up mirrors (Figure 3).

Figure 4 shows a mirror curved the other way (**convex**). You can use this sort of mirror as a car driving mirror. It gives the driver a wide field of view. As you can see more, the image is smaller.

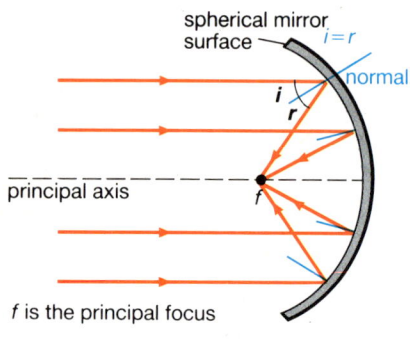

spherical mirror surface

$i = r$

normal

i
r

principal axis

f

f is the principal focus

Figure 2
A concave or converging mirror

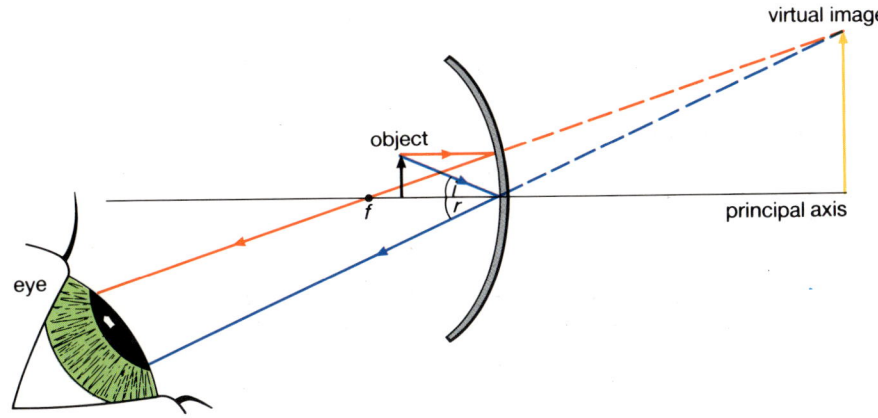

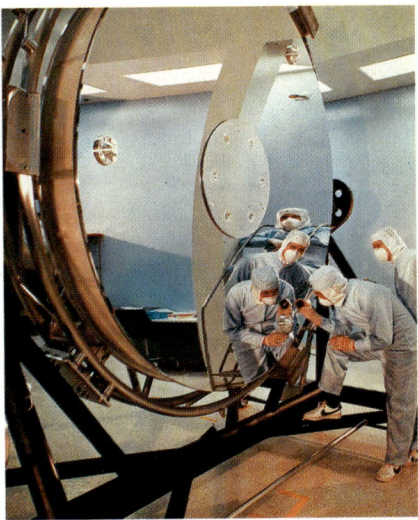

Figure 3
A concave mirror can be used as a shaving or make-up mirror. By putting your face within the focal length of the mirror you can see a magnified image of yourself

This 2.4 metre mirror is being prepared for NASA's space telescope, which will be in orbit round the Earth in the 1990s. Astronomers will be able to see further into space without our cloudy and dusty atmosphere getting in the way

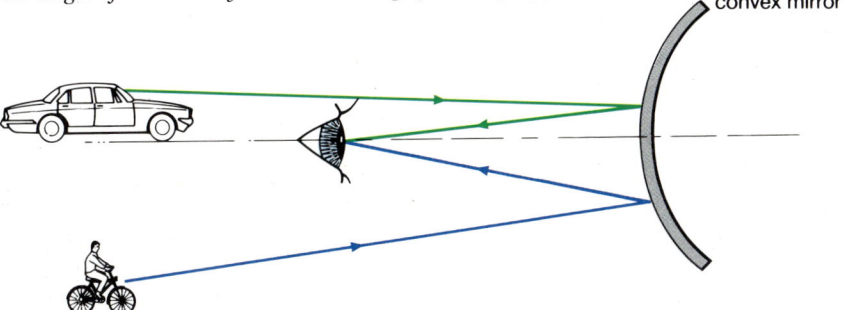

Figure 4
A convex mirror used as a driving mirror

Questions

1 (a) You want to make a telescope with a large magnification. What sort of lenses do you choose? An objective lens with (i) a long or (ii) a short focal length; and an eye lens with (iii) a long or (iv) a short focal length?
(b) Use the data in the table below to show that the magnification of a telescope does not depend on the diameter of the objective lens.

(c) Large diameter lenses do not help to produce a big magnification, so what is the point of having them in telescopes?
(d) Use the data to produce a formula that relates the magnification of a telescope to the focal lengths of the objective and eye lenses.
2 The diagram shows a dual-beam headlight for a car. Filament X is at the

principal focus of the reflector mirror.
(a) Which filament produces the full beam, and which filament is for 'dipped' headlights?
(b) Copy the diagram below and draw in rays to show the path of light from each filament (remember $i = r$).

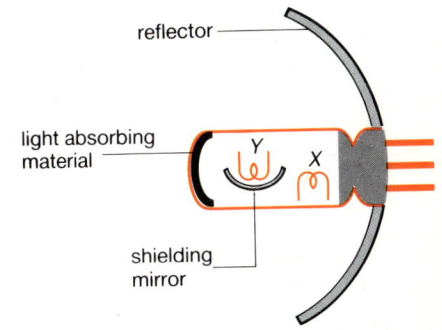

Diameter of objective lens (cm)	Focal length of objective (cm)	Focal length of eyepiece (cm)	Magnification
10	100	5	× 20
10	50	10	× 5
20	100	5	× 20
20	300	3	× 100
40	100	5	× 20
40	200	4	× 50

9 Colour

Raindrops split sunlight into separate colours to produce a rainbow

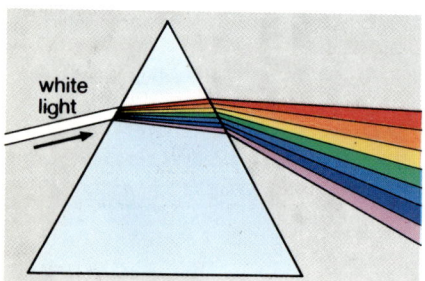

Figure 2
A prism can produce a spectrum from white light

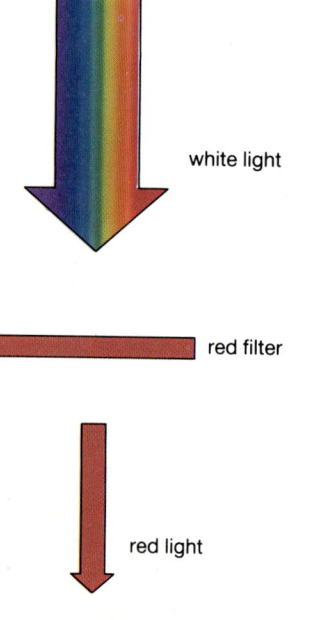

Figure 3

If you look into the sky when it has been raining (and if it is sunny or reasonably bright), you are quite likely to see a rainbow. White light from the sun is a mixture of many different colours. When sunlight passes through raindrops in the sky it is split up into its separate colours (Figure 1).

You can produce your own 'rainbow' or **spectrum** of colours by usng a *prism*. In Figure 2 a ray of white light is split into its separate colours. Blue light is bent more by the prism than the red light. This tells us that blue light is slowed down more by the glass than the red light.

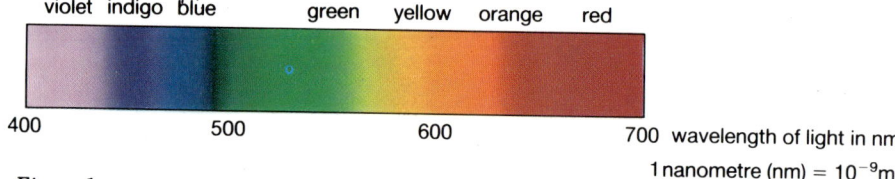

Figure 1
The colours of the rainbow. Different colours have different wavelengths

Making and mixing coloured lights

In this disco the mixing of four coloured lights has produced a white spot on the dancefloor

In your eyes there are cells called cones that are sensitive to colours. There are three types of cone cell; one that detects red light, one that detects green light and one that detects blue light. Red, green and blue are called the **three primary colours**. They cannot be made by mixing together other colours. Most other colours *can* be made by mixing together various amounts of the primary colours.

You can make red, green or blue light with a filter (Figure 3). A red filter, for example, allows red light to pass through it but absorbs all the other colours of light. The red light would be absorbed by a green filter. So if you looked at white light through a green and a red filter, you would see nothing. Figure 4 shows you the sort of effect that you will see if you mix together light of the three primary colours.

You can do this by using three slide projectors, one with a red filter, one with a green filter and one with a blue filter. Where the three colours overlap in the middle, white light is produced. At other places the **secondary colours** are produced. Cyan is a mixture of green and blue; magenta is a mixture of red and blue; yellow is a mixture of red and green.

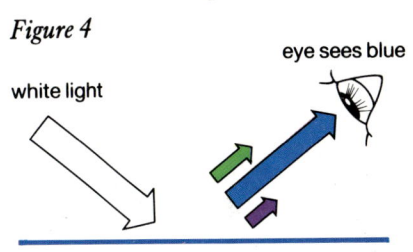

Figure 4

Making colours by reflection

Things appear coloured because they absorb some colours of light and reflect others. For example a shirt may look blue because it reflects blue light and absorbs the other colours. In fact, it is not quite as simple as that. The blue shirt will probably reflect a little green and violet light as well, but the shirt looks blue to the eye. Other colours behave in the same way (Figure 5). The blue shirt will look blue as long as you look at it in white or blue light. However, if you shine red light on the shirt it will look black, because there is no blue light for it to reflect.

What colour shirt would you get if you dyed it in a mixture of yellow and blue dyes? The answer is green. The yellow dye absorbs blue and violet, but reflects green, yellow and red. The blue dye absorbs yellow, orange and red, but reflects violet and green. The only colour that is *not* absorbed is green. Mixing paints and dyes gives a different effect from mixing light. If you mix green and red light you get yellow. If you mix green and red paint you get a nasty dark mess!

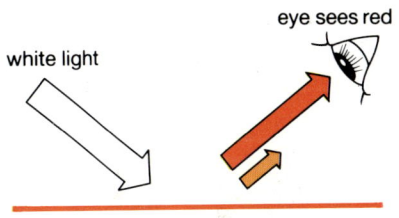

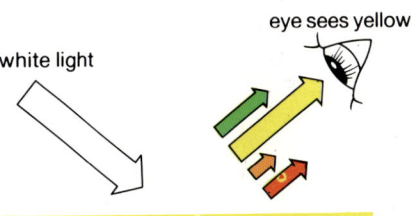

Figure 5

Questions

1 Inspector Grappler of the vice-squad is on patrol in the red light district of New York. He is in disguise wearing a red cap, blue shirt and green trousers. Copy and fill in the table below to show what colour he looks in red and yellow light.

Colour of			
light	cap	shirt	trousers
Red			
Yellow			

2 What colour of paint will you get if you mix yellow and cyan paints?

3 The graph (right) shows how the brightness of light, passing through two filters A and B, depends on the wavelength of light.
(a) What colour will you see if you look at a white light bulb through (i) filter A (ii) filter B? (You will find Figure 3 useful)
(b) What colour will you see if you make the light pass through both filters?
(c) What colour will a red car look if you view it through: (i) filter A, (ii) filter B?

10 Light in the Night Sky

The article below is about astronomy. Read it carefully and then answer the questions that follow. This is an opportunity for you to apply your knowledge and understanding of Physics to an unfamiliar situation.

Many years ago astronomers were puzzled by strange fuzzy objects that were visible in the night sky. Some of these objects can be seen by eye, others were only visible in their telescopes. They called the objects nebulas; the word nebulous means hazy or indistinct. The nature of the nebulas remained a mystery for a long time because no telescope was powerful enough to show any details of any nebula.

In 1947 the Hale telescope on Mount Palomar was completed. This telescope is the world's largest. Its mirror has a diameter of 5 m. The mirror alone weighs over twenty tonnes. The focal length of the mirror is about 20 m. When astronomers used this telescope to look at nebulas the results were truly amazing. Some of the nebulas turned out to be clouds of gas and stars at distances of only a few hundred light years. But some nebulas were found to be whole galaxies of stars like our own Milky Way.

Discoveries with the Hale telescope came thick and fast. It is now thought that the universe contains some thousand million (10^9) galaxies. On average each galaxy contains about a hundred thousand million (10^{11}) stars. The furthest galaxies from us are enormous distances away. The Hale telescope has photographed galaxies at distances of 5000 million light years. Light from these galaxies started its journey towards us before the Earth had been formed.

Distance finding
● Parallax. When you go for a walk through a wood you will notice that the trees nearest to you appear to move relative to the more distant trees. The same effect happens with stars. Those stars near to us appear to move relative to the more distant stars as we go around the sun. The bigger the apparent movement of the star the closer it is to us. Figure 1 shows how a star near to us makes a different angle with a distant star during the course of a year.

If the parallax angle is 1 second of arc then the star is said to be 1 parsec away. (1 parsec is about 3.3 light years). If the angle is 1/10 of a second of arc then the star is 10 parsecs away. The smaller the angle, the bigger the distance. Parallax angles are very small and hard to measure. Only distances as far as 200 light years or so can be measured by parallax.

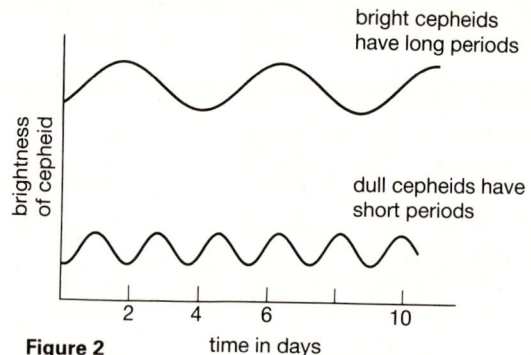

Figure 2

● Cepheid variables are stars whose brightness varies over a period of a few days. The bigger and brighter the star is the longer it takes for its brightness to go through one full cycle (Figure 2). The important point is that any two stars, which take the same time for one cycle (3 days for example), are equally bright. So if you see two cepheids with the same *period* but different apparent *brightnesses* you can conclude that the duller one is further away.

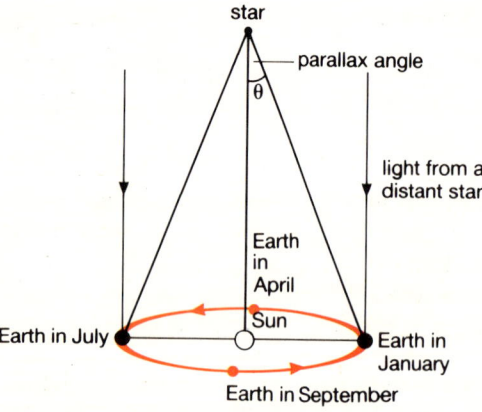

Figure 1.   Measuring star distances by parallax

The Andromeda galaxy and its two dwarf companion galaxies. This galaxy is some 2 million light years away from our galaxy, the Milky Way. It is the largest of the nearby galaxies and measures some 17000 light years across. It is visible as a faint patch with the naked eye

Questions

1 In the article the author mentions two types of nebula. What are these two types?

2 Use the information in the text to estimate the total number of stars in the universe.

3 (a) A star has a parallax angle of 1/30 second. How far away is it in: (i) parsecs, (ii) light years?

(b) A light year is the distance that light will travel in a year. Work out the answer to part (a) in km. (Light travels at a speed of 300 000 km/s; there are about 3×10^7 s in a year).

4 The Andromeda Nebula and the Magellanic Clouds are two galaxies which can be seen by eye. Clarissa, an astronomer, knows that the Magellanic Clouds are about 180 000 light years away from us. She does not know how far away the Andromeda Nebula is but she wants to work it out. In the next column you can see some data she found by looking at some cepheid variables in the two galaxies.

(a) Which galaxy is nearer to us?

(b) Compare cepheids with the same period from the two galaxies. Roughly how many times brighter are the cepheids in the Magellanic Clouds?

(c) What relative brightness does a cepheid of period 0.7 days have, in the Andromeda Nebula?

(d) Now use the rule shown in the diagram, and your answer to part (b), to calculate how far the Andromeda Nebula is away from us.

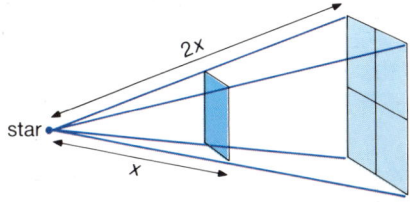

Galaxy	Period of cepheid in days	Relative brightness of cepheid
Andromeda Nebula	1.7	13
	2.4	19
	3.8	30
Magellanic Clouds	0.7	490
	1.7	1280
	3.8	3100
	4.9	4080

1 The periscope shown in the diagram below contains two mirrors. It has

A a plane mirror at (1) and a concave mirror at (2).
B a concave mirror at (1) and a plane mirror at (2).
C plane mirrors at both (1) and (2).
D concave mirrors at both (1) and (2).
E a concave mirror at (1) and a convex mirror at (2).

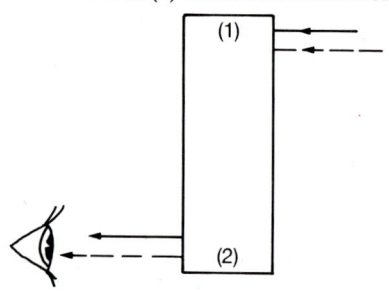

NI

2 The image observed in the periscope is
A upright and virtual.
B upright and real.
C enlarged and real.
D inverted and real.
E inverted and virtual.

NI

3

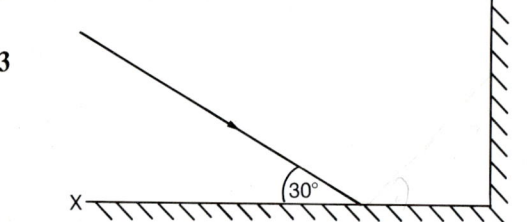

Two mirrors X and Y are shown inclined to each other at 90 degrees. A ray of light falls upon mirror X as shown. The value in degrees of the angle of incidence upon mirror Y is
A 15   **B** 30   **C** 45   **D** 60   **E** 90.

NI

4 An image of a lamp is formed on the screen of a pinhole camera as shown below.

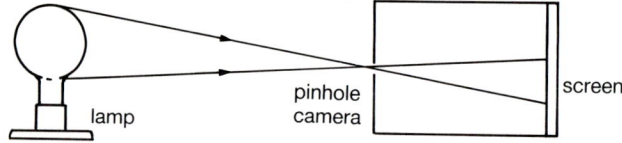

Which of the following changes, made on its own, would make the image smaller?
A moving the camera nearer to the lamp
B moving the camera away from the lamp
C increasing the diameter of the pinhole
D placing the diverging lens over the pinhole
E making the camera longer

MEG

5 In the diagram a ray of light is shown approaching an air/water boundary. Five lines upon which arrows have been drawn are also shown.

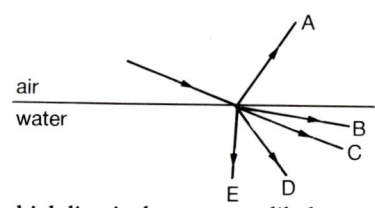

Along which line is the ray most likely to travel after reaching the boundary?

LEAG

6 Which diagram correctly shows the path of a ray of light through the glass block?

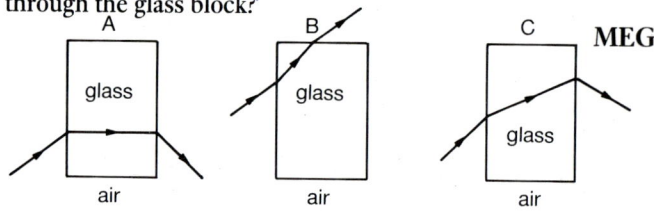

MEG

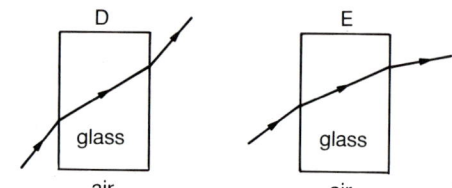

7 A hollow glass sphere is held under water. A beam of light from a torch shines on the sphere. Does the light beam converge or diverge? Explain your answer.

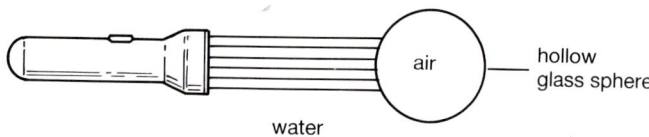

8 The diagram below shows the position of the image of a very distant object produced by an eye.

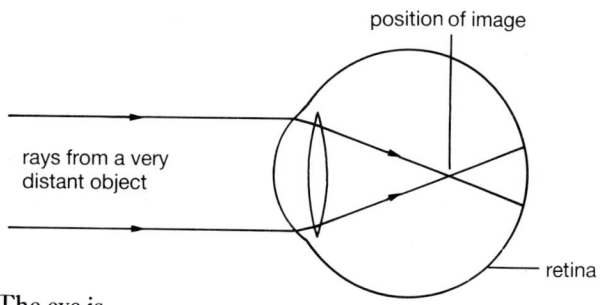

The eye is
A short sighted and can be corrected using a converging lens.
B short sighted and can be corrected using a diverging lens.
C normal.
D long sighted and can be corrected using a converging lens.
E long sighted and can be corrected using a diverging lens.

NEA

9 Red, green and blue lamps are arranged as shown. A piece of metal is held between the lamps and the white screen.

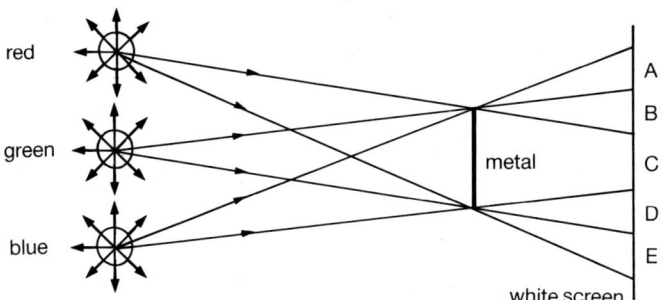

Which of the places on the screen, **A, B, C, D** or **E**, will be yellow?

MEG

10 Which of the shaped pieces of glass, **A, B, C, D** or **E**, shown below has the greatest power?

A   B   C   D   E

MEG

11 Diagram 1 shows a simple diagram of the eye.

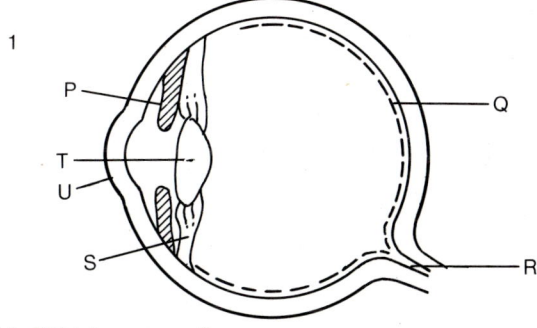

(a) Which arrow refers to
(i) the retina   (ii) the cornea?
(b) A person enters a brightly lit room from a dark corridor.
(i) State the effect on the pupil of the eye.
(ii) How does this affect the amount of light entering the eye?
(c) A colour television screen consists of many thousands of dots arranged in threes as shown in diagram 2. When the dots are lit, one produces red light, one produces green light and the third produces blue light.

 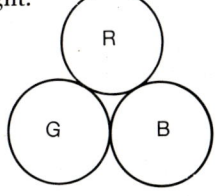

(i) Which dots are lit if the television screen is
1. magenta,   2. yellow,   3. white?
(ii) Explain how it might be possible to make the screen orange.

NEA

12 (a) Copy and complete the diagram below to show refraction and dispersion by a prism. (The emerging red ray is shown).

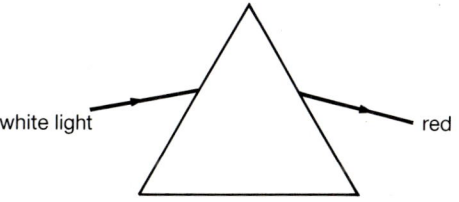

(b) Diagrams 2, 3 and 4 show two semicircular glass blocks and a triangular prism which are made of the same material. The material has a critical angle of incidence of 40°. Measure the angles of incidence at points *A*, *B* and *C*, then copy and complete the diagrams to show the approximate path of the ray of light in each case.

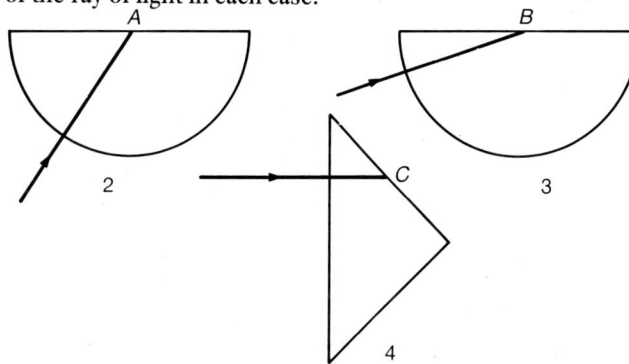

(c) State **one** practical use of the arrangement shown in diagram 4.

LEAG

13 Copy the diagram, which shows an object O placed between the principal focus F and the optical centre of a converging (convex) lens.

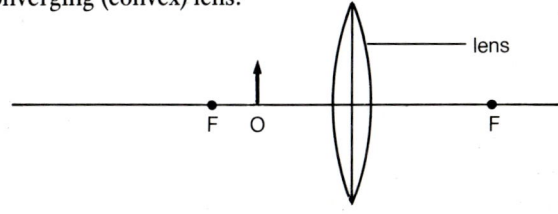

(a) By drawing two rays of light from the object, complete the diagram to show the image of O which is formed by the lens.
(b) The converging lens is now held near a window and an image of some distant houses is formed on a screen.
(i) State **three** properties of the image on the screen.
(ii) The distance between the lens and the screen is 0.15 m. State and explain what can be deduced from this information.

NEA

STUDY QUESTIONS

14 The three diagrams below show rays of light travelling from water to air.

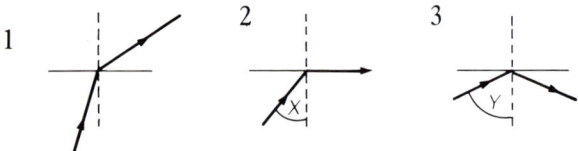

(i)   Why does the ray bend as shown in diagram 1?
(ii) What name is given to the angle marked X in diagram 2?
(iii) What name is given to the bending of the ray shown in diagram 3, and why does it bend as shown?

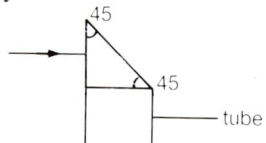

(b) A prism periscope consists of a tube and two identical prisms. Copy and complete the diagram above to show the periscope turning a ray of light through 180 degrees.

NEA

15 Quasars are some of the most distant objects in the universe. Astronomers have observed quasars at distances of about 10 000 million light years. However, some quasars appear brighter than expected. One theory to explain this is the idea of a gravitational lens. According to Einstein's general theory of relativity, light is bent when it passes close to a massive object. A few galaxies have masses equal to that of 10 000 billion (10^{13}) suns. Such a large galaxy acts as a gravitational lens. The diagram shows the idea. Light from the quasar is bent, when it passes close to the large galaxy. The light is focussed by the gravitational lens. An image of the quasar is now formed nearer to the Earth. The lens makes the quasar look brighter than it would do without it.

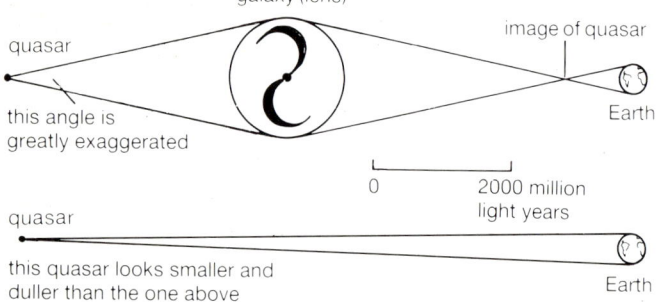

(a) Explain why the quasar would look duller without the gravitational lens.
(b) Use the diagram to measure the distance from (i) the quasar to the galaxy (ii) the galaxy to the image.
(c) Draw a ray diagram to show that the focal length of the gravitational lens is 2000 million light years.
(d) Explain why the quasar and its image are the same size.
(e) Now explain why the lens also makes the quasar look bigger than other quasars seen without the lens.

16 The diagram below shows light from the Sun falling on to a lens.
(a) Explain why light rays coming from the Sun are parallel.
(b) Sketch the drawing below and complete it to show the position of the Sun's image. The focal length of the lens is 10 cm.

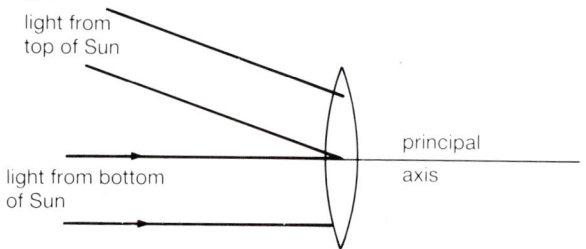

(c) Using the same scale make a second sketch to show the size and position of the Sun's image produced by a lens of focal length 20 cm.
(d) Use the data below to select the best lens for each of these jobs: (i) burning a piece of paper with the Sun's rays (ii) projecting an image of the Sun to look for sun spots.

Lens	Diameter (cm)	Focal length (cm)
1	5	10
2	10	20
3	10	10
4	20	50

17 The diagram shows a **cassegrain reflecting telescope**. C is a curved mirror which reflects light from a star. f is the point where the image of the star would be, if the plane mirror P were removed.

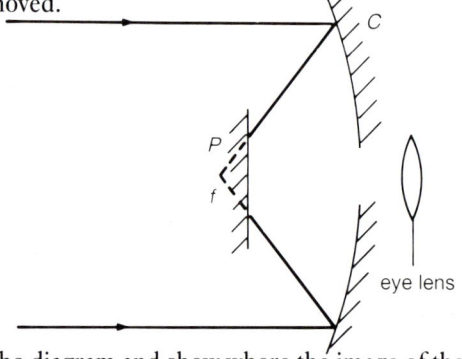

(a) Copy the diagram and show where the image of the star is, after the rays are reflected from P.
(b) Continue the rays to show them passing through the eye lens. What sort of image is formed by the lens?
(c) What would be the effect of doubling the diameter of the mirror C? (Assume that C keeps the same amount of curvature).
(d) What are the advantages of using a reflecting telescope, rather than a telescope which uses lenses?

SECTION 1
Electricity

An engineer repairing some overhead lines. In the background you can see the huge cooling towers of the Connah's Quay power station, and the tall pylons which carry electricity to the national grid

1 Introducing Electricity

Birmingham at night. Major cities use an enormous amount of electricity to provide light

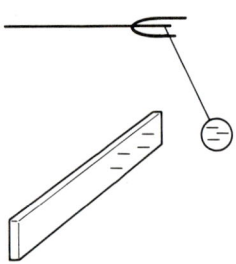

Figure 1
Attraction or repulsion?
(a) Like charges repel

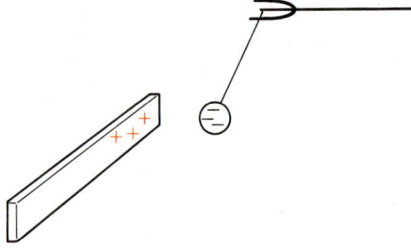

(b) Unlike charges attract

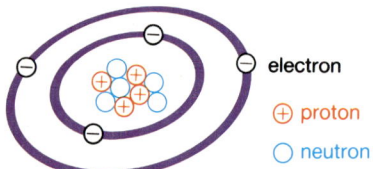

Figure 2
This beryllium atom is neutral. Four
negatively-charged electrons balance four
positively-charged protons

We all take electricity for granted. At home you can turn on a light or a fire at the flick of a switch. You may very well take a snack out of the fridge or deep freeze and cook it in your microwave oven, before sitting down in front of your favourite television programme. Without electricity, our lives would be completely different and less comfortable.

Electrical charge

Electricity was discovered a long time ago when the effects of rubbing materials together was noticed. You will have seen these effects for yourself. If you take a shirt off in a dark room you can hear the shirt crackle and you may also see some sparks. A well-known trick at children's parties is to rub a balloon and stick it on to the ceiling. You have probably felt an electrical shock after walking across a nylon carpet. In these examples, you, or the balloon, have become charged as a result of **friction** (rubbing).

There are two types of electrical charge, positive and negative. A **positive** charge is produced on a perspex ruler when it is rubbed with a woollen duster. You can put **negative** charge onto a plastic comb by combing it through your hair.

Some simple experiments show us that like charges repel each other, and unlike charges attract each other. These experiments are shown in Figure 1.

Where do charges come from?

There are three types of small particles inside atoms. There is a very small centre of the atom called the **nucleus**. Inside the nucleus there are **protons** and **neutrons**. Protons have a positive charge but neutrons have no charge. The **electrons** carry a negative charge and they move around the nucleus. The size of the charge on an electron and on a proton is the same. There are as many electrons as protons. This means that the positive charge of the protons is balanced by the negative charge of the electrons. So the atom is neutral or uncharged (Figure 2).

When you rub a perspex ruler with a duster some electrons are removed from the atoms in the ruler and are put onto the atoms in the duster. As a result the ruler has fewer electrons than protons and so it is positively charged. But the duster has more electrons than protons and is negatively charged (Figure 3).

Picking up litter?

You may know that you can use your comb to pick up small pieces of paper, but it is quite difficult to explain it. Figure 4 shows the idea. The comb is negatively charged, after combing your hair, so when it is placed close to the paper, **electrons** in the paper are pushed to the bottom or **repelled**. The top of the paper becomes positively charged and the bottom negatively charged. The negative charges on the comb attract the top of the paper upwards. The same charges repel the bottom, but the positive charges at the top of the paper are closer to the comb. As a result the upwards force is bigger than the repulsive downwards force and the piece of paper is picked up.

(a)

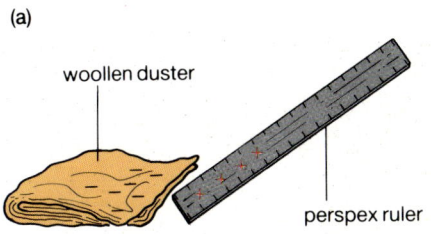

(b)

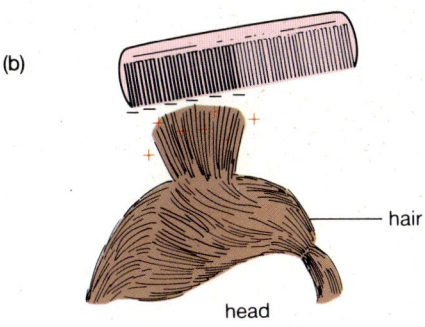

Figure 3
Charging by friction

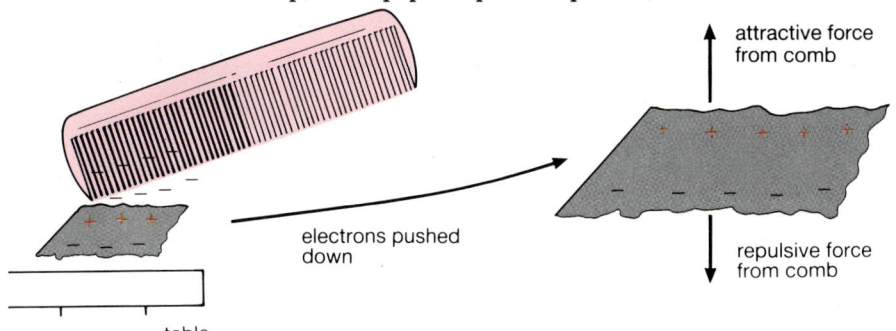

Figure 4
Your comb can pick up small pieces of paper

In some parts of the world, there is no electricity supply, and people do without many of the things that we take for granted

Questions

1 (a) List 4 machines, at home, that work on electricity.
(b) How would you manage without these machines in a power cut that lasts for 3 days?
2 The diagram below shows 2 small plastic balls. The balls are charged. What can you say about these charges on the balls?

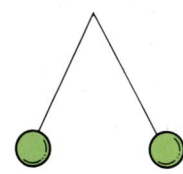

2 Electrostatics at Work

(a) Jack is positively charged. His charge attracts electrons on to the parts of the ship near to him

(b) Jack touches the ship with his hand. Electrons flow on to him to cancel out his positive charge. The spark could cause an explosion

Figure 1

The conducting tail on this car prevents static electricity from building up. When could such a build-up be dangerous?

An accidental spark on an oil tanker could be fatal. Great care is taken on these ships to avoid the build up of electrical charge

Spark hazards

Jack is a sailor who works on board an oil tanker. Jack has to wear shoes that conduct electricity. We say a material conducts electricity, if charge can flow through it. Metals are conductors of electricity. Materials like plastic and rubber do not conduct electricity; they are called insulators. Wearing shoes with rubber soles on board a tanker could be extremely dangerous. When Jack moves around the ship in rubber shoes, charges can build up on him as he works. Then when he touches the ship, there will be a small spark as charge flows away from his body. Such a spark could ignite oil fumes and cause an explosion. Some very large explosions have destroyed tankers in the past. So Jack wears shoes with soles that conduct. Now any charge on him flows away and he cannot make a spark.

Sparks are most likely to ignite the oil when it is being unloaded. To avoid this, the surface of the oil is covered with a 'blanket' of nitrogen. This gas does not burn, so a spark will not cause an explosion.

Electrostatic precipitation

Most of our power stations still burn coal to produce electrical energy. When coal is burnt a lot of soot is produced. It is important to remove this soot before it gets into the atmosphere. One way of doing this is to use an **electrostatic precipitator**. Inside the precipitator there are some wires which carry a large negative charge. As the soot passes close to these wires the soot particles become negatively charged. Now these particles are repelled away from the negative wires and are attached to some positively charged plates. The soot sticks to the plates, and can be removed later. Some large precipitators, in power stations, remove 30 or 40 tonnes of soot per hour.

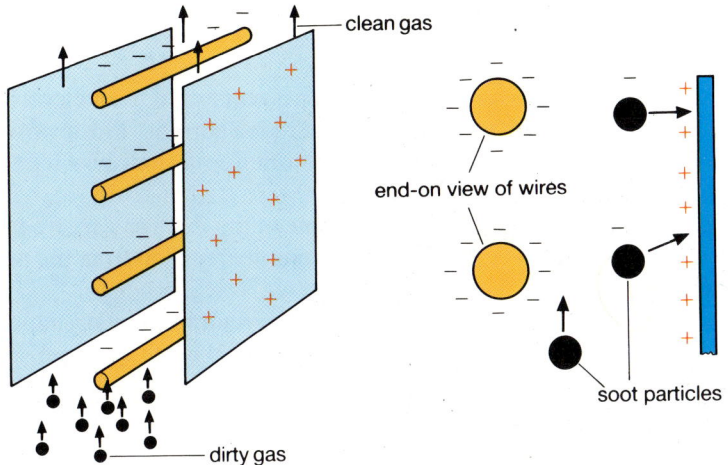

Figure 2
Soot particles in dirty fumes are removed in an electrostatic precipitator

Photocopying

Nowadays, most offices have a photocopying machine. The key to photocopying is a plate that is affected by light. When the plate is in the dark its surface is positively charged. When the plate is in the light it is uncharged.

An image of the document to be copied is projected on to the plate (Figure 3(a)). The dark parts of the plate become charged. Now the plate is covered with a dark powder, called **toner**. The particles in the toner have been negatively charged (Figure 3(b)). So the toner sticks to the dark parts of the plate, leaving a dark image. Next, a piece of paper is pressed on to the plate. This paper is positively charged, so the toner is attracted to it (Figure 3(c)).

Finally, the paper is heated. The toner melts and sticks to the paper, making the photocopy of the document (Figure 3(d)). In modern photocopiers the whole process takes two or three seconds.

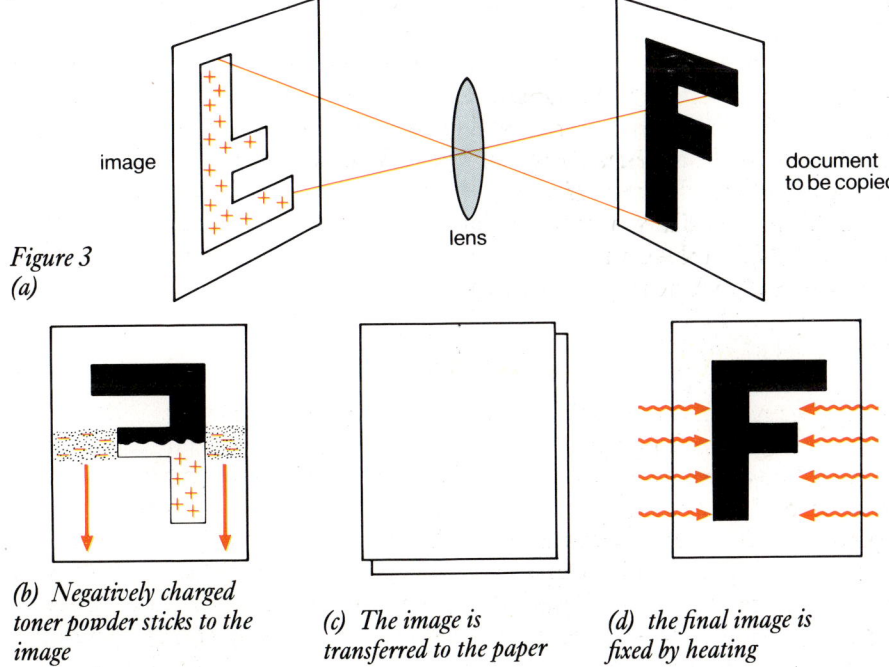

Figure 3
(a)

(b) Negatively charged toner powder sticks to the image

(c) The image is transferred to the paper

(d) the final image is fixed by heating

Questions

1 The tyres on aircraft are made from special rubber that conducts electricity. Explain why.

2 (a) Explain why a photocopier needs toner powder. Why does the powder need to be charged?
(b) When you get your photocopy out of the copier it is usually warm. Why?
(c) The lens in Figure 3(a) produces an image of the document, which is upside-down and back-to-front. Explain why the final image is the right way round.

3 (a) Explain this: When you polish a window using a dry cloth on a dry day it soon becomes dusty. Why does this not happen on wet days?
(b) Cling film is a thin plastic material that is used for wrapping up food. When you peel the film off the roll it sticks to itself. Can you suggest why this happens?

3 Electric Current

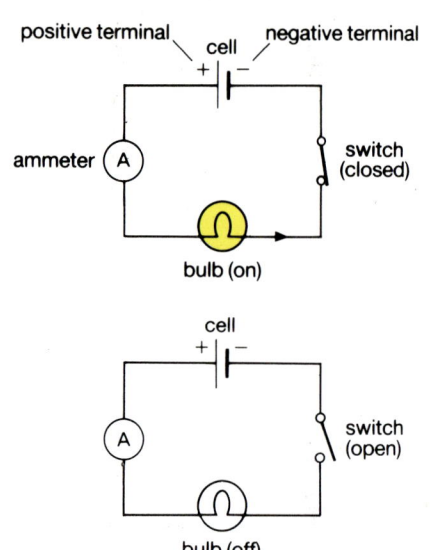

Figure 1
An electrical circuit

Circuits

Figure 1 shows an electrical **circuit**. The bulb is connected to an **electric cell** by copper wires. When the lamp lights we say there is a current. The **ammeter** measures the current. The needle on the meter moves to show how big the current is.

A current is a flow of charge. In copper some of the electrons are free to move. When a current goes through the wire, electrons are repelled from the negative terminal (Figure 2).

A current flows only when there is a complete circuit without any gaps. The same idea works when water flows around your central heating system. You must have a complete pathway (or circuit), so the water goes from the boiler out to the radiators, and back to the boiler to be warmed up again.

In Figure 1 there is a switch. The simplest switch can be made out of a springy piece of copper. When the switch is open there is no conducting path for the electrons to flow round. Then the bulb is off.

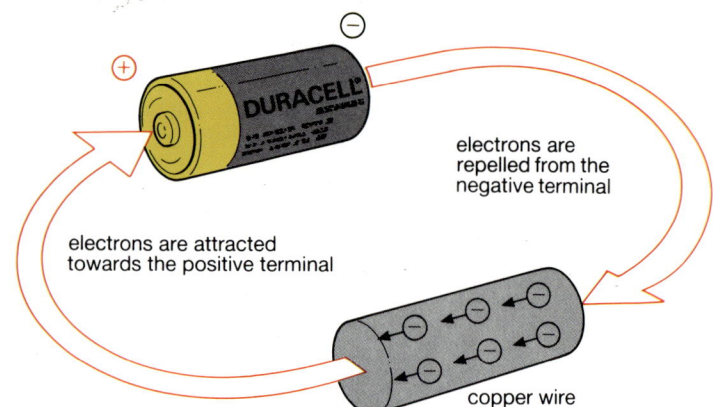

Figure 2
A current is a flow of charge

Which materials conduct?

The table (left) shows good and bad conductors, and insulators. The best conductors are metals; they contain electrons that are free to move. A current can also be carried by **ions**. The word ion is used to describe an atom (or molecule) that has lost or gained an electron. When an atom loses an electron it is a positive ion. When it gains an electron it is a negative ion.

Conductors	Insulators
Good	rubber
metals e.g.	plastics e.g.
copper	polythene
silver	PVC
mercury	perspex
aluminium	china
steel	air
Moderate	
carbon	
silicon	
germanium	
Poor	
water	
humans	

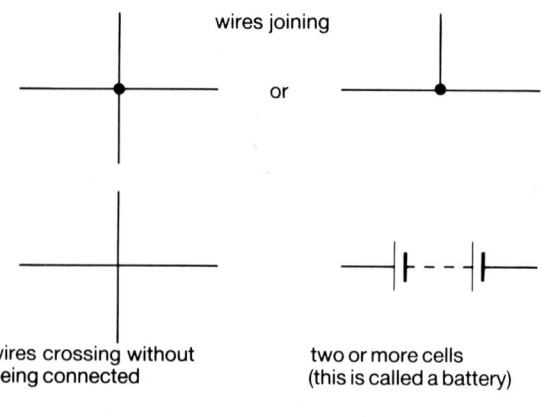

Figure 3
Circuit symbols for wires and cells

Your body is full of ions, so you conduct electricity. Electrical signals from your brain travel along nerves to give instructions to your muscles. Because you are a conductor, you can get a shock from the electricity mains supply.

Which way does current flow?

Figure 4 shows three examples of charge flowing. In all three cases the current is flowing to the right. The current is either carried by positive particles moving to the right, or negative particles moving to the left (or by both). It is easiest to say that current flows in the direction that positive charge moves in. So we say that current flows from the positive terminal of a cell to the negative terminal.

Measuring charge and current

We measure the current in a circuit using an ammeter. The unit of current is the **ampere** (A), though most of us call it an amp.

When the current is big (10 A), the charge moves round the circuit quickly. When the current is small (0.001 A) the charge moves round the circuit slowly. Current is the rate at which charge flows round a circuit.

$$\text{Current } (I) = \frac{\text{charge flowing } (Q)}{\text{time } (t)}$$

$$I = \frac{Q}{t}$$

We could measure charge by counting the number of electrons flowing. However, electrons have only a very small charge. Our unit of charge is the **coulomb** (C). 1 coulomb is equivalent to the charge on six million million million (6×10^{18}) electrons.

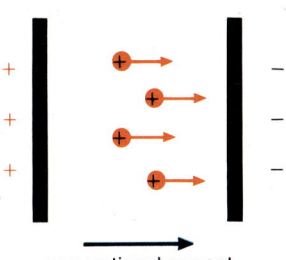

Figure 4
(a) Positive particles in a semiconductor

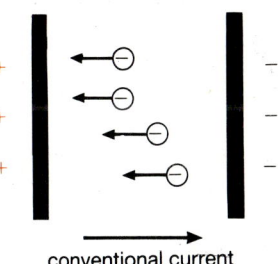

(b) Electrons in a metal

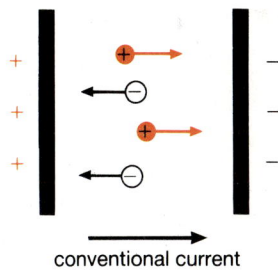

(c) Positive and negative ions in a solution

Questions

1 (a) In circuit (a) (below) which of the switches must be closed for the bulb to light?

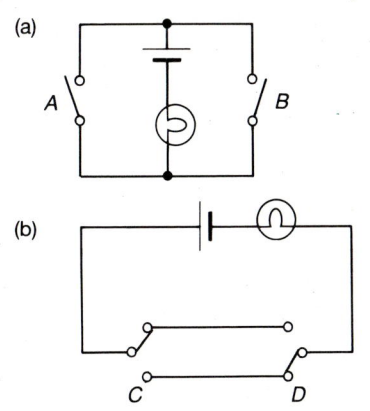

(b) Copy this table and fill in the third column to show if the bulb is on or off.

Switch C	Switch D	Lamp (on or off)
Up	Up	
Up	Down	
Down	Up	
Down	Down	

(c) What is the difference between the switches in the two circuits?
(d) Which circuit could be used for turning a light on or off at the top or bottom of a staircase?

2 Copy the circuit in the next column and mark in the direction of the current. Also show the direction in which the electrons are moving round the circuit.

3 In this circuit the ammeter A_1 reads 0.5 A. Which of the following statements is true: (i) A_2 reads 0.5 A, (ii) the current through the cell is zero, (iii) in 10 minutes, 5 C of charge flow through the bulb?

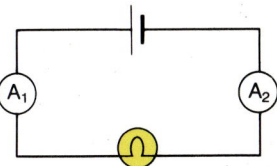

4 The bottom of a thundercloud stores about 10 C of charge. A flash of lightning lasts for about 0.001 s. Work out the current that flows between the Earth and the cloud during a flash of lightning.

4 More about Circuits

The thick cables on overhead pylons are made from aluminium. This transmission line uses a voltage of 440 kilovolts. To minimise heat loss, high voltages such as this are always used for the long distance transmission of electricity. In this case only 3% of the energy carried by these wires is wasted as heat

Splitting the current

We cannot lose current as it goes round a circuit (see Figure 1). The same current flows through bulbs *B* and *C* (and through the cell). The bulbs are **in series**. You can measure the current going into the bulbs and out of the bulbs with two ammeters. Both ammeters read 1 A. If they did not read the same, we would have lost some electrons.

Exactly the same idea applies if we have a circuit with branches in it. In Figure 2 a current of 2 A goes through bulb *D*. Then the current splits to go through bulbs *E* and *F*; each bulb gets a current of 1 A. But then 2 A flows back to the negative terminal of the cell.

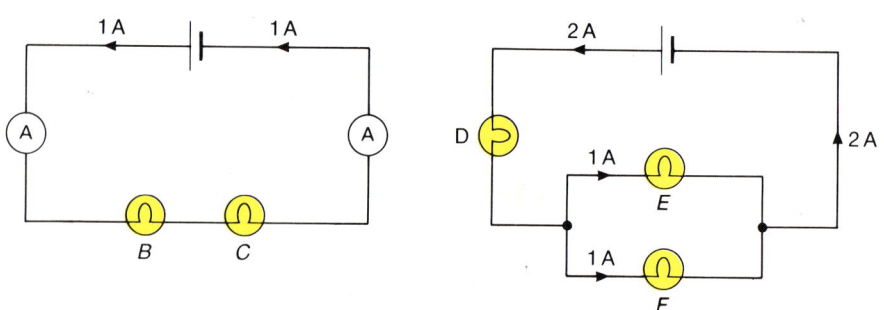

Figure 1
An electrical circuit

Figure 2
A current is a flow of charge

The currents do not always split equally. More current goes through the easier path. In Figure 3 it is easier for the current to go through bulb *I*, than it is to go through bulbs *G* and *H*. So 2 A goes through bulb *I*, and 1 A through the other two bulbs.

Figure 4 shows a **short circuit**. Bulb *J* has been shorted out by some copper wire. The current takes the easy path through the wire. No current goes through the bulb *J* so it is off.

CURRENT RULES
- All parts of a series circuit receive the same current
- The same current goes in as comes out
- More current goes through the easier path

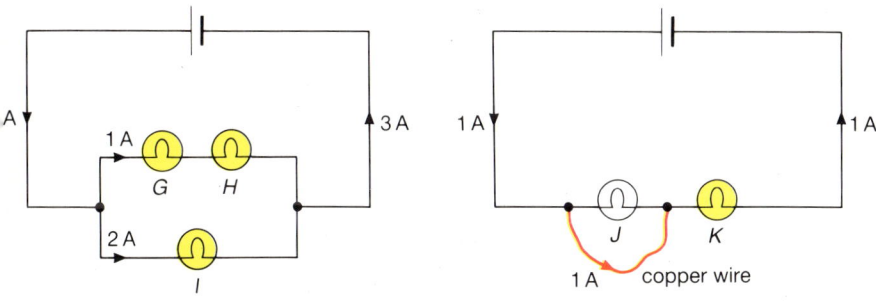

Figure 3

Figure 4

Voltage

Look at the circuit in Figure 5. Will the bulb work if you take the cell away? Of course it would not. The cell provides the energy for the bulb to light. The cell turns chemical energy (from the substances inside it) into electrical energy. The cell gives energy to the electrons to move around the circuit. When the electrons reach the bulb their energy is turned into heat and light energy.

There is an electrical energy difference between the positive and negative terminals in the cell. This energy difference is measured by the cell's **voltage**. (Voltage is also known as **potential difference** or **p.d.**) We use a voltmeter to measure voltages. Notice that the voltmeter is connected across the cell (**in parallel**), not in series with it.

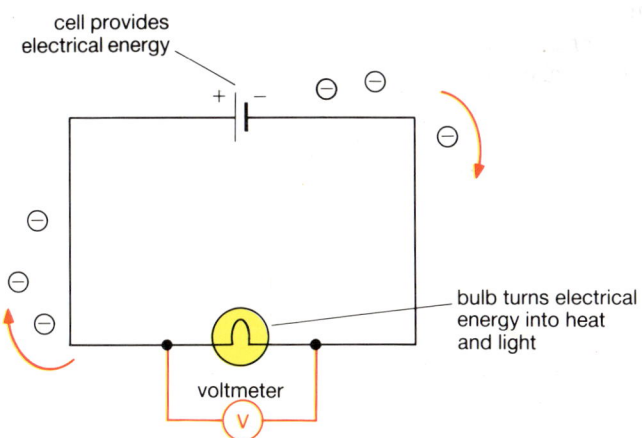

Figure 5

High voltages can kill!

The more energy a cell or battery can provide, the larger its voltage. A transistor radio uses a 9 volt (9 V) battery. If you put your tongue across the terminals of a 9 V battery you can feel a little tingle. (DON'T use a battery of more than 9 V). Mains electricity is supplied at 240 V. A shock from the mains gives your body a lot of energy. A mains shock is very painful and could kill you.

Figure 6 shows a circuit to supply two headlamps from a car battery. The bulbs are in parallel and each gets the 12 V from the battery.

If you put the two bulbs in series, they are dimmer. Each one only gets half the battery voltage (Figure 7). The battery voltage does not always split equally. A large bulb uses more energy. So it gets more of the voltage (Figure 8). Notice that the sum of the voltages across the bulbs is always equal to the battery voltage. This is because the energy provided by the battery equals the energy used by the bulbs.

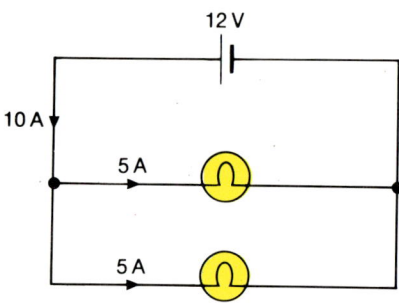

Figure 6
A circuit to supply two headlamps from a car battery

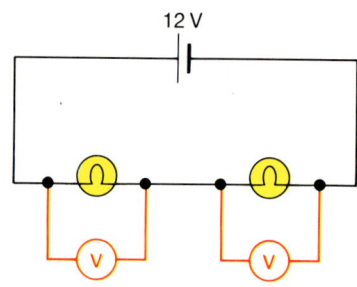

Figure 7
Each voltmeter reads 6 V

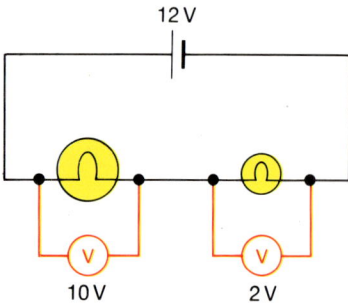

Figure 8
The larger bulb gets the larger voltage

Questions

1 Which current in the diagram below is the largest, *p*, *q* or *r*? Which current is the smallest?

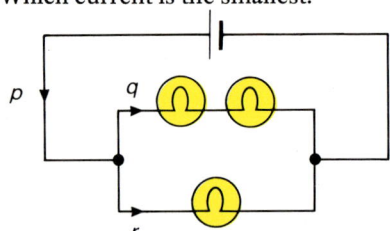

2 Bulbs *X* and *Y* (below) are identical. A_3 reads 0.2 A
What do ammeters A_1 and A_2 read?

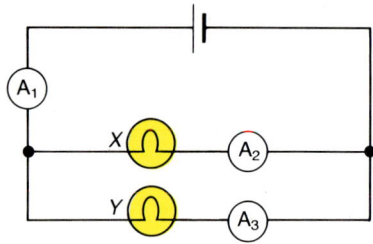

3 Copy the diagrams below and mark in the missing values of current.

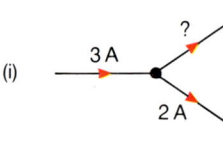

(i)

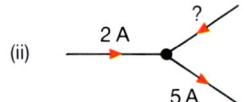

(ii)

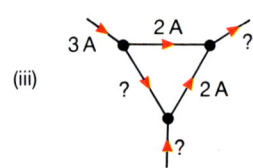

(iii)

4 Some industrial machines use a 415 V supply. Explain why no machines use a 415 V supply in your home.

5 Explain carefully the energy changes that occur when a cell is connected to light a bulb.

6 In the circuit below, a voltmeter connected across *AB* measures 6 V. What does the voltmeter measure across:
(i) *CD*, (ii) *GH*, (iii) *FI*, (iv) *EI*, (v) *DI*, (vi) *DH* (quite hard)?

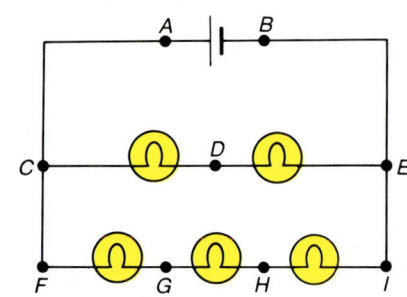

5 Resistance

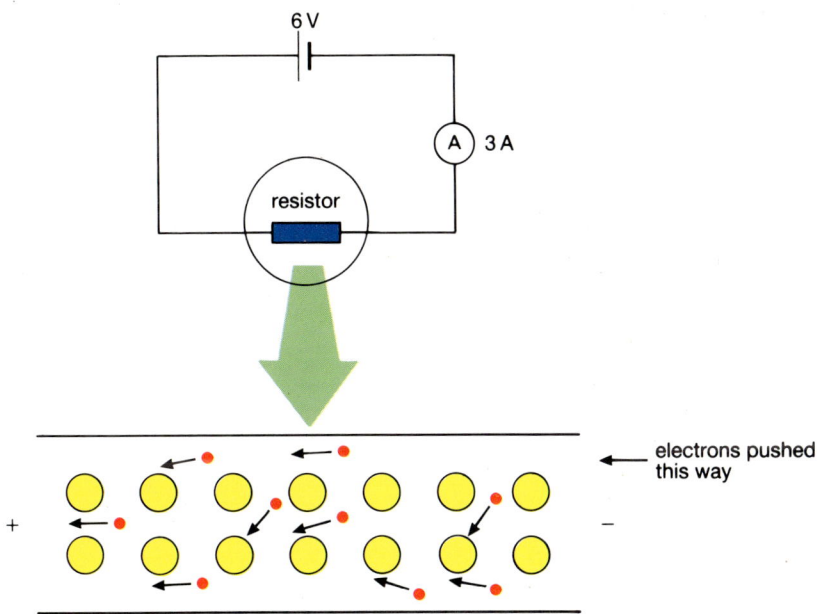

Porcelain is an excellent electrical insulator, so it is used to isolate electrical cables from metal pylons

Figure 1
Moving through a resistor, electrons keep bumping into atoms. This makes the resistor hotter

Figure 1 shows a resistor in a circuit. If we want to calculate the current flowing through it we need to know how much **resistance** it has. A resistor that has a large resistance only allows a small current through it. A small resistance allows a large current through.

$$\text{Resistance } (R) = \frac{\text{p.d. across resistor, in volts } (V)}{\text{current flowing through it, in amperes } (I)}$$

$$R = \frac{V}{I}$$

The unit of resistance is the ohm, symbol Ω. Large resistances are measured in thousands or millions of ohms. So, $1000\,\Omega = 1\,k\Omega$ (1 kilohm); and $1\,000\,000\,\Omega = 1\,M\Omega$ (1 megohm).

In an electrical circuit it is the resistors, not the wires connecting them to the battery, that get hot. The enlarged picture of a resistor in Figure 1 helps to show you why. Electrons are being pushed around the circuit by the battery and they collide with the atoms in the resistor. Energy is given to the atoms so that they vibrate faster – this means the resistor is getting hotter.

Measuring resistance

The simplest way for you to measure resistance is shown in Figure 2. The voltage across the resistor is measured using the voltmeter. The current flowing through the resistor can be read from the ammeter. Then the resistance can be calculated using the formula: $R = V/I$. A voltmeter has a large resistance so hardly any current flows through it. If current does flow through the voltmeter, the ammeter measures the extra current, and this will give you the wrong answer.

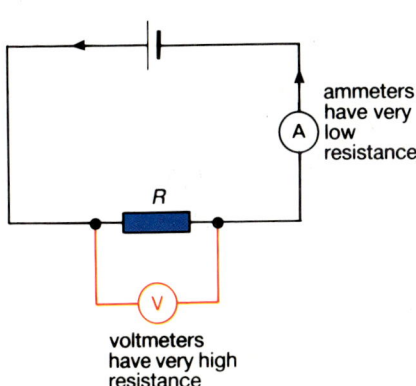

Figure 2

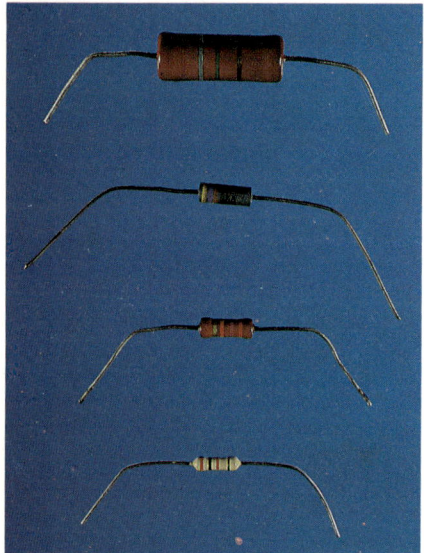

Some practical resistors. The various coloured bands show the value of the resistor and the accuracy with which it is made

You can increase the current flowing through your resistor by increasing the number of cells. You can then plot a graph of the current against the voltage. If your resistor is made from a metal or from carbon, your graph will look like Figure 3. The graph shows us that the resistance of the resistor does not change as the current increases. Such a resistor is said to be **ohmic** – one that obeys **Ohm's Law**.

Ohm's Law states that for some conductors, the current flowing is proportional to the voltage, provided the temperature does not change.

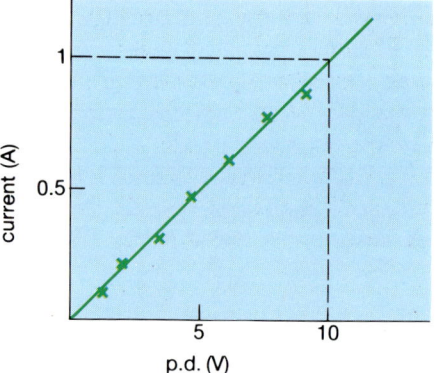

Figure 3

You can use the circuit in Figure 4 to see how the resistance of a wire is proportional to its length. A resistance wire is stretched between *AB* and a steady current of 1A flows through it. A voltmeter is attached to one end of the wire at *A*, and the other end of the voltmeter can be connected to any point on the wire. Table 1 shows some typical results. You can see that double the length of wire gives double the resistance.

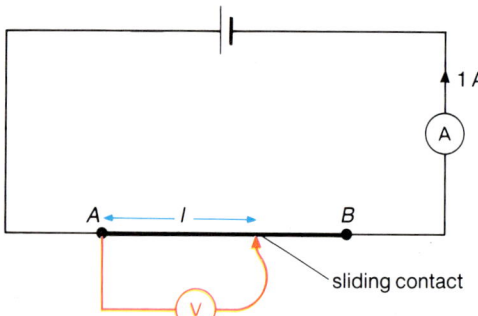

Figure 4
The resistance of a wire becomes bigger, the longer it is

Voltage (V)	Current (A)	Resistance (Ω)	Wire length (cm)
2.0	1.0	2.0	15
4.0	1.0	4.0	30
8.0	1.0	8.0	60
12.0	1.0	12.0	90

Table 1

Combination of resistors

- **Series**. The last experiment with a wire makes it easy to make a rule to work out the total resistance of two or more resistors in series. If the length of the wire is doubled the resistance is doubled. So when resistors are in series we simply **add** them to work out the total resistance. You can see in Figure 5 that the resistance between *X* and *Y* is 20 Ω.

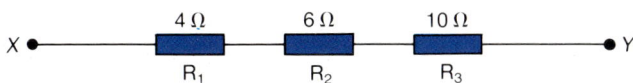

Figure 5
Resistors in series: $R = R_1 + R_2 + R_3$

- **Parallel**. Working out how to combine resistors in parallel is harder. Figure 6 shows you an example. The most important point you should realise is that the resistance between X and Y must be *less* than either of the 3 Ω or 6 Ω resistors. When you put a resistor in parallel with another, you increase the *total* current, therefore you have made the resistance *smaller*.

   Using the equation $I = V/R$ you can see that a 6 V battery will drive 1 A through the 6 Ω resistor, and 2 A through the 3 Ω. So the *total* current flowing out of the battery is 3 A. You can now calculate the combined resistance of the two resistors.

$$R = \frac{V}{I} = \frac{6\,\text{V}}{3\,\text{A}} = 2\,\Omega$$

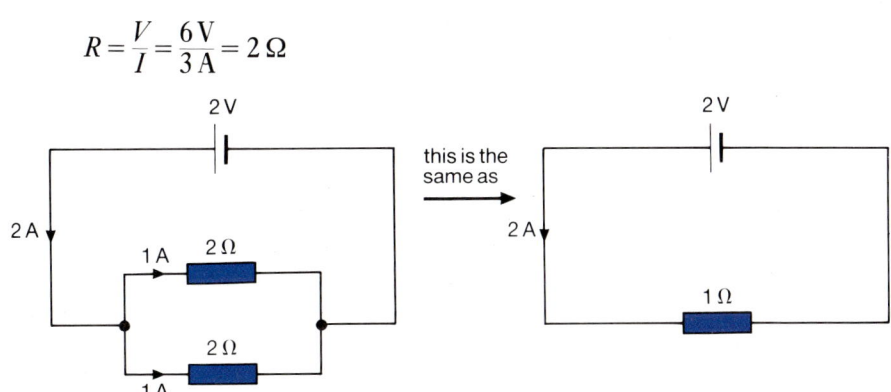

this is the same as →

Figure 7
Putting two resistors of the same value parallel halves the resistance

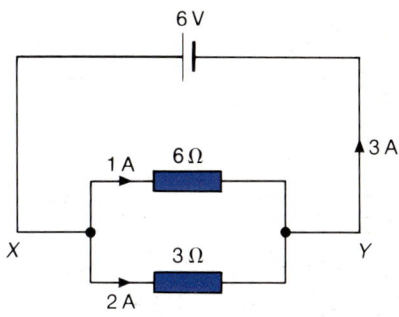

(a) Resistors in parallel

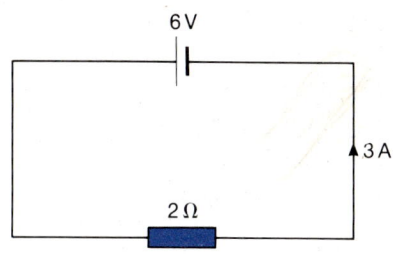

(b) The total resistance is smaller; in this example the two resistors in (a) are the same as one 2 Ω resistor

Figure 6

Questions

1 Use the data in Table 1 to calculate the resistance of 1 m of the wire.
2 Figure 3 shows an I/V graph for a resistor. Calculate its resistance.
3 Calculate the resistance of the following combination of resistors.

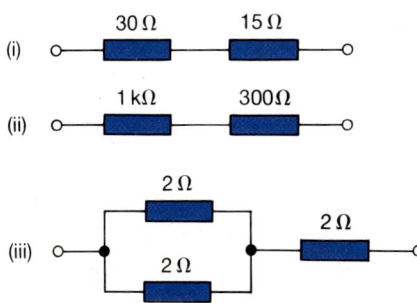

4 (a) Work out the current flowing through: (i) the 12 Ω resistor, (ii) the 4 Ω resistor, (iii) the battery.
(b) What single resistor has the same value as these two resistors which are connected in parallel?

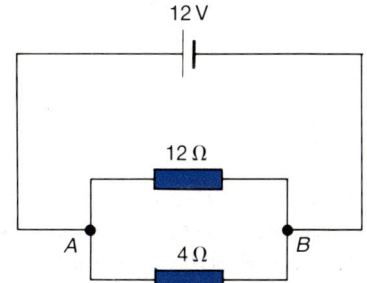

6 Changing Resistance

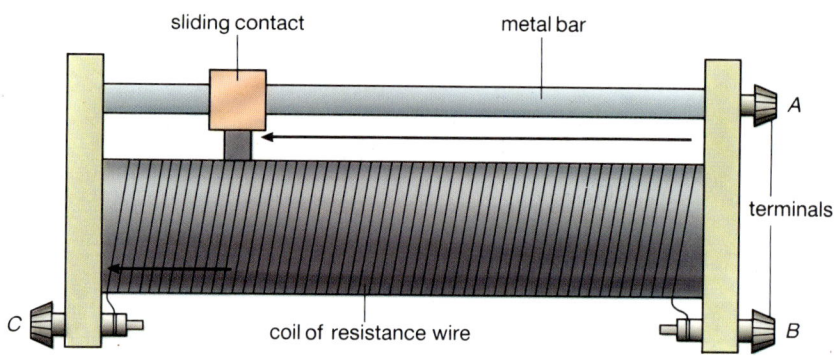

Figure 1
A variable resistor

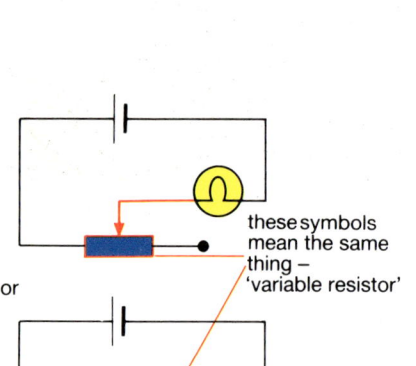

Figure 2
These circuits show the symbols for a
variable resistor

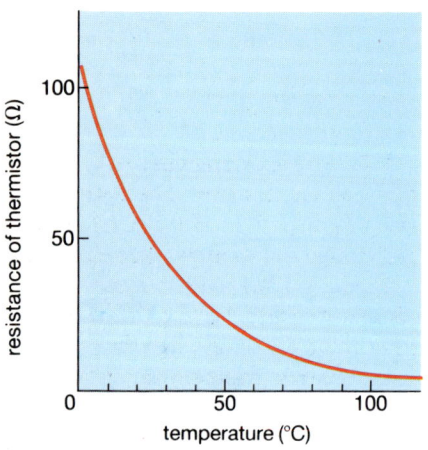

Figure 3
The resistance of a thermistor changes with
temperature

Variable resistors

In Figure 1 you can see one type of **variable resistor**. It can be used as a fixed resistor by using terminals B and C, when the current goes through the full length of resistance wire. Or, it can be used as a variable resistor, when you use terminals C and A. The current now flows along the thick metal bar, then down through the sliding contact and along the coil of resistance wire to C. The metal bar has a very small resistance, so the further the contact slides towards C, the less the resistance becomes. When the sliding contact is half way down the bar, the resistance between A and C is about half of the total resistance of the coil between B and C. Figure 2 shows the circuit symbols for a variable resistor.

The thermistor

A **thermistor** is a resistor whose resistance changes considerably with temperature (Figure 3). At low temperatures a thermistor's resistance is high, but as the temperature rises the resistance becomes less. At low temperatures, electrons are fixed on to atoms and so cannot move. As the electrons get hotter they receive enough energy to escape from their atoms, so the thermistor becomes a better conductor. Materials whose resistances change in this way are known as semiconductors.

Resistance of a light filament

The filament of a light bulb also changes with temperature, but its resistance gets bigger as the temperature rises. In this case the number of electrons carrying the current remains constant as the temperature rises. However, the increased vibrations of the atoms as they get hotter makes it harder for the electrons to pass through the filament, so the resistance increases.

You can investigate the resistance of the light filament using the circuit in Figure 4(a). The variable resistor allows you to control the brightness of the bulb. Moving the slider towards Y reduces the resistance, R, so more current flows, making the bulb brighter. Moving the slide towards X dims the bulb. You can show your results by plotting a graph of the current against voltage across the bulb (see Figure 4(b)).

You can see that the current is *not* proportional to the voltage, so the light filament does not obey Ohm's Law. Some simple calculations shown on the graph show you how the resistance changes. At point A, where the current is only 0.2 A and the bulb is quite cool, the resistance is 5 Ω. At B the bulb is hot, having a current of 0.4 A going through it, and the resistance is now 15 Ω.

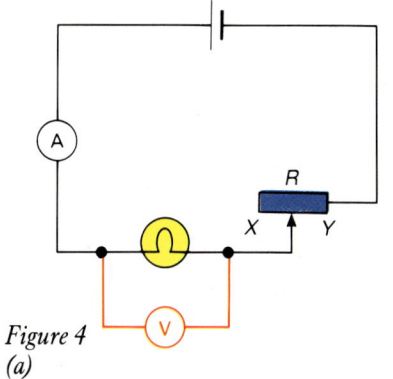

Figure 4
(a)

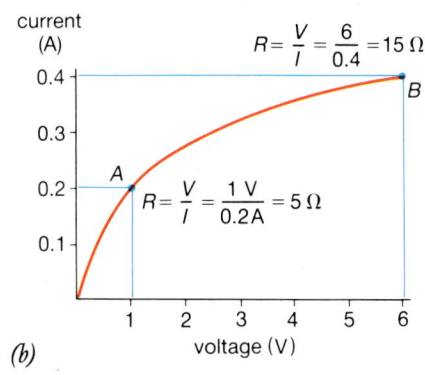

(b)

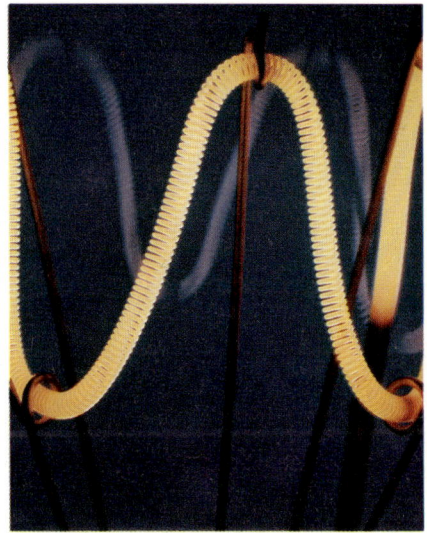

A close-up of the tightly coiled tungsten filament in a light bulb

Manufacturing light bulbs

Making a light bulb is quite a difficult problem. The wire used to make the filament is *tungsten*, which has a melting point of 3400°C. When the filament is working, its temperature is about 2500°C. At that temperature most other metals have melted. For the light bulb to work on the mains voltage of 240 V, the filament needs a high resistance. This is usually about 1000 Ω. As you know, metals are good conductors, so metal wires have low resistances. 1 m of tungsen wire with a diameter of 1 mm has a resistance of 90 Ω (at 2500°C). To make a filament have a larger resistance it must be made of very fine wire (thinner than your hair). The 1 m length of filament is coiled tightly, and then coiled again to fit into the bulb, as shown in the photograph.

You might have thought that the problems were over now. The next difficulty is that tungsten at high temperatures would oxidise in air, so the filament is placed in a mixture of argon and nitrogen gases. Even so, the filament's life is limited because at such high working temperatures it tends to evaporate. Next time a light bulb blows at home, look at it. You will see that the glass bulb has blackened due to the evaporated tungsten. Nevertheless the average light bulb has a lifetime of about 1000 hours. That is excellent value for money, at a cost of about 40p!

Questions

1 Figure 1 shows a variable resistor. The resistance of the coil between B and C is about 100 Ω. Estimate what resistance you would find if you used terminals: (i) A and B, (ii) A and C.
2 Use the graph in Figure 3 to estimate the resistance of the thermistor at: (i) 30°C, (ii) −10°C.
3 (a) The graph shows I/V graphs for three electrical components X, Y, Z. (i) Which one obeys Ohm's law? (ii) Which one could be a light bulb? Explain your answers.

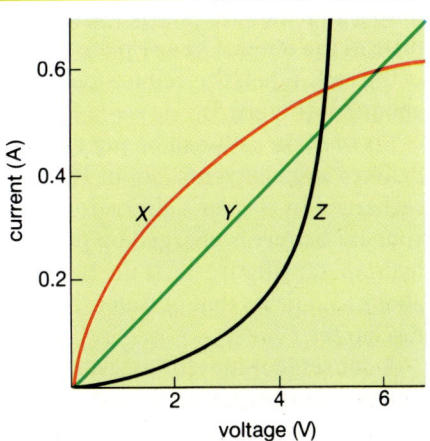

(b) The electrical components X, Y, Z are now all put in series. The current flowing through them is 0.2 A. (i) What is the voltage across each component of the circuit? (ii) Now work out the resistance between p and q. (iii) Will the resistance be the same when the current is 0.4 A? Explain your answer.

7 Electrical Cells

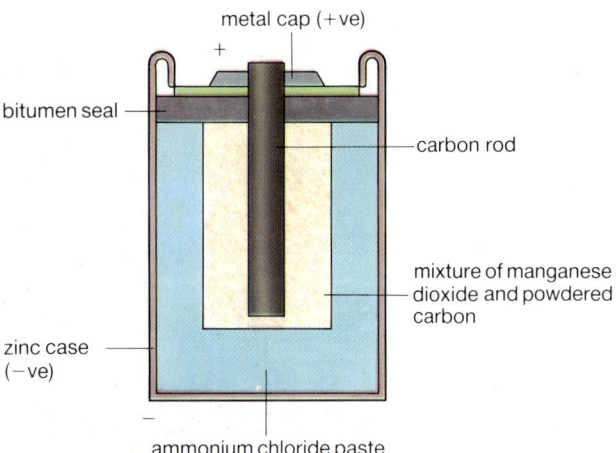

Figure 1
A primary cell. The diagram shows the structure of a Leclanche dry cell. This cell produces a voltage of about 1.5 V

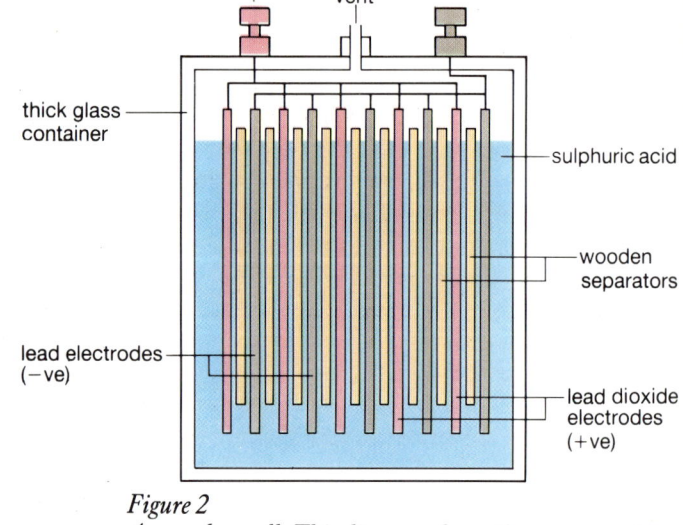

Figure 2
A secondary cell. This diagram shows the structure of a lead-acid cell. This cell produces a voltage of about 2 V

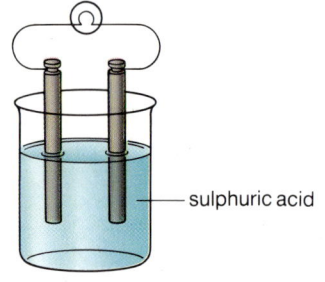

Figure 3
A simple lead-acid cell
(a) Two lead plates in sulphuric acid – no voltage

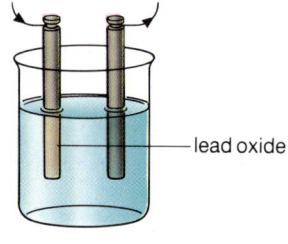

(b) A current passes through the cell and energy is stored in it

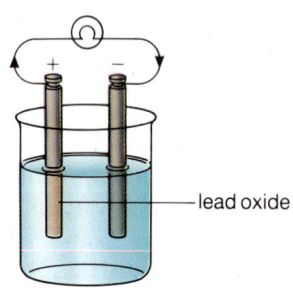

(c) Energy can now be released. Voltage is about 2 V

Primary cells

Sometimes when various substances are mixed together, chemical reactions produce electrical energy. These chemical reactions form the basis for electrical cells. Figure 1 shows the principle of the **Leclanche dry cell**. You will probably just know it as a torch battery. The reaction between ammonium chloride and zinc makes the zinc negative with respect to the carbon. The carbon is the positive terminal of the cell and the zinc the negative terminal. The manganese dioxide helps to prevent hydrogen gathering near the carbon electrode which would stop the reaction.

This type of cell is called a primary cell. Once the chemicals have been used up the cell is finished and you throw it away.

Secondary cells

As you may know, torch batteries are quite expensive. If you need a lot of electrical energy it may be better to buy a **rechargeable** battery. Rechargeable batteries are more expensive to buy in the first place, but you can use them over and over again.

Figure 2 shows a rechargeable lead–acid battery, the sort that is used in cars. Two lead plates are put into sulphuric acid, and a current is then passed from one plate to the other. On one plate a coating of lead oxide forms and energy is stored in the cell. When the cell has been charged up in this way it produces a voltage of about 2 V (Figure 3).

Six of these cells can be put together in series to form a car battery. This delivers large currents (about 100 A) to drive the starter motor. The battery is recharged by the car's alternator when the car is in motion. You can tell whether your car battery is charged up properly by measuring the density of the acid with a hydrometer (Figure 4). If the battery is fully charged the density of the acid is about 1.25 times that of water, but it falls to about 1.11 times that of water as it discharges.

A convenient unit for measuring the charge stored in a battery is the **ampere-hour** (**Ah**). If a battery is rated at 4 ampere-hours, it means that it can deliver a current of 1 A for 4 hours, or 0.5 A for 8 hours, and so on.

Internal resistance and e.m.f.

The voltage that is produced across the terminals of a cell is called the **electromotive force** of the cell (**e.m.f.**). This is the maximum possible voltage that can be produced by the cell. As soon as the cell delivers a current, the voltage that you measure across the terminals of the cell drops. The problem is that the current has to pass through the cell itself. In the Leclanche dry cell the current has to pass from the zinc outside through the chemicals and to the carbon electrode. The electrical resistance of this path is about 0.5 Ω; we call this the internal resistance of the cell. The internal resistance of a car battery is very low indeed. It is usually about 0.01 Ω.

The circuit in Figure 5 shows you a way of measuring the internal resistance of torch cell. With the switch open the voltmeter reads 1.5 V. This is the cell's e.m.f. When the switch is closed the voltmeter reads 1.2 V. This means that of the cell's 1.5 V, 0.3 V are used across the internal resistance, r, and 1.2 V across the resistance R. The calculations show that in this case, $r = 0.5\,\Omega$.

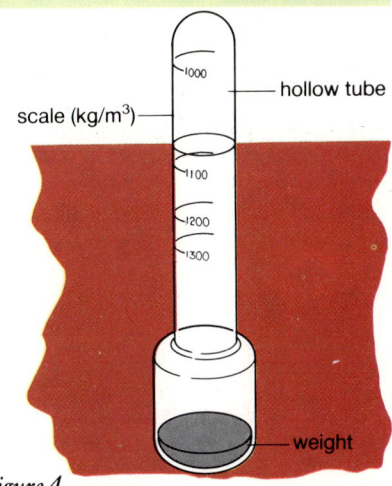

Figure 4
A battery hydrometer. When the acid is more dense, the hydrometer floats higher

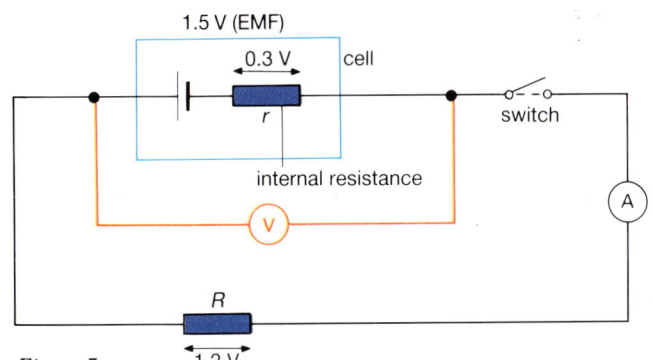

Figure 5
A circuit to measure internal resistance, with results and calculations

	Voltmeter	Ammeter
Switch open	1.5 V	0
Switch closed	1.2 V	0.6 A

$$\text{Internal resistance, } r = \frac{V}{I}$$

$$= \frac{0.3\,\text{V}}{0.6\,\text{A}}$$

$$= 0.5\,\Omega$$

Energy from cells

On page 187 you read that voltage is a measure of energy. If a cell has a voltage of 1 V, it means that each coulomb of charge that leaves the cell has one joule of electrical energy.

$$\text{volts} = \frac{\text{joules}}{\text{coulombs}}$$

$$V = \frac{J}{C}$$

Questions

1 What advantage does a secondary cell have over a primary cell?

2 Why do you not use a lead–acid battery to power your portable radio?

3 An electronic wrist watch is powered by a small mercury cell. The cell lasts for 3 years before it needs replacing. The data below contains the manufacturer's information for the cell, and other information you need.

(a) How much charge (in coulombs) does the cell store?

(b) What average current does the watch use? Give your answer in μA.

(c) How much energy does the cell store? (Hint: use the formula:
energy = volts × charge)

- voltage = 1.35 V
- height = 3.4 mm
- diameter = 11.6 mm
- charge stored = 0.08 Ah
- cost = £1.00
- 3 years = 100 million seconds
- $1\,\mu\text{A} = \dfrac{1}{1\,000\,000}\,\text{A}$

(d) Work out how many joules of energy you buy for 1p.

(e) The electricity board sells energy at a cost of 600 000 J for 1p. How many times more expensive is the energy from the cell?

(f) Why is energy so expensive from this cell?

4 A car battery has an e.m.f. of 12 V and an internal resistance of 0.01 Ω. Its terminals are shorted out. What current flows through the battery?

8 Electrical Power

Milk floats such as this one can deliver milk within an 11 kilometre radius of the depot. They have a top speed of just over 20 kilometres per hour. Some 15 batteries are used to provide power and after a typical day's use it takes around 10 hours to recharge them. Why are there so few electric cars around?

When you switch on a fire to warm up a room, electrical energy is changed into heat energy. If you want to know how quickly the room is going to warm up you need to know the rate at which electrical energy is used, or the power used by the fire. **Power used = Energy used/second** (see page 74).

You can work out the electrical power, *P*, used in a circuit, if you know the voltage of the supply, *V*, and the current, *I*, by using the formula:

$$P = V \times I$$

This gives us the power because

$$\text{volts} \times \text{amps} = \frac{\text{joules}}{\text{coulombs}} \times \frac{\text{coulombs}}{\text{seconds}} = \frac{\text{joules}}{\text{seconds}}$$

Power is measured in **joules per second** (J/s) or **watts** (W).

Often power used comes in rather large units. For example 1 bar of an electric fire uses about 1000 W; this is called a **kilowatt (1 kW)**. If you want to talk about power stations you will need to use **megawatts**. 1 megawatt is 1 000 000 W. A large power station can produce 2000 MW of electrical power. In the cold spell of January 1987, Britain's peak power usage was 48 300 MW. Table 1 shows you the power used, and the operating currents and voltages of several electrical devices.

Device	Power (W)	Operating voltage (V)	Current (A)
Torch bulb	0.9	3	0.3
Mains filament bulb	100	240	0.4
Electric kettle	3000	240	12.5
Iron	1000	240	4
TV set	60	240	0.25
Car starter motor	1200	12	100
Pocket calculator	0.0003	3	0.0001
Milk float	3600	72	50

Table 1.

Electricity bills

What is it that you pay the electricity board for? You pay for the energy that you use and it costs you about 6p for 1 unit. The unit that the electricity board uses is the **kilowatt hour, (kWh)**. If you leave a 1 kW fire on for 1 hour then you have used 1 kWh of energy.

The kilowatt hour does not look like a unit of energy; but it is. We can turn a kilowatt hour into joules like this:

$$\text{Energy used} = \text{Power} \times \text{time}$$
$$= 1000\,\text{W} \times 3600\,\text{s}$$
$$= 3\,600\,000\,\text{J}$$

Example. What does it cost to cook a turkey for 6 hours using a 2 kW electric oven?

$$\text{Energy used} = 2\,\text{kW} \times 6\,\text{h}$$
$$= 12\,\text{kWh}$$

Each kWh costs 6p, so total cost = 12 × 6p = 72p

Running a large company can be an expensive business

Questions

1 (a) A mains bulb is marked 240 V, 40 W and a torch bulb is marked 6 V, 3 W. Explain what the markings mean.
(b) Calculate the current that flows through each bulb when they are working normally.

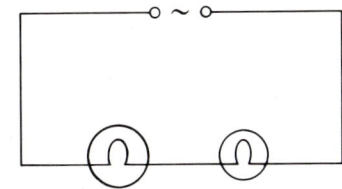

(c) Calculate the normal working resistance of each bulb.

(d) Explain what will happen if both bulbs are put in series across the mains as shown.

2 (a) Calculate the current flowing in this circuit.

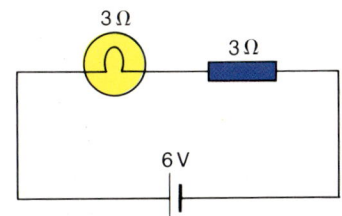

(b) What power does the battery deliver to the circuit?

(c) How much power is the light bulb using?

3 An electric kettle has an element of resistance 24 Ω. Calculate how much power it uses when it is connected to the mains (supply voltage 240 V).

4 A houseowner goes away for two weeks holiday. To deter burglars she leaves on two 100 W light bulbs in the house while she is away. If 1 kWh of electricity costs 6p, how much do these bulbs cost her over the two weeks?

5 What does it cost to use a 1 kW iron for 30 minutes?

9 Electricity in the Home

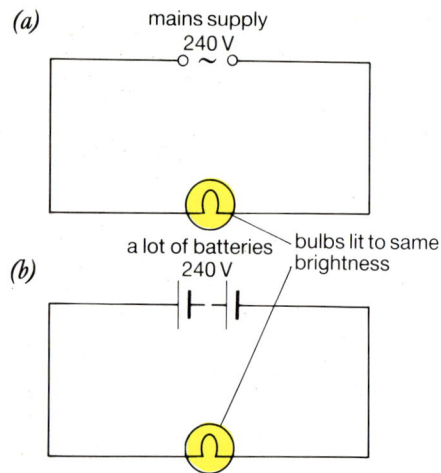

(a)

mains supply
240 V
~

a lot of batteries
240 V

bulbs lit to same
brightness

(b)

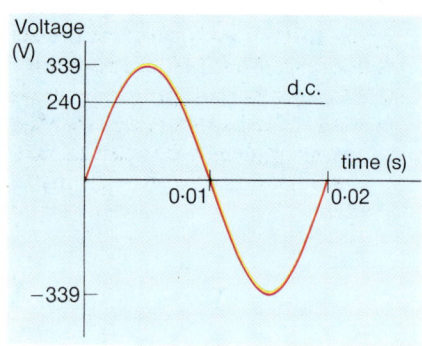

(c)

Figure 1

Alternating current

The electricity supply to your home is an **alternating current** supply (**a.c.**). The voltage of the supply is 240 V. Figures 1(a) and (b) show two light bulbs, one on a 240 V mains a.c. supply and the other on a 240 V d.c. supply. They both have the same brightness, so each supply delivers the same power to the light. Figure 1(c) shows how the voltage across each lamp varies with time. For the d.c. supply the voltage is constant at 240 V. However, for the light powered by the a.c. supply, the voltage across it changes from $+339$ V to -339 V. When the voltage becomes negative the lamp still works. It simply means that the current is flowing the other way round – just like turning the battery around in a circuit. 339 V is the peak voltage of the mains supply. The reason that the peak voltage is bigger than 240 V is that for part of the time the mains voltage is near to zero, and at those times little power is given to the lamp. So the peak voltage has to be more than 240 V, to make up for those times when little power is given to the light bulb.

We call 240 V the **root mean square** (**r.m.s.**) **value** of the mains supply. In general

$$\text{r.m.s. voltage} = \frac{\text{peak voltage}}{\sqrt{2}}$$

The frequency of the mains supply is 50 Hz, which means that there are 50 complete voltage cycles per second.

Mains supply to your house

The mains electricity supply comes into your house on two cables called the **live** and **neutral**. In any circuit, you must have two wires. The current comes into the house on one wire and returns to your local substation on the other. The live wire is the dangerous wire – the voltage on this wire changes between $+339$ V and -339 V. The voltage on the neutral wire is close to zero. Figure 2 illustrates how your cooker is wired to the mains. Notice how the direction of current flow reverses during one voltage cycle.

Figure 3 shows a plan of your house wiring. The supply comes through a main fuse and the electricity meter, and then to the fuse box. The fuse box is the distribution point for your house's electricity supply. In the box there are about six fuses which lead to different circuits in the house. The size of fuse depends on the current that flows in the circuit. Cookers, immersion heaters or electric shower heaters use large currents and so have their own circuit.

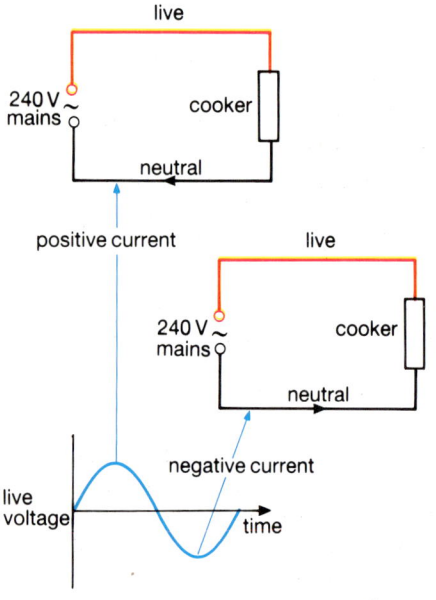

Figure 2

- Your house will have two or three **ring main circuits**, which supply all the wall sockets. On each ring main there are usually about 10 sockets. Notice that all the sockets are in parallel, so the full mains voltage is supplied to everything that is plugged into a socket. The advantage of using a ring main circuit is that current can flow two ways to a particular socket. So the connecting wires can be thinner, because they carry a smaller current than they would do otherwise. In addition to the live and neutral wires, the ring main circuit carries an **earth wire**. The earth wire is there for reasons of safety (see page 201).

- The lights for your house have their own circuit. Again each light fitting is in parallel, so that each light bulb recieves the mains voltage of 240 V. Light bulbs draw a small current (about 0.4 A for a 100 W bulb) so about 10 lights can be safely run through a 5 A fuse.

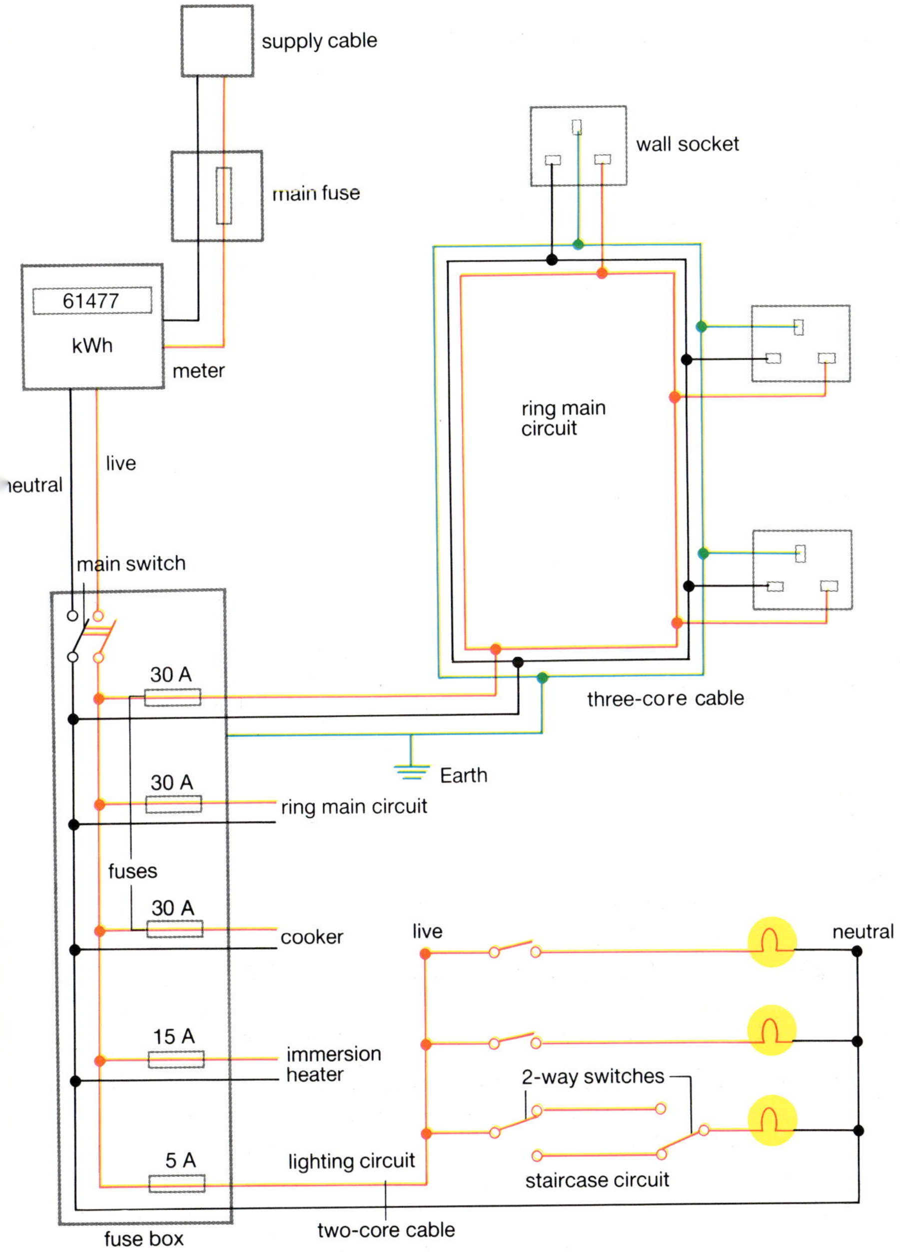

supply cable

main fuse

61477

kWh

meter

neutral

live

main switch

wall socket

ring main circuit

three-core cable

30 A

30 A

ring main circuit

fuses

30 A

cooker

Earth

15 A

immersion heater

5 A

lighting circuit

fuse box

two-core cable

live

2-way switches

staircase circuit

neutral

Figure 3
A house electricity supply

Circuits in buildings are usually protected by fuses, which 'blow' if too much current flows through them. This fuse box in an office building has fuses for a large number of circuits. Circuit breakers are now often used instead of fuses. It is important that the supply is turned off before fuses are changed

Questions

1 Explain what is meant by an alternating current.

2 Why do we say that the mains voltage is 240 V, when its maximum value is 339 V?

3 What is the advantage of using a ring main circuit?

4 What is the greatest number of 60 W bulbs that can be run off the mains, if you are not going to overload a 5 A fuse?

5 A hot water tank containing $0.2\,m^3$ of water (200 kg) is to be heated from 15°C to 40°C, by a heater drawing a 15 A current from the mains supply (240 V). (i) What is the power of the heater? (ii) What is the resistance of the heater? (iii) How much energy is used to warm up the tank?. (1 kg of water needs 4200 J to warm it up by 1°C), (iv) How long will it take to warm up the tank, (v) How much does it cost to warm up the tank, if 1 kWh costs 6p?

10 Electrical Safety

Although electricity is very useful to us it can also be very dangerous. Electrical faults can cause fires which damage property and sometimes result in injury or death. You can also give yourself a very painful electric shock from the mains, and electric shocks do occasionally kill people. The purpose of this unit is to teach you how to keep your home safe.

Fuses

The main purpose of a fuse is to prevent a fire, although a fuse can also prevent damage to expensive electrical equipment. In electrical cables that are bent and twisted it is quite common for the electrical insulation to crack and break. For example this could easily happen with your electric lawn mower or an iron. As a result of the broken insulation, the live wire could come directly into contact with the neutral wire. This would short-circuit the electrical device (Figure 1).

This adaptor is being used dangerously; far too many devices are plugged into it. Question 3 opposite shows the problems that can occur

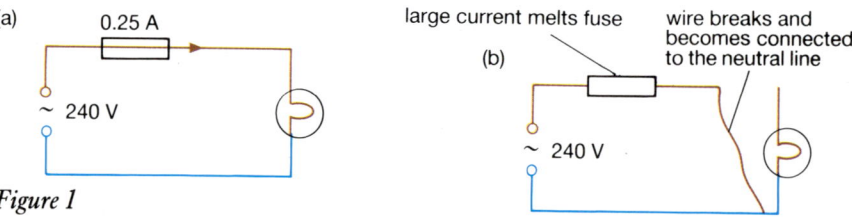
Figure 1

Normally the current flowing through the connecting wires is quite small: 0.25 A in our example. By contrast, if there is a short circuit, the resistance in the circuit is so small that the current becomes very large: 100 A or more. Such a large current would cause enormous heating and a fire is the likely result. However, you are protected by a fuse. A fuse contains a thin piece of resistance wire which melts easily if the current becomes too large. So when a fault happens, the fuse melts and the circuit is broken (Figure 1(b)).

The fuse is put into the *live* wire. Then if a fault occurs and the fuse melts, the live (dangerous) wire is disconnected. Switches are also put into the live wire for the same reason. If a switch is put into the neutral wire, you could get a shock from an electric fire element even though it is switched off (Figure 2).

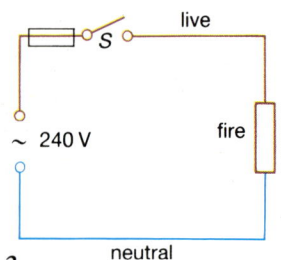
Figure 2
Switches, as well as fuses, are put into the live wire

All devices are protected by a fuse in the plug, and if that fails, the main fuse box in the house also has a fuse for each circuit. Nowadays, it is quite common to find **circuit breakers** in houses. If the current becomes too big the circuit breaker throws open a switch which can be reset when the fault has been corrected.

Choosing a fuse

When wiring up a plug you should always choose the right fuse for the device you are going to use. For example, what fuse should you use for a 1000 W iron?

$$P = V \times I$$

$$I = \frac{P}{V} = \frac{1000\,\text{W}}{240\,\text{V}} = 4\,\text{A}$$

The most commonly used fuses in the home are 3 A, 5 A, and 13 A. So you should choose the 5 A for your iron, then the fuse will blow if any small extra current flows due to a fault. Table 1 shows some appliances and their recommended fuses.

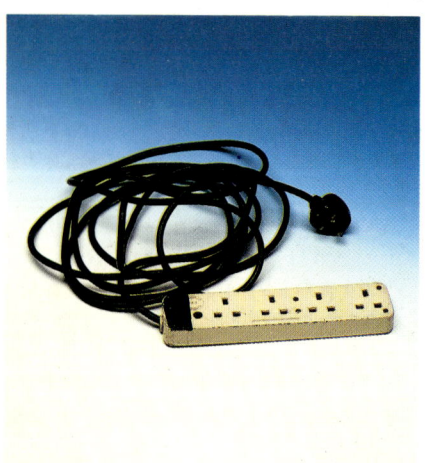
This extension lead has a 13 A fuse. This is a safe way of plugging many things into one socket

Appliance	Power (W)	Normal Current (A)	Recommended Fuse (A)
Lamp	100	0.4	3
TV set	70	0.3	3
Hair dryer	500	2.1	5
Toaster	1200	5.0	13
Kettle	2750	11.5	13

Table 1

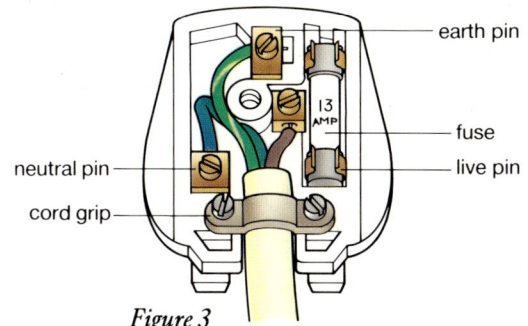

Figure 3
A correctly wired plug

Wiring a plug

You will often want to wire up a new plug to some electrical device. Don't rush at it; it will take you at least 15 minutes to do it well. Figure 3 shows you what it should look like when finished. These are the points to watch for:

- Make sure the wires are in the right place; earth is green/yellow, live is brown and neutral is blue.
- Do not strip the insulation back too far.
- Make sure the main thick cable is held in the cord grip.
- Do the screws up well.
- Choose the right fuse.

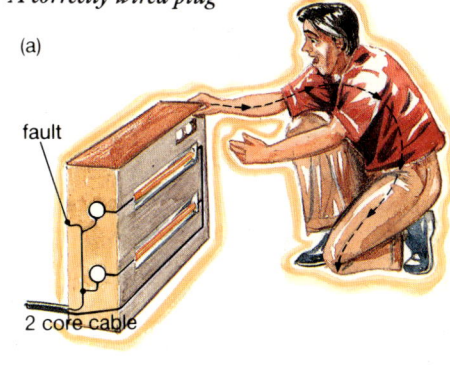

Earthing appliances

The pictures in Figure 4 tell a story. In (a) a fault has developed in an electric fire and the live wire has come into contact with the metal casing. The man gets a shock. In (b) the casing was attached to earth, so as soon as the live wire touched the case, a large current flowed and the fuse blew. In the second case the man does not get a shock.

A lot of modern appliances are made out of plastic, such as the bases of electric lamps. In these cases there is no need for an earth wire since plastic does not conduct electricity. This method of protection is called **double insulation** and it is shown by the sign ▣ .

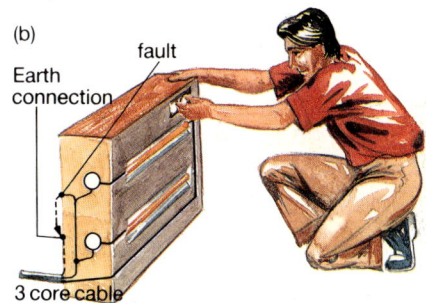

Figure 4

Questions

1 (a) Explain why we need fuses and how they work.
(b) Why is it dangerous to put a fuse into the neutral wire?
(c) Why are the outer metal casings of electrical devices earthed?

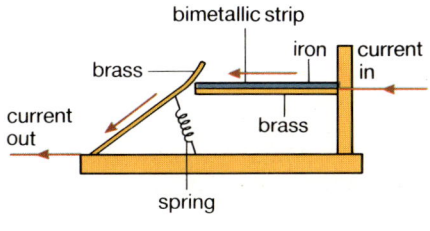

2 (a) The diagram shows a thermal circuit breaker. Explain how it works.
(b) The circuit breaker is designed to trip for currents greater than 2 A. When the current is 3 A the breaker takes 2 minutes to trip.
When the current is 10 A the breaker takes 5 seconds to trip.
Account for these times.
3 The devices below are all drawing current through an adaptor plugged into a single socket.
(a) Copy the table and fill in the column to show how much current each device takes.

(b) What fuse should be fitted for each device?
(c) What is the total current that goes through the adaptor when all three devices are plugged in? Comment on the size of the current.
(d) Will this current blow any of the fuses in the plugs?

Device	Power	Voltage	Current
Kettle	3 kW	240 V	
Fan heater	2 kW	240 V	
Iron	1 kW	240 V	

11 Electron Beam Tubes

Thermionic emission

When a current goes through a wire, the wire gets hot. At very high temperatures some electrons have enough energy to escape from the wire. If the wire is charged negatively the electrons are repelled and can be collected by a positive plate (Figure 1). The positive plate is called an **anode** and the negative wire the **cathode**. When this effect (thermionic emission) was first discovered, the emitted electrons were called **cathode rays**, because they were emitted from the cathode.

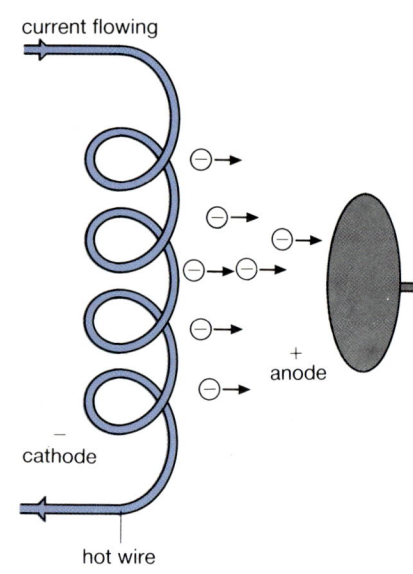

Figure 1
Thermionic emission

The Maltese Cross tube

Figure 2 shows how a **'Maltese Cross'** tube works. A 6 V battery is connected to the filament so that it is hot enough to give out electrons. A 4000 V power supply is then connected across the cathode and anode. The negative electrons are attracted towards the positive anode. Some of the electrons hit the Maltese Cross and go back to the power supply. Others carry on, and hit the fluorescent screen. When an electron hits the screen, its kinetic energy is used to make the fluorescent screen give out some light.

You can control the brightness of the fluorescent screen in two ways. First the voltage of the supply to the anode can be increased. This makes the electrons travel faster, so they will have more kinetic energy when they hit the screen. Secondly the filament can be made hotter by making a bigger current flow through it. Then more electrons are given out, so more electrons hit the screen.

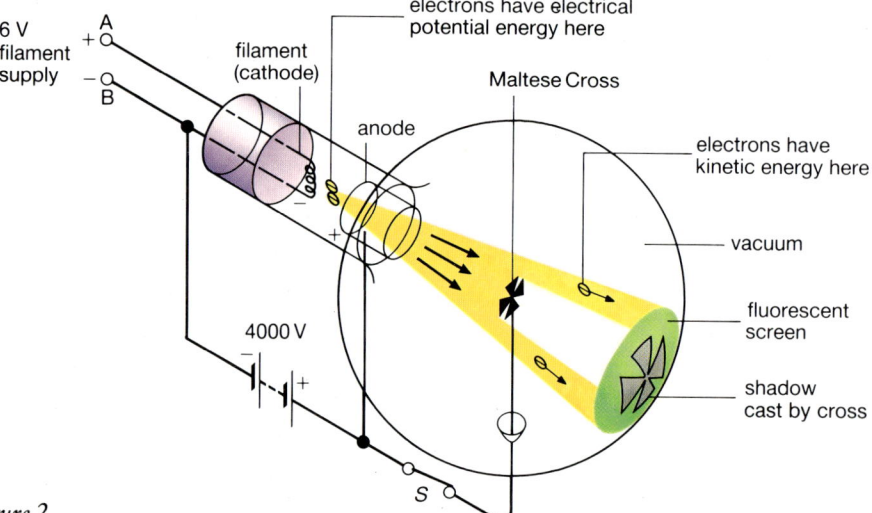

Figure 2
A Maltese Cross tube. A metal cross lies in the path of electrons, so a shadow is cast on the fluorescent screen

The cathode ray oscilloscope

Figure 3 shows a **cathode ray oscilloscope** (CRO). Electrons are accelerated away from a heated filament by the pull of a positive anode. Then the electrons travel through a vacuum until they hit a fluorescent screen. Unlike the Maltese Cross tube the CRO produces a very fine beam of electrons, which produces a small spot of light on the screen. The beam is focused by the two anodes. The brightness of the spot is controlled by the grid. If the grid is charged slightly

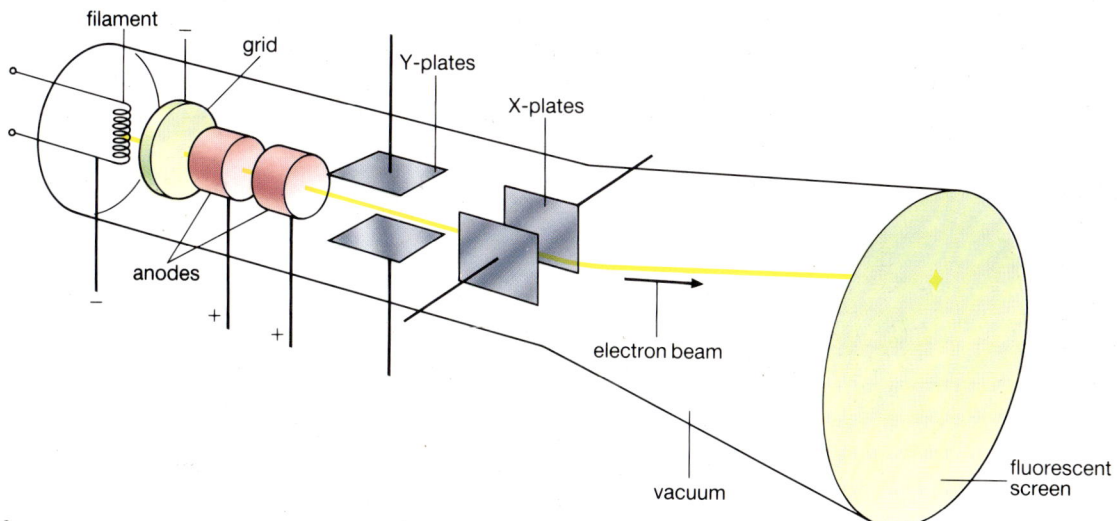

Figure 3
A cathode ray oscilloscope (CRO)

negative, electrons from the filament are repelled so that fewer of them hit the screen. This makes a smaller, duller spot.

The electron beam can be deflected vertically or horizontally by charging the Y-plates or the X-plates. Figure 4 illustrates how the X and Y plates may be charged to deflect the beam. In this diagram you are looking down the length of the oscilloscope, through the screen towards the deflection plates. Notice that when alternating current is applied to the Y-plates a vertical line is produced. As a result of the alternating current, the charge on the Y-plates is switching around very rapidly. Sometimes the top plate is positive, but a fraction of a second later it is negative. This makes the electron beam switch quickly from being deflected upwards to being deflected downwards. The switching occurs so rapidly that the spot on the screen has no time to fade, and we therefore see a continuous straight line.

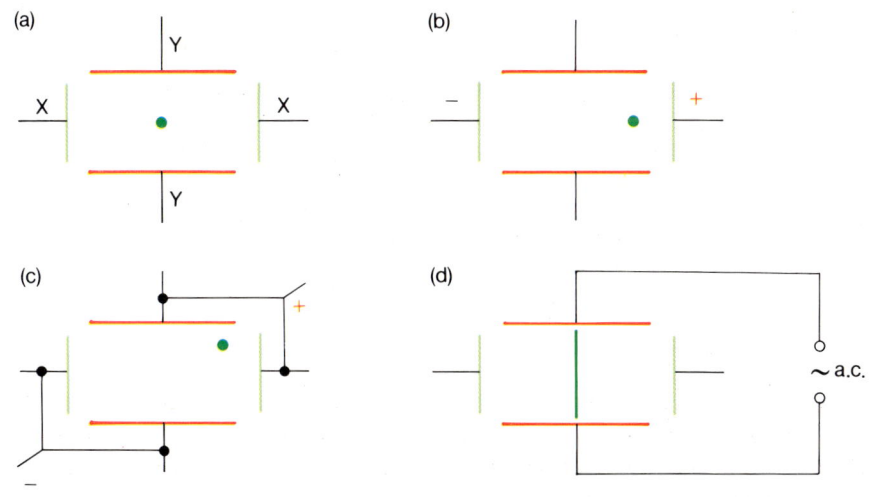

Figure 4

Questions

1 Look carefully at Figure 2.
(a) (i) Does it matter which way round the 6 V battery is connected to the filament?   (ii) Does it matter which way round the 4000 V supply is connected? Explain your answers.
(b) What change would you see if you reduced the voltage between the cathode and the anode to 2000 V?
(c) Why do we have to keep a vacuum in the tube?
(d) The diagrams below show the shadow at the end of the Maltese Cross tube. In one of them the switch S is open, in the other it is closed. Can you explain their shapes?

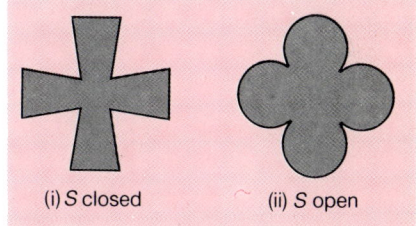

(i) *S* closed   (ii) *S* open

2 Draw diagrams similar to those in Figure 4 to show the deflection of the spot in an oscilloscope in the following cases:
(a) top Y-plate negatively charged, bottom Y-plate positively charged;
(b) alternating current supplied to both Y and X plates together.

12 Using the Cathode Ray Tube

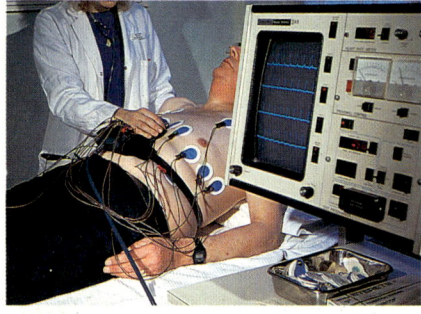

Cathode ray tubes are used in electrocardiograph (ECG) machines which monitor the electrical activity of the heart. This photograph shows an ECG machine being used to investigate whether the patient has suffered a heart attack.

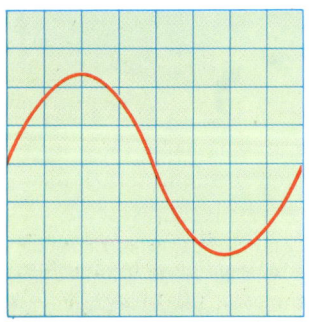

Figure 1
How the voltage applied to the X-plates varies with time

potential difference applied to X-plates (volts)

spot is moving across the screen

time

spot flies back again

TV studios use Cathode ray tubes as monitors. This picture shows the presentation control room during transmission. From here, the transmission controller keeps a close watch on the schedule and on the technical quality of the programmes

The time base

In the last unit you read how the X and Y plates in an oscilloscope can be used to deflect the electron beam. What really makes the oscilloscope useful to us is its time base. The **time base circuit** works like this. A changing voltage is applied to the X-plates of the CRO, so that the spot moves across the screen from left to right. The moment the spot reaches the right-hand side of the screen it returns immediately to the other side to start its journey across the screen again. Figure 1 shows how the voltage applied to the X plates varies with time.

The time base allows us to look at changing voltages (*signals*). The CRO plots a graph of voltage (y-axis) against time (x-axis). We can apply a signal to the Y-plates so that the spot is moving up and down. At the same time the time base makes the spot move sideways. A typical signal is shown in Figure 2. In this diagram the right-hand dial shows you that each centimetre in a vertical direction means a voltage of 2 V. So the peak voltage is 4 V. The left-hand dial shows you that the beam crosses the screen horizontally, taking 2 ms (0.002 s) for each centimetre. One complete cycle of the voltage waveform takes 4 squares of the grid, or 8 ms. The time period of the cycle is 8/1000 s. Therefore the frequency is $1000/8 = 125$ cycles per second or 125 Hz.

Oscilloscopes are widely used in industry to monitor and display changing signals. They are also useful in hospitals where they are used to show the strength and frequency of a patient's heart beat, or to build up a picture after an X-ray or ultrasound scan.

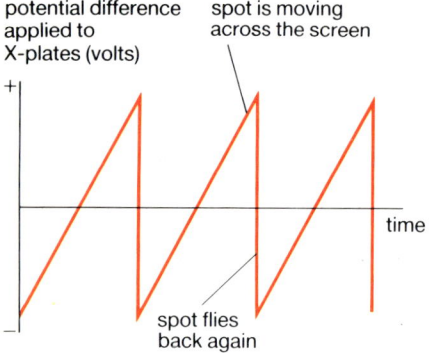

Figure 2
A typical CRO signal

Televisions

Getting our television pictures is a complicated process, but the television set itself is rather like a CRO. The photograph opposite shows a magnified picture of a TV screen. You can see that it is made up of lots of small dots which are blue, green or red. These are the three primary colours, and if they are mixed together they make white. Various combinations of the colours can make the other colours. Table 1 gives some examples.

To make a colour picture the TV set needs three electron beams, as shown in Figure 3. A beam of electrons leaves the 'green' gun and passes through a hole in the shadow mask. Then the beam hits a small fluorescent spot that gives out green light. Similarly electron beams leave the other two guns and hit fluorescent spots that give out blue and red light. The purpose of the shadow mask is to make sure the beams hit the right part of the screen. The small holes in it only allow the beam to hit a small part of the screen.

Like an oscilloscope, the television has a time base. The whole television picture is built up by the beams sweeping backwards and forwards across the screen very rapidly indeed. The beams go from top to bottom of the screen in about 0.025 seconds. During this time they have crossed the screen about 625 times.

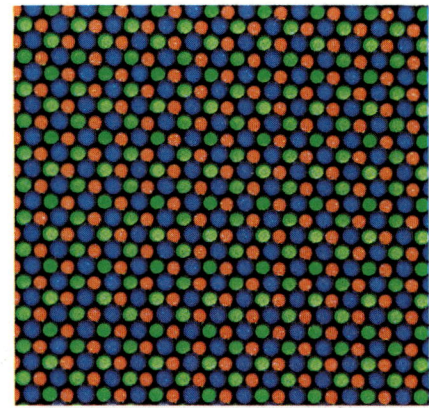

A close-up of a TV screen, showing the tiny dots that seem to blend together at a distance to make up a smooth picture

Blue	Green	Red	Mixture	
●	●	●	○	white
●	●		●	cyan
	●	●	●	yellow
●		●	●	magenta

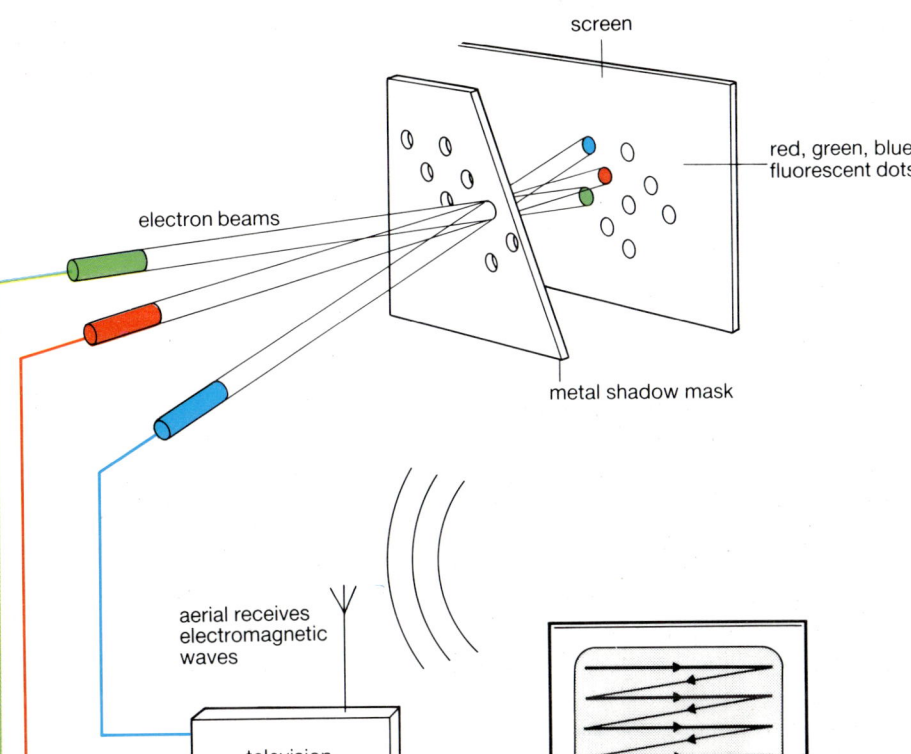

Figure 3
The principle of the colour television

Figure 4
The time taken for the beam to travel from top to bottom of a screen is 0.025 seconds

Questions

1 Look at the signal in Figure 2. Draw another diagram to show how the signal looks when you make these two separate changes: (i) increase the Y-gain to 1 V/cm, (ii) increase the time base speed to 1 ms/cm.

2 The diagram shows a waveform on an oscilloscope screen; the time base is set to 1 ms/cm and the Y-gain control to 2 V/cm. Calculate the peak voltage and frequency of the a.c. supply.

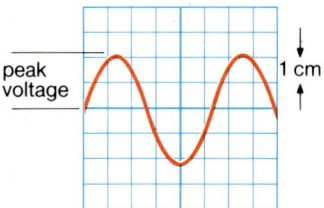

3 Explain why a colour television needs three electron beams. What colour do you see on the screen when two red spots are illuminated for every blue and green spot?

13 Thunder and Lightning

Bolts of forked lightning during an electrical storm

Even though scientists now understand thunderstorms, we still find them both spectacular and frightening. A flash of lightning occurs when static electric charge on the bottom of a thundercloud suddenly flows towards the Earth. Inside clouds, strong convection currents of air carry small ice crystals upwards. These crystals collide with large hailstones that are falling downwards. As a result of these collisions, the hailstones become negatively charged and the ice crystals positively charged (Figure 1). This means that the top of the cloud becomes positively charged, and the bottom negatively charged.

In between the ground and the bottom of the cloud there is now a space in which very strong electric forces act. These forces can be so strong that electrons are pulled out of air molecules, so that charged particles (or ions) are formed. With ions present, air becomes a conductor and a large amount of charge leaves the bottom of the thundercloud – we see this as a flash of lightning. The lightning flash heats the air to several thousand degrees centigrade which causes the air to expand very quickly. This rapid expansion makes the noise we call thunder.

ice crystal after collision

falling hailstone

ice crystal before collision

Figure 1
Charging by friction in a thundercloud

A special test rig being used to create artificial lightning by the Central Electricity Generating Board

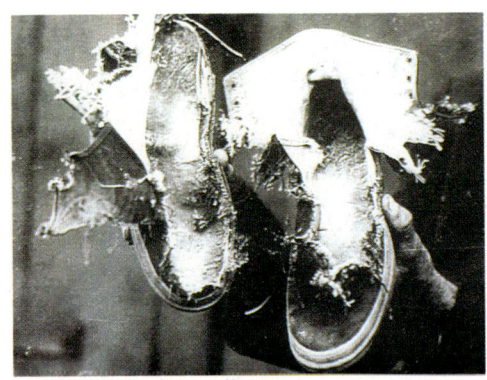

(left) These shoes were blasted off an unfortunate man's feet by lightning. What steps might he have taken to avoid this accident?

(right) This tree was damaged by lightning

Reducing the damage

When lightning strikes a tree or building the results can be devastating. A flash of lightning carries a lot of energy. The base of a thundercloud stores about 10 C of charge. The voltage between the Earth and the thundercloud can be about 100 million volts. This means that when the cloud loses all of its charge, about 1000 million joules (10^9 J) of energy are released. Two thousand cars travelling down the motorway have about the same amount of kinetic energy.

To reduce the damage to buildings during thunderstorms, we use **lightning conductors**. A lightning conductor is made from a long metal rod. One end must be fixed into the ground near to the base of the building. The other end of the rod should be pointed, and must be the highest point on the building. When a thundercloud passes overhead a large amount of charge builds up near the point of the conductor. The electric forces near to the point are now so large that negative and positive ions are produced. The positive ions are repelled away from the conductor, and are attracted towards the negative thunder cloud. So the positive charge from the conductor neutralises some of the cloud's negative charge. Now the lightning strike is less likely.

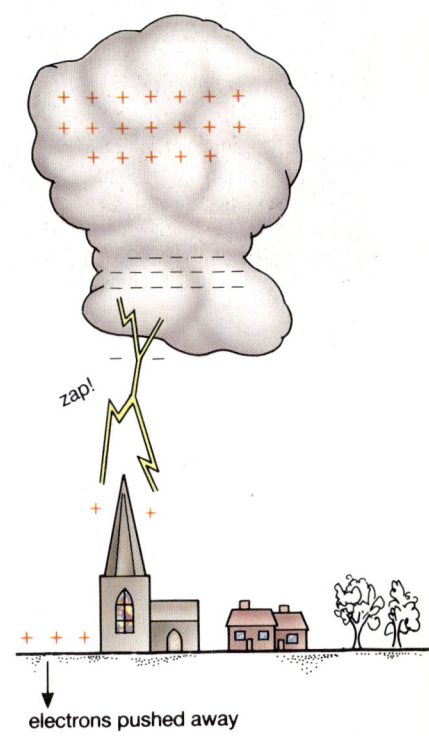

Figure 2
Lightning strikes tall buildings as these are closest to the thundercloud

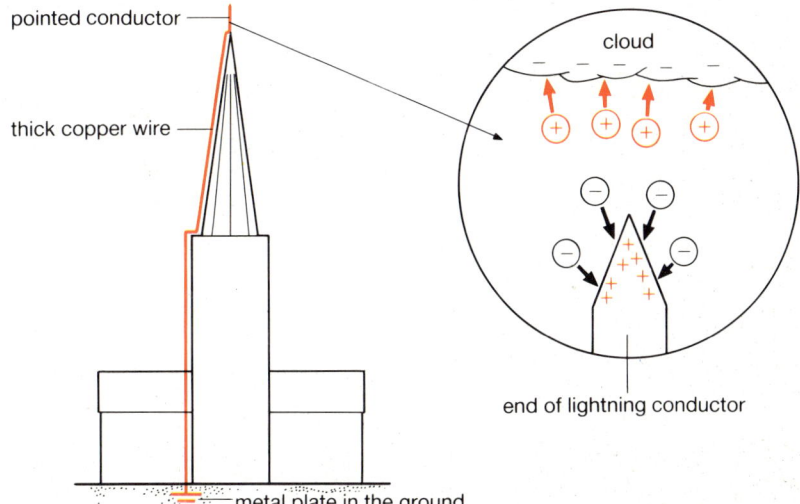

Figure 3
The principle of a lightning conductor

Questions

1 When you see a flash of lightning you often have to wait 5 or 10 seconds before you hear the thunder. Explain why.

2 Explain why it is dangerous to stand under a tree during a thunderstorm.

3 Why should lightning conductors be: (i) fixed to the ground, (ii) made of metal, (iii) pointed, (iv) the highest point on a building?

4 In the text the author wrote that a thundercloud stored about 1000 million joules. Why does this follow from the information given in the previous two sentences?

SECTION I: *STUDY QUESTIONS*

1 This diagram shows the screen, Y-gain and time-base controls from a typical oscilloscope displaying a waveform.

screen

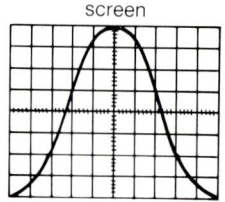

The graticule has a 1 cm grid

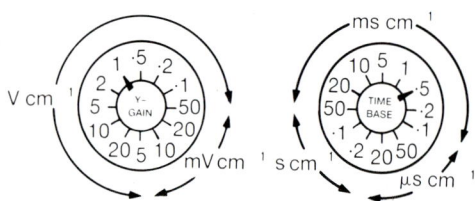

(a) What is the setting of the Y-gain control?
(b) What is the peak voltage of the waveform?
(c) What is the time base setting?
(d) What is the period of the trace?
(e) What is the frequency of the waveform?
(f) If the time base setting is altered to 1 ms cm^{-1} and the Y-gain to 2 V cm^{-1}, what would the resultant trace look like?

NEA

2 Diagram 1 shows the inside of a mains-operated hair-dryer. The fan can blow either hot or cold air.
Diagram 2 is a circuit diagram of the same dryer showing how it is wired up.

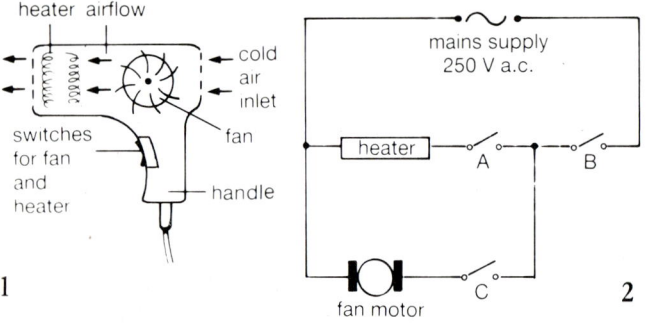

1   2

(a) Copy the table below and show by placing ticks in the boxes, which switches need to be ON to get the result shown. *(You may use each switch more than once, once or not at all.)*

Result	Switch A	Switch B	Switch C
a blow of cold air			
a blow of hot air			

(b) The heater must not be on without the fan.
(i) Which of the switches A, B, C must always be on to achieve this?
(ii) Explain carefully what you would expect to happen if the fan failed to work when the heater was on.

(c) The manufacturer wishes to include a two-speed fan. This could be done by connecting a suitable resistor across one of the switches as follows:

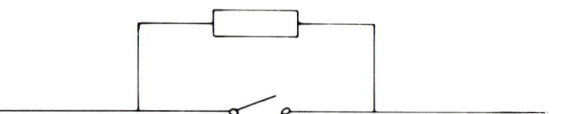

(i) Copy diagram 2 and draw a resistor across the correct switch in order to make a two-speed fan.
(ii) When this switch is open, will this give a fast or slow speed? Explain your answer.
(d) The details on the hair-dryer are 250 V 500 W. Calculate the current from the mains supply when the drier is working at the stated power.
(e) Fuses for the mains of 3A, 5A and 13A are available.
(i) Which fuse would you choose for use in the plug attached to the drier?
(ii) Which wire in the mains cable should be connected to the fuse?
(f) A girl needs to use the dryer for 10 minutes. Calculate the energy converted during this time.
(g) The manufacturer makes a different hair-dryer which will work from a 12 V car battery. You are required to find the power taken by the 12 V dryer. Copy and complete the circuit diagram below to show how you would connect up an ammeter and voltmeter to do this.

LEAG

12 V car battery

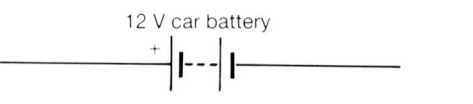

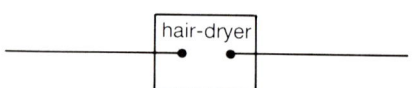

3 A battery and three similar lamps are connected as shown.

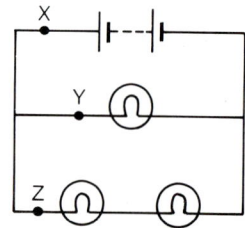

Which of the following statements about the currents at X, Y and Z is correct?
A The current at Z is greater than that at Y.
B The current at Y is greater than that at Z.
C The current at X equals the current at Y.
D The current at X equals the current at Z.
E The current at Z equals the current at Y.

MEG

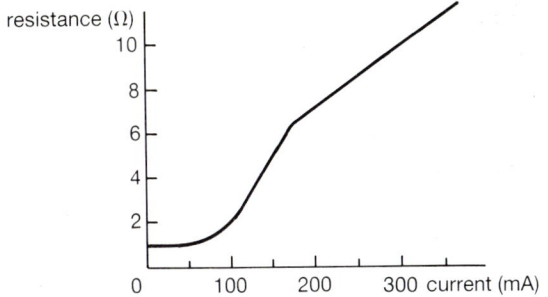

4 The graph shows how the resistance of a bulb varies with the current through it.
(a) What is the value of the resistance of the bulb when it is not switched on?
(b) The bulb can be used as a simple overload protection and warning device for use with model railway controllers. The bulb is placed in one of the wires leading to the track, as in the circuit diagram below.

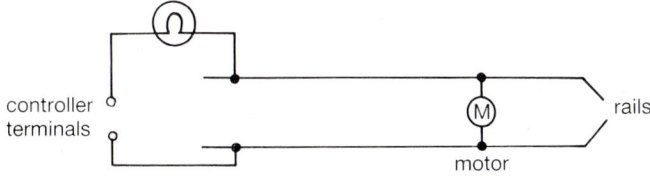

When operating normally the current through a model railway locomotive is about 250 mA.
(i) What is the potential difference across the bulb when 250 mA passes through it?
(ii) If the output voltage of the controller is 12 V, what is the potential difference across the motor when 250 mA passes? (Assume the resistance of the rails is negligible.)
The train derails and short-circuits the two rails (i.e. it joins the two rails together by a good conductor).
(iii) State and explain what happens to the bulb when this occurs.
(iv) How does the bulb provide overload protection for the control unit?

MEG

5 When a 1.5 V, 6.0 W bulb is connected to a 1.5 V cell, the number of coulombs passing through the bulb per second is
A 0.25   B 1.5   C 4.0   D 6.0   E 9.0

NI

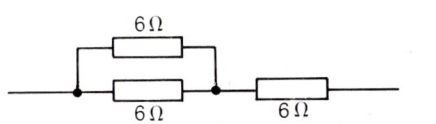

6 The diagram above shows part of a circuit. If this arrangement of three resistors was to be replaced by a single resistor, what should its resistance be?
A 2 Ω   B 4 Ω   C 6 Ω   D 9 Ω   E 18 Ω

LEAG

7 (a) The diagram below shows part of an electrical circuit. What is the reading on the ammeter (i) at S? (ii) at T?

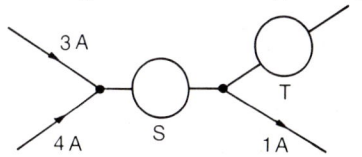

(b) This diagram shows three 2-ohm resistors connected in different ways.

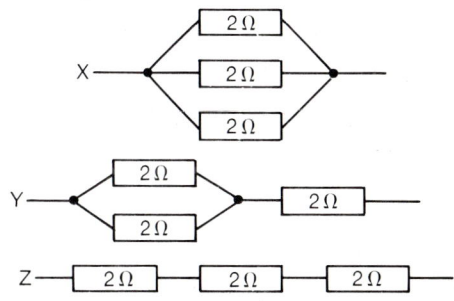

(i) Calculate the combined resistance of each of the three arrangements, X, Y and Z. Show your working.
(ii) The voltage applied across each arrangement is 2 volts. Which arrangement would allow the highest current to pass? Calculate this current.
(c) By using the appropriate symbols for a 12 V battery, an ammeter and a voltmeter, copy and complete the circuit below to show how you would measure:
(i) the current through the resistor marked P.
(ii) the voltage across the resistor marked P.

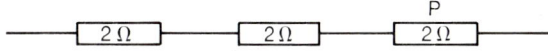

(d) A consumer buys two appliances, an electric iron marked 1000 watts and a desk light marked 100 watts. They are to be used on the mains 250 volts a.c.
(i) What would be the current passing through the two appliances? Show your working.
PATTERN: Power = Current × Voltage
(ii) If 2A, 5A and 13A fuses are available, which fuse should be used with: 1. the electric iron   2. the desk light?
(iii) Explain why it is important to include a fuse in the circuit.

LEAG

8 An electric heater is rated at 2 kW. Electrical energy costs 6 p per kWh. What is the cost of using the heater for 3 hours?
A 4 p   B 6 p   C 12 p   D 18 p   E 36 p

MEG

9 For safety reasons, where in a house are mains 13 A sockets **not** allowed?
A bathroom   B bedroom   C dining room
D kitchen   E staircase

MEG

STUDY QUESTIONS

10 Which of the following would be the best arrangement of an ammeter A and a voltmeter V to obtain readings to find the power of the lamp?

MEG

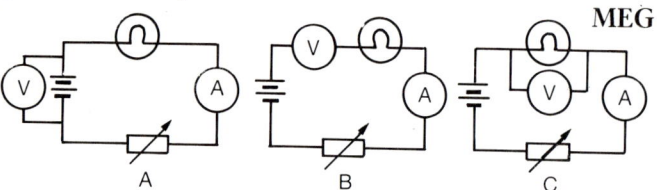

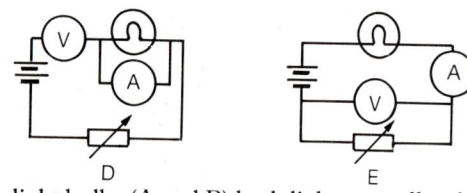

11 Two light bulbs (A and B) both light normally when connected to a 5 V battery. The graph shows how the current varies with the applied voltage (potential difference) for each bulb.

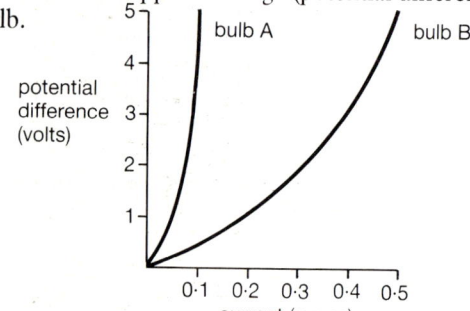

(a) When they are working normally what is the resistance of each bulb?
(b) A 5 V battery is connected across *XY* in diagram *1*.
(i) Which bulb will be brighter?
(ii) How much power is used by each bulb?

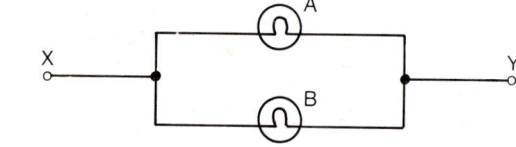

(c) The bulbs are now connected in series as shown in Diagram 2. Describe carefully what happens as the voltage across *PQ* is increased slowly from 0 to 5 V.

12 The diagram shows **a gold leaf electroscope**. This can be used for detecting charges. When the electroscope is uncharged the leaf hangs down. But when a charged rod is brought close to the plate at the top, the leaf rises. This is because the positive charged rod attracts the electrons to the top. The leaf and stem are now left positively charged so the leaf is pushed away from the stem.

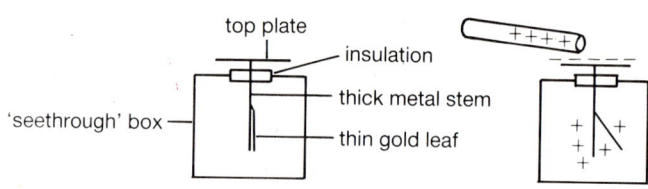

Negative charge is now placed on the electroscope. The leaf rises as shown in *A*.

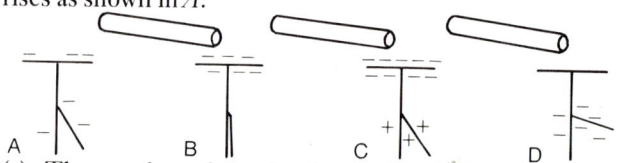

(a) Three rods are brought close to the electroscope. What can you say about the sign of the charges on *B, C* and *D*?
(b) What can you say about the size of the charges on *B* and *C*?

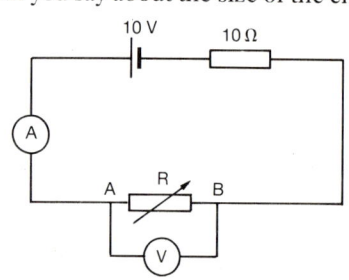

13 In the circuit R is a variable resistor. The table below shows six different values of *R*. For two of these values, the current flowing, the voltage (potential difference) across R and the power used by R have been calculated.

R (Ω)	Current (A)	Voltage across *AB* (V)	Power used in R (W)
2	$\frac{10}{12}$	$\frac{10}{12} \times 2$	1.4
5			
10			
15	$\frac{10}{25}$	$\frac{10}{25} \times 15$	2.4
20			
40			

(a) Explain how these values were calculated.
(b) Copy the table and fill in the missing values of current, voltage and power.
(c) Plot a graph of power (*y*-axis) against *R* (*x*-axis). For what value of *R* is the power used by R the greatest?
(d) What can we say about the power used by R when *R* is:
(i) very small — nearly zero.
(ii) very large — about 10 000 Ω.
Explain your answers.

SECTION J

Magnetism and Electromagnets

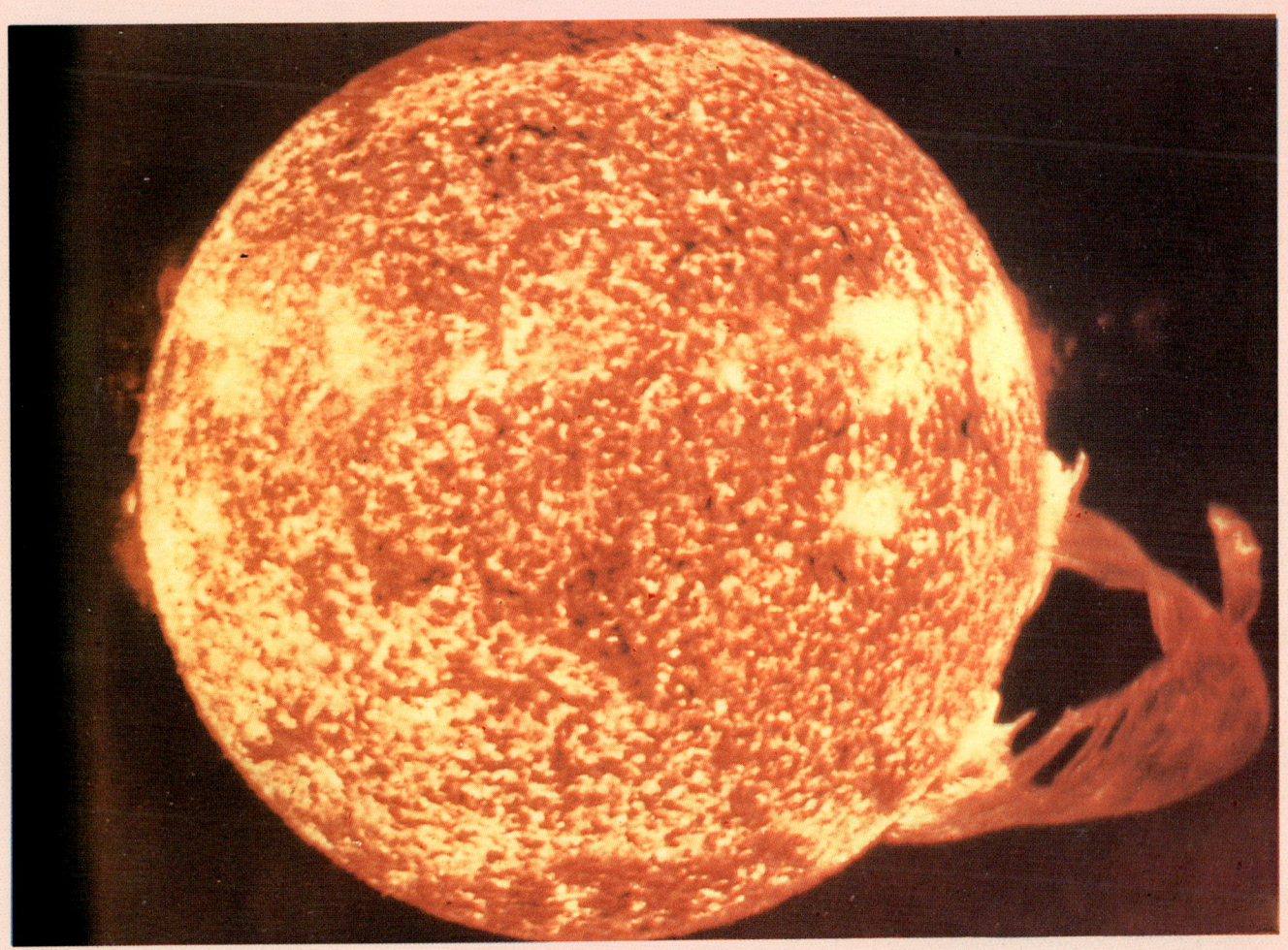

This photograph of the Sun, taken in ultraviolet light, shows a loop of material above the Sun's surface. Loops like this are called prominences. They are held up by the Sun's magnetic field, and are colossal. The Earth could fit many times into the prominence shown here

1 Magnets

Some metals, for example iron, cobalt and nickel, are **magnetic**. A magnet will attract them. If you drop a lot of pins on to the floor the easiest way to pick them all up again is to use a magnet.

Poles

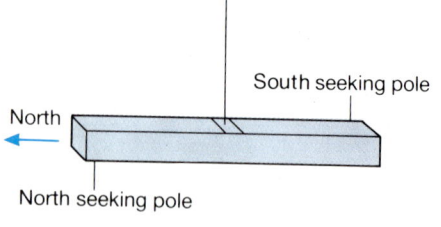

Figure 1

In Figure 1 you can see a bar magnet which is hanging from a fine thread. When it is left for a while, one end always points north. This end of the magnet is called the **north-seeking pole**. The other end of the magnet is the **south-seeking pole**. We usually refer to these poles as the north and south poles of the magnet.

The forces on pins, iron filings and other magnetic objects are always greatest when they are near the poles of a magnet. Every magnet has two poles which are equally strong.

When you hold two magnets together you find that two north poles (or two south poles) repel each other, but a south pole attracts a north pole (Figure 2).

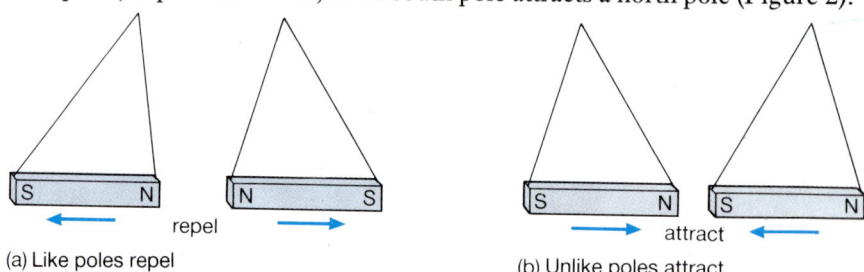

Figure 2

Magnetic fields

There is a magnetic field in the area around a magnet. In this area there is a force on a magnetic object. If the field is strong the force is big. In a weak field the force is small.

The direction of a magnetic field can be found by using a small plotting compass. The compass needle always lies along the direction of the field. Figure 3 shows how you can investigate the field near to a bar magnet, using a compass. We use magnetic field lines to represent a magnetic field. Magnetic field lines always start at a north pole and finish on a south pole. When the field lines are close together then the field is strong. The further apart the lines are the weaker the field is. (Magnetic field lines are not real, but they make a useful model that helps us to understand magnetic fields.)

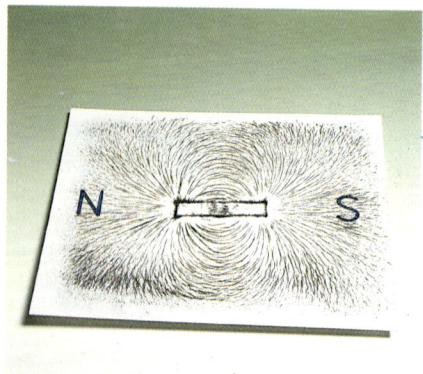

You can see the shape of a magnetic field by using iron filings as in this photograph

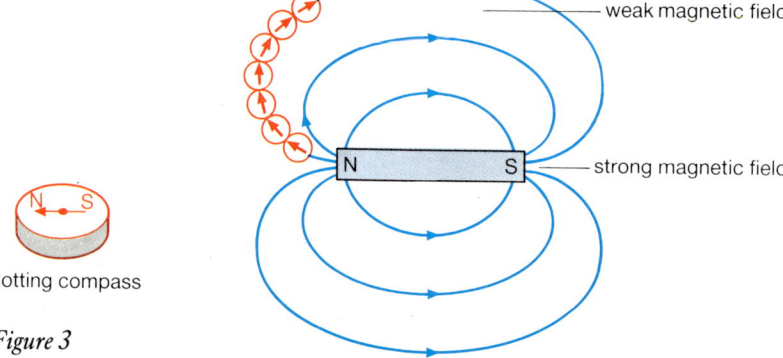

Figure 3

Combining magnetic fields

The pattern of magnetic field lines close to two (or more) magnets becomes complicated. The magnetic fields from the two magnets combine. The field lines from the two magnets *never* cross. If field lines did cross, it would mean that a compass would have to point in two directions at once.

Figure 4 shows the sort of pattern when two north poles are near each other. The field lines repel. At the point X there are no field lines. The two magnetic fields cancel out. X is called a neutral point.

Figure 5 shows the pattern produced by a north and a south pole. Notice that there is an area where field lines are equally spaced and all point in the same direction. This is called a uniform field.

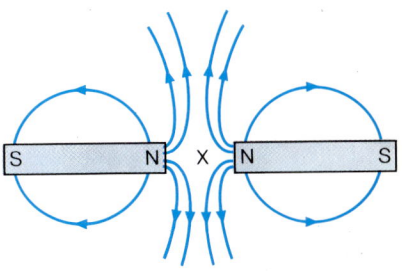

Figure 4
X is a neutral point; a compass placed here can point in any direction

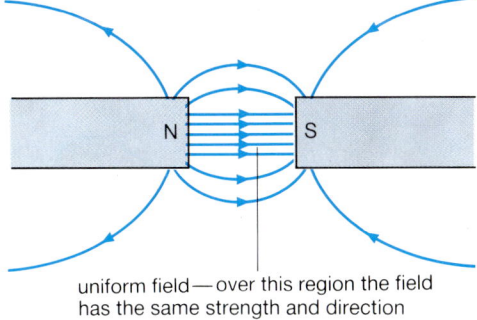

uniform field—over this region the field has the same strength and direction

Figure 5

The Earth's magnetic field

Figure 6 shows the shape of the Earth's magnetic field. Magnetic north is not in the same place as the geographic North Pole. At the moment magnetic north is in the sea north of Canada! Over a period of centuries the direction of the field alters.

The north (seeking) pole of a compass points towards magnetic north. Unlike poles attract. This means that magnetic north behaves like a south-seeking magnetic pole.

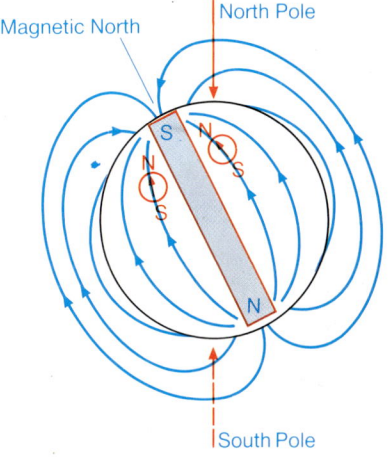

Figure 6
The Earth's magnetic field

Questions

1 (a) The diagram shows a bar magnet surrounded by four plotting compasses. Copy the diagram and mark in the direction of the compass needle for each of the cases *B, C, D*.

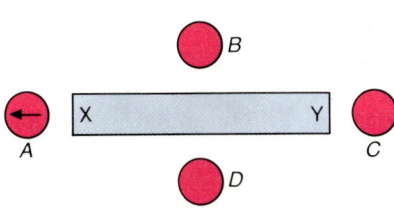

(b) Which is a north pole, X or Y?

2 Draw carefully the shape of the magnetic field surrounding these magnets. Mark in any neutral points.

3 Two bar magnets have been hidden in a box. Use the information in the diagram below to suggest how they have been placed inside the box.

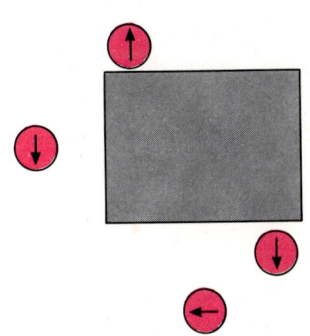

2 Currents and Magnetism

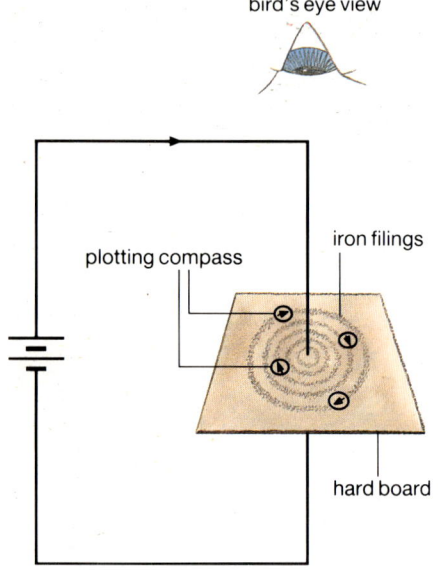

bird's eye view

plotting compass

iron filings

hard board

Figure 1
This experiment shows there is a magnetic field around a current-carrying wire

The field near a straight wire

In Figure 1, a long straight wire, carrying an electrical current, is placed vertically so that it passes through a horizontal piece of hardboard. Iron fillings have been sprinkled onto the board to show the shape of the field. Below are summarised the important points of the experiment:

- If the current is small, the field is weak. But if a large current is used (30 A) the iron fillings show a circular magnetic field pattern.
- The magnetic field gets weaker further away from the wire.
- The direction of the magnetic field can be found using a compass. If the current direction is reversed, the direction of the magnetic field is reversed.

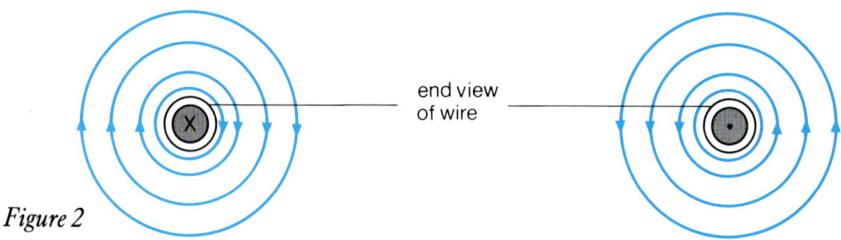

end view of wire

Figure 2

Figure 2 shows the pattern of magnetic field lines surrounding a wire. When the current flows into the paper (shown $\otimes$) the field lines point in a clockwise direction around the wire. When the current flows out of the paper (shown $\odot$) the field lines point anticlockwise. The **right-hand grip rule** will help you to remember this. Put the thumb of your right hand along a wire in the direction of the current. Now your fingers point in the direction of the magnetic field.

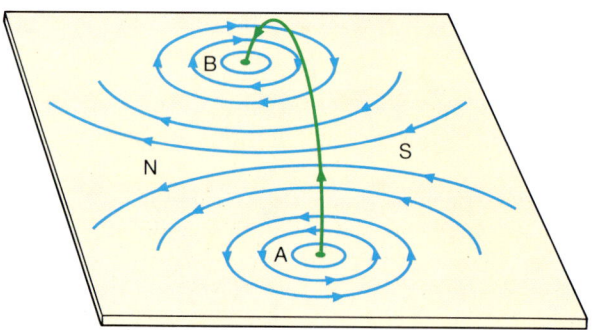

Figure 3
The magnetic field near a single loop of wire

The field near coils of wire

Figure 3 shows the magnetic field around a single loop of wire, which carries a current. You can use the right-hand grip rule to work out the field near to each part of the loop. Near A the field lines point anticlockwise as you look at them, and near B the lines point clockwise. In the middle, the fields from each part of the loop combine to produce a magnetic field running from right to left. This loop of wire is like a very short bar magnet. Magnetic field lines come out of the left-hand side (north pole) and go back into the right-hand side (south pole). Figure 4 shows the sort of magnetic field that is produced by a current flowing through a long coil or **solenoid**. The magnet field from each loop of wire adds on to the next. The result is a magnetic field that is like a long bar magnet's field.

The iron filings show the shape of the magnetic field around a solenoid

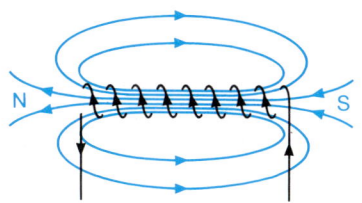

Figure 4
The magnetic field near a long solenoid

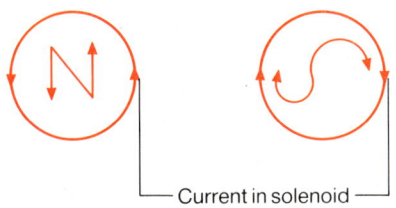

Current in solenoid

Figure 5
This is a good way to work out the polarity of the end of a solenoid

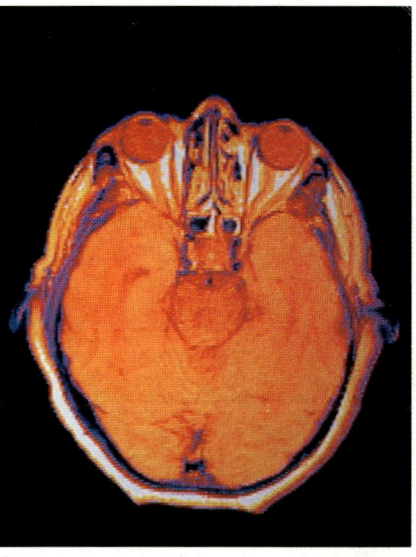

Strong magnetic fields are used by doctors to examine soft body tissue using a technique called nuclear magnetic resonance (NMR). This photograph shows a section through a brain. NMR is a very safe technique because unlike X-rays it does not damage the body

Producing large magnetic fields

The strength of the magnetic field produced by a solenoid can be increased by:
- using a larger current.
- using more turns of wire.
- putting some iron into the middle of the solenoid.

But there is a limit to how strong you can make a magnetic field. If the current is made too large the solenoid will get very hot and start to melt. For this reason large solenoids must be cooled by water. There is a limit to the number of turns of wire you can put into a space. Eventually iron becomes magnetically 'saturated' and its magnetisation gets no stronger.

Nowadays, the world's strongest magnetic fields are produced by **superconducting magnets**. At very low temperatures (about 4 K) some materials, such as nobium and lead, become superconductors. A superconductor has no electrical resistance. This means that a current can flow without causing any heating effect. So a large magnetic field can be produced by making enormous currents (5000 A) flow through a solenoid which is kept cold in liquid helium.

Questions

1 The diagram below shows two plotting compasses, one above and one below a wire. Draw diagrams to show the position of the needles when: (i) there is no current, (ii) the current is very large (30 A), (iii) the current is small (1 A).

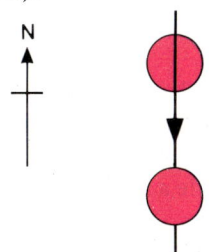

2 Diagram (i) below shows an electron moving around the nucleus of an atom.
(a) Explain why this atom is magnetic.
(b) Sketch the shape of the magnetic field near the atom.

(c) What happens to the magnetic field if the electron moves the opposite way.
(d) What can you say about the atom in diagram (ii)? Is it more or less magnetic than the other atom?

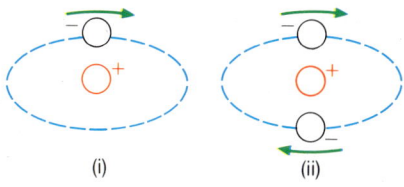

(i)

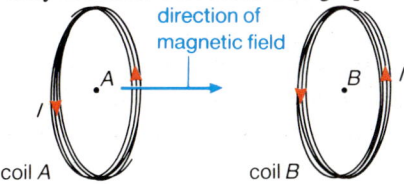

(ii)

3 The diagram shows two coils which carry the same current *I*. The graph

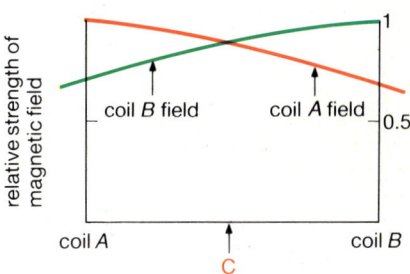

direction of magnetic field

coil A

coil B

shows how the strengths of the magnetic fields, from each coil, change along the line *AB*.

(a) Use the information in the graph to plot another graph to show the resultant strength of both fields added together.
(b) What is the strength of the magnetic field at *C* when these separate changes are made? (i) The current in coil *A* is reversed, (ii) The current in coil *B* is doubled.

3 Magnetising

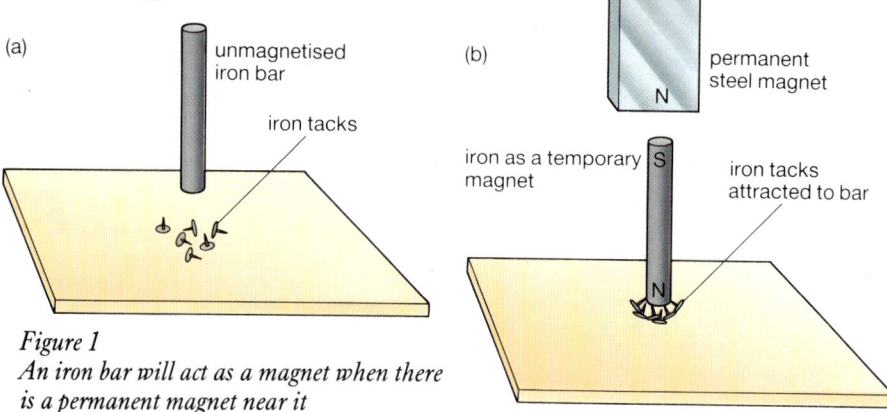

Figure 1
An iron bar will act as a magnet when there is a permanent magnet near it

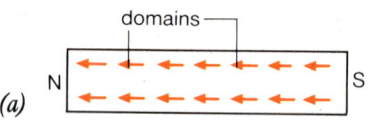

(a)

(b)

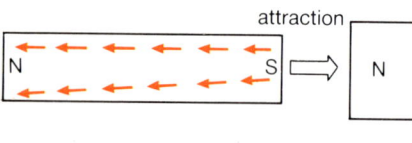

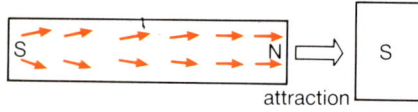

(c)

Figure 2

Magnetic domains

Steel can be a permanent magnet; once a piece of steel has been magnetised it remains a magnet. Iron can be a temporary magnet; iron is only magnetised when in a magnetic field. This field can be provided by a magnet or by a solenoid (Figure 1).

We think that the insides of magnetic materials are split up into small regions, which we call **domains**. Each domain acts like a very small magnet. In any iron or steel bar there are thousands of domains. This idea helps us to understand permanent and temporary magnets.

In a steel bar, once the domains have been lined up, they stay pointing in one direction (Figure 2(a)).

In an iron bar the domains are jumbled up when there is no magnet near (Figure 2(b)). But as soon as a magnet is put near to an iron bar the domains are made to line up. So an iron bar will be attracted to either a north or south pole of a permanent magnet (Figure 2(c)).

Testing for a magnet

When two like poles of permanent magnets meet, they repel. When two unlike poles meet, they attract each other. If you want to find out whether a metal rod is magnetised you must check if it can be repelled by a known permanent magnet. Repulsion proves that the rod is a magnet, but attraction does not. An unmagnetised iron bar is attracted to *both* poles of a permanent magnet (see Figure 2(c)).

Magnetising

One way to make a magnet out of an unmagnetised steel bar is to stroke it with a permanent magnet. The movement of the magnet along the steel bar is enough to make the domains line up (Figure 3(a)).

A magnet can also be made by putting a steel bar inside a solenoid. A short but very large pulse of current through the solenoid produces a strong magnetic field. This magnetises the bar.

The apparatus shown in Figure 3(b) can also be used to make an **electromagnet**. However an iron bar is used instead of a steel bar in the solenoid. When a current is switched on the iron becomes magnetised; when the current is switched off the iron is no longer magnetised. Electromagnets are widely used in industry.

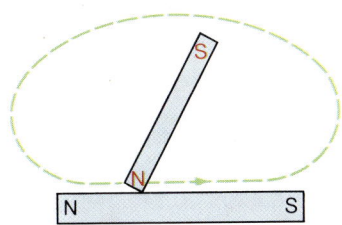

Figure 3
(a) To magnetise a steel bar stroke it 20 times like this

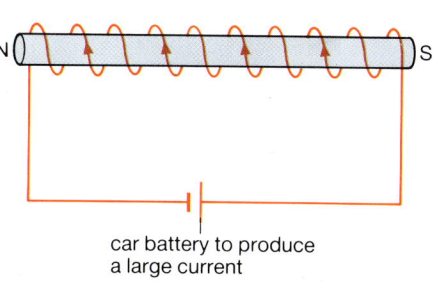

car battery to produce
a large current

(b) Magnetisation using a solenoid

A pair of electromagnets lifting railway lines in a stockyard

Demagnetising

To make a steel bar lose its magnetism you need to jumble up the domains inside the bar. There are three ways to do this:

- Hit the bar with a hammer.
- Put the bar inside a solenoid that has an alternating current supply. The alternating current produces a magnetic field that switches backwards and forwards rapidly. The domains are left jumbled up after the current has been reduced gradually to zero.
- Heat the bar to about 700°C. At low temperatures the atoms inside the steel line up to magnetise each domain. At a very high temperature the atoms vibrate at random so much that each domain is no longer magnetised.

Questions

1 (a) Tracey has magnetised a needle as shown below. She then cuts the needle into four smaller bits as shown. Copy the diagram and label the poles *A–H*.

(b) Lee says 'this experiment helps to show that there are magnetic domains in the needle'. Comment on this observation.

2 The diagram below shows an electromagnet. As the current increases the magnet can lift a larger load, because the domains in the iron core line up more and more.

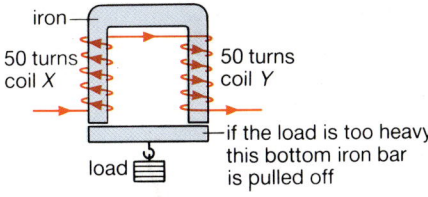

iron

50 turns
coil *X*

50 turns
coil *Y*

if the load is too heavy this bottom iron bar is pulled off

load

(a) Use the data provided to draw a graph of the load supported (*y*-axis) against current (*x*-axis).
(b) Use the graph to predict the load supported when the current is:
(i) 1.5 A, (ii) 6.0 A.
(c) Sketch on your graph how the load varies with current when each of the coils *X* and *Y* has: (i) 25 (ii) 100 turns.

(d) What happens when you reverse the windings on coil *X*?
(e) Make sketches to show how the domains are lined up when the current (in the original diagram) is: (i) 0, (ii) 0.5 A, (iii) 5 A.
(f) Use your answer to part (e) to explain why there is a maximum load that an electromagnet can lift.

Load (N)	Current (A)
0	0
5	0.5
8.5	1.0
12.0	2.0
14.0	3.0
14.8	4.0
15.0	5.0

4 The Motor Effect

Aluminium foil carrying a current is pushed out of a magnetic field

In the photograph you can see a piece of aluminium foil that has been fixed between the poles of a strong magnet. When there is a current through the foil it is pushed upwards away from the magnet. This is called the **motor effect**. It happens because of an interaction between the two magnetic fields, one from the magnet and one from the current.

In Figure 1 below you can see the way in which the two fields combine. By itself the field between the poles of the magnet would be nearly uniform. The current through the foil produces a circular magnetic field. The magnetic field from the current squashes the field between the poles of a magnet. It is the squashing of the field that catapults the foil upwards.

The force acting on the foil is proportional to:
- the strength of the magnetic field between the poles.
- the current.
- the length of the foil between the poles.

Fleming's left-hand rule

To predict the direction in which a wire moves in a magnetic field you can use the **left-hand rule** (Figure 1). Spread out the first two fingers and the thumb of your left hand so that they are at right angles to each other. Let your first finger point along the direction of the magnet's field, and your second finger point in the direction of the current. Your thumb then points in the direction that the wire moves.

This rule works when the field and the current are at right angles to each other. When the field and current are parallel to each other, there is no force on the wire and it stays where it is.

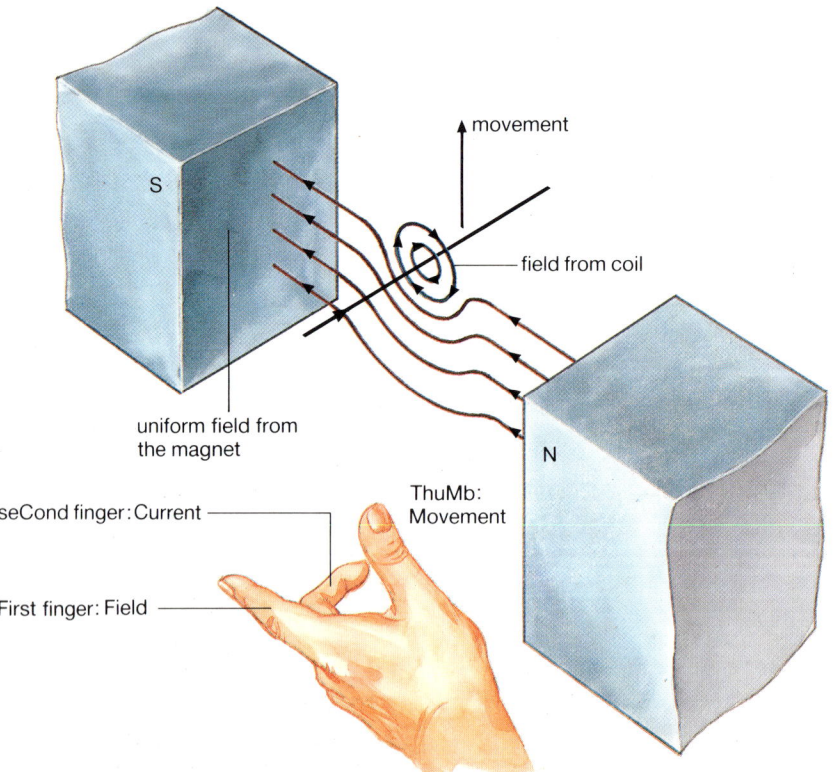

Figure 1 Left-hand rule

Ammeters

Figures 2 and 3 show the idea behind a moving coil ammeter. A loop of wire has been pivoted on an axle between the poles of a magnet. When a current is switched on, the left-hand side of the loop moves downwards and the right-hand side moves upwards. (Use the left-hand rule to check this). If nothing stops the loop it turns until side *DC* is at the top and *AB* is at the bottom. However, when a spring is attached to the loop it only turns a little way. When you pass a larger current through the loop, the force is larger. This will stretch the spring more and so the loop turns further.

The photograph in the questions section shows how a model ammeter can be made using a coil of wire, a spring and some magnets.

We say that an ammeter is sensitive if it turns a long way when a small current flows through it. The sensitivity of an ammeter will be large when:

- a large number of turns is used on the coil.
- strong magnets are used.
- weak springs are used.

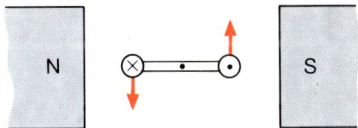

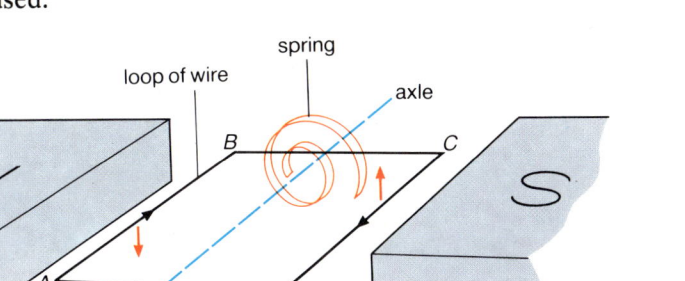

Figure 2
The principle of the moving-coil ammeter

Figure 3
(a) The end-on view of the wire loop in Figure 2. The turning effect is maximum in this position

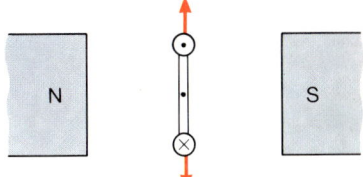

(b) In this position the forces acting on the coil produce no turning effect

Questions

1 (a) This question is about the photograph on page 218. Draw the magnetic fields near to: (i) the two magnetic poles shown, (ii) the wire carrying a current into the paper.
(b) Draw the combined field when the wire is placed between the magnetic poles.

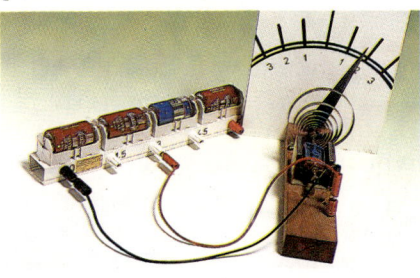

(c) Which way does the wire move when: (i) the current is reversed, (ii) the north and south poles are changed round?

2 This question is about the model ammeter shown in the photograph. The graph shows the angle that the coil turned through, against the current flowing through the coil.
(a) Use the graph to work out the current when the angle is: (i) 30°, (ii) 50°
(b) Make a sketch of the graph. Add to it further graphs to show the deflection for different currents when the ammeter is made: (i) with stronger springs, (ii) with twice the number of turns on its coil.

(c) Could this ammeter be used to measure alternating currents? Explain.
(d) (Quite hard) Why is the graph not a straight line?

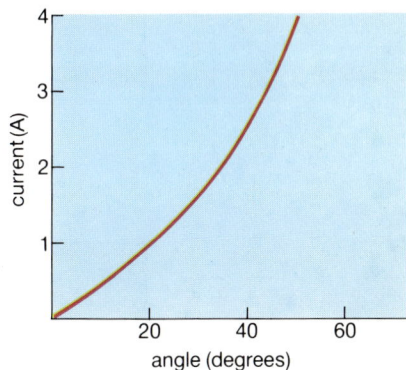

5 Electric Motors

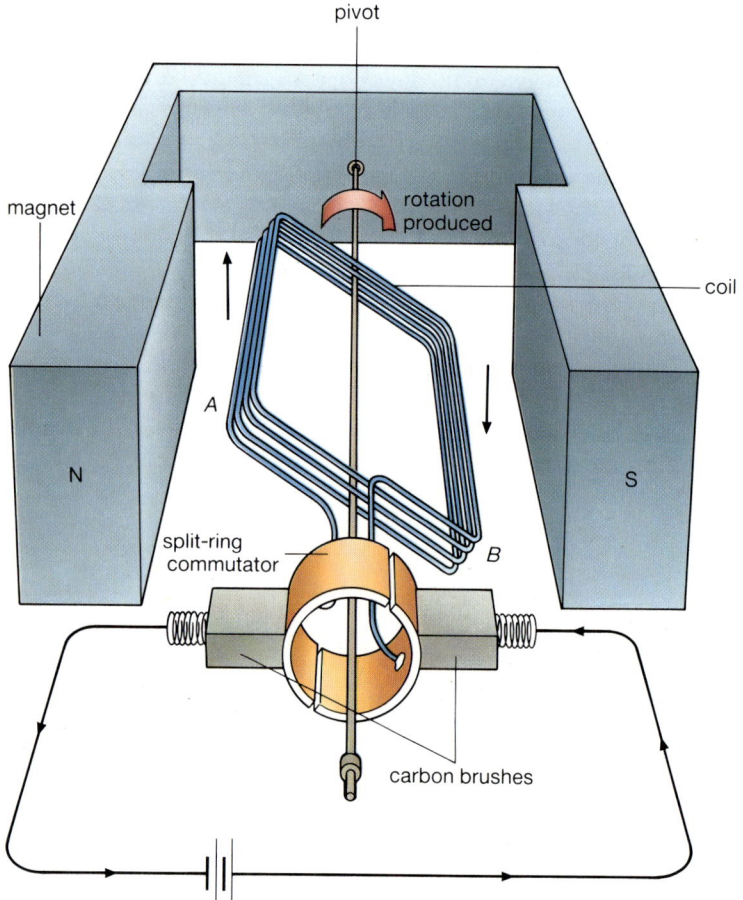

Figure 1
A design for a simple motor

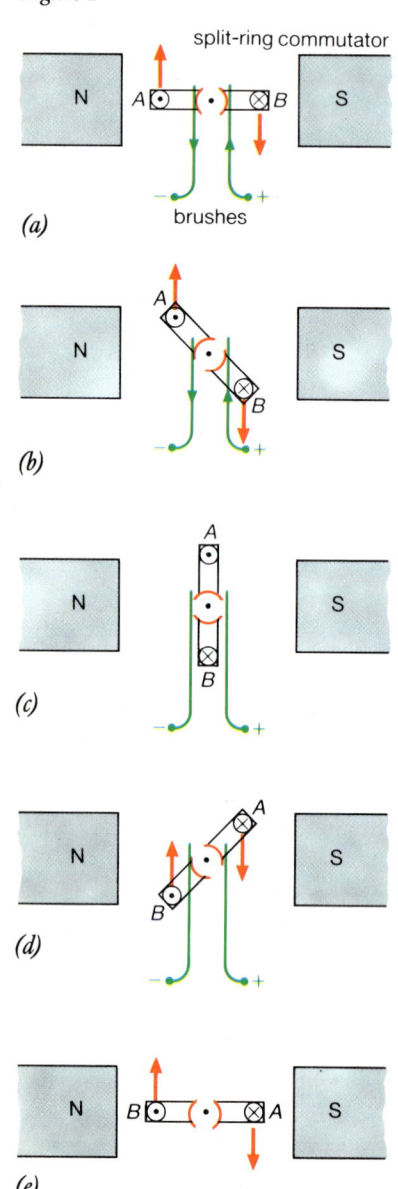

Figure 2

In the last unit you learnt that a coil carrying a current rotates, when it is in a magnetic field. However, the coil can only rotate through 90° and then it gets stuck. This is not good for making a motor. We need to make a motor rotate all the time.

Figure 1 shows the design of a simple motor that you can make for yourself. A coil carrying a current rotates between the poles of a magnet. The coil is kept rotating continuously by the use of a **split-ring commutator**. This causes the direction of the current in the coil to reverse, so that forces continue to act on the coil to keep it turning. Figure 2 explains the action of the commutator.

A current flows into the coil through the commutator so that there is an upwards push on side A, a downwards push on B (Figure 2(a)). The coil rotates in a clockwise direction (Figure 2(b)).

When the coil reaches the vertical position no current flows through the coil. The coil continues to rotate past the vertical due to its own momentum (Figure 2(c)).

Now side A is on the right-hand side. The direction of current flowing into the coil has been reversed (Figure 2(d)).

Side A is pushed down and side B is pushed upwards. The coil continues to rotate in a clockwise direction (Figure 2(e)).

Commercial motors

You use electric motors every day. Every time you put your washing into the washing machine or use a vacuum cleaner, you are using an electric motor. Your car also uses an electric motor to get it started. For these sort of uses, motors need to be made as powerful as possible.

Here are some of the ways that can be used to make a motor more powerful (Figure 3):

- A large current should be used.
- As many turns as possible should be put on the coil.
- More than one coil can be put on the rotor. Each coil then experiences a force, so the total force turning the motor is larger.
- The coils can be wound on an iron core to increase the magnetic field.
- Curved pole pieces make sure that the magnetic field is always at right angles to the coils. This gives the greatest turning effect.

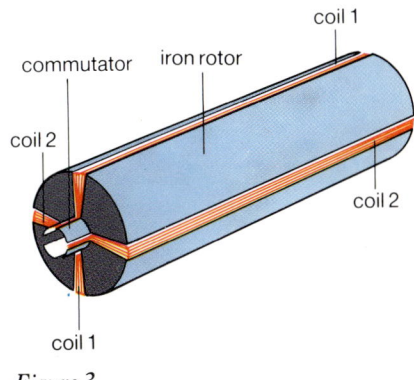

Figure 3
(a)

A cut-away view of a Black and Decker drill, showing the electric motor and gears inside it

(b)

Questions

1 Name three machines (other than those mentioned in the text) that use electric motors.

2 What two properties of carbon make it a good material to use for motor brushes?

3 (a) Look at Figure 2. Which position is the motor likely to stick in?
(b) Explain why the motor shown in Figure 3 is less likely to stick.

4 Copy Figure 3(b) and draw in some field lines to show the shape of the magnetic field.

5 The motor shown to the right is used to lift up a load.
(a) Use the information in the diagram to calculate the following: (i) the work done in lifting the load, (ii) the power output of the motor while lifting the

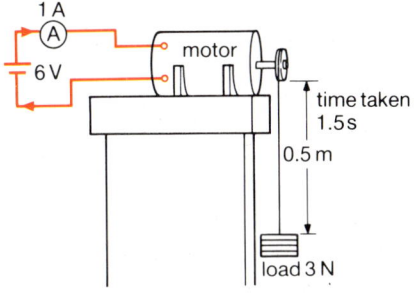

load, (iii) the electrical power input to the motor.
(b) Calculate the efficiency of the motor.
(c) Explain carefully the energy changes that occur during the lifting process.

6 The diagram below shows how an electric motor can be made using an electromagnet. You can see the forces acting on the coil when a current flows.
(a) The battery is now turned round. What effect does that have on: (i) the polarity of the magnet? (ii) the direction of current in the coil? (iii) the forces acting on the coil?
(b) Would this motor work with an a.c. supply? Explain.

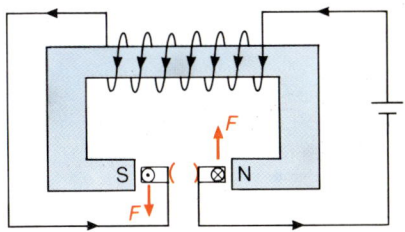

6 Electromagnetic Devices

Reed switches

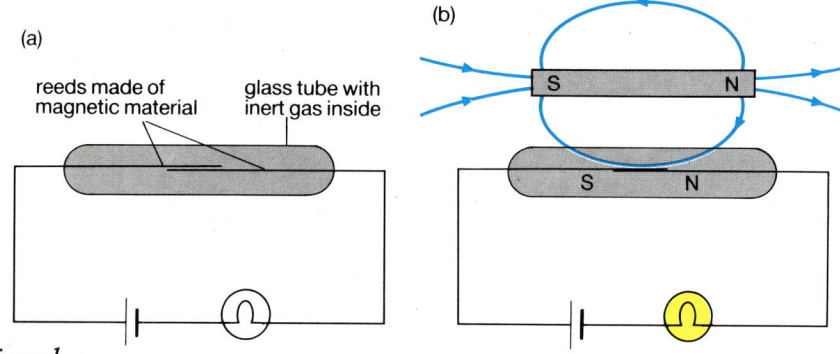

Figure 1

Two reeds made of magnetic materials are enclosed inside a glass tube. Normally these reeds are not in contact so the switch is open (Figure 1(a)). When a magnet is brought close to the reeds they become magnetised so that they attract each other. The switch is now closed (Figure 1(b)). These switches can be used to work circuits using a magnetic field. This field can be supplied using a permanent magnet or a solenoid (Figure 2).

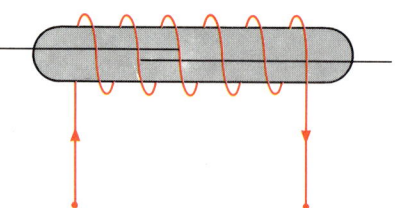

Figure 2
Reed switches can also be opened and closed using the magnetic field from a solenoid. This device is sometimes called a reed relay

Relays

A car starter motor needs a very large current of about 100 A to make it turn round. Switching large currents on and off needs a special heavy-duty switch. If you had such a large switch inside the car it would be a nuisance since it would take up a lot of space. The switch would spark and it would be unpleasant and dangerous. A way round this problem is to use a relay.

Figures 3(a) and 3(b) show how a car starter relay works.

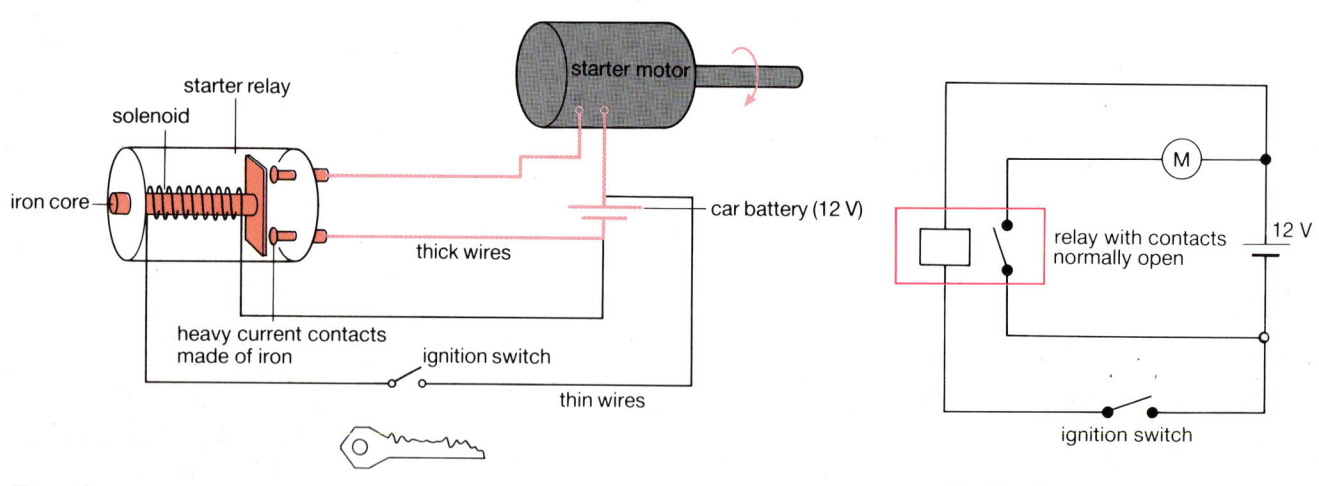

Figure 3
(a) A car starter relay

(b) The circuit diagram for the starter motor

Inside the relay a solenoid is wound round an iron core. When the car ignition is turned, a small current magnetises the solenoid and its iron core. The solenoid is attracted towards the heavy-duty electrical contacts, which are also made of iron. Now current can flow from the battery to the starter motor. The advantage of this system is that the car engine can be started by turning a key at a safe distance!

Energy changes

In Figure 2(a) a current is being produced by moving a magnet into a solenoid. When the current flows a compass needle is attracted towards end Y. So end Y behaves as a south pole and end X as a north pole. This means that as the magnet moves towards the solenoid there is a magnetic force that repels it. There is a force acting against the magnet, so you have to do some work to push it into the solenoid. The work done pushing the magnet produces the electrical energy.

In Figure 2(b) the magnet is being pulled out of the solenoid. The direction of the current is reversed and now there is an attractive force acting on the magnet. The hand pulling the magnet still does work to produce electrical energy. When the switch S is opened, there is no current. So the magnet can move in and out of the coil without any repulsion or attraction.

When a current is produced by electromagnetic induction, energy is always used to create the electrical energy. In the example described in Figure 2 the energy originally came from the muscles pushing the magnet.

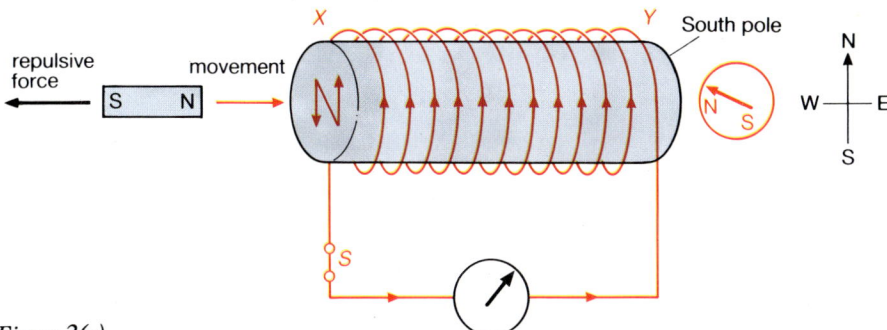

Figure 2(a)

Lenz's law of electromagnetic induction says: 'When a current is induced it always opposes the change in magnetic field that caused it.' The end X of the solenoid is a north pole. So the effect of the induced current is to push the magnet back. This opposes the motion of the magnet, and agrees with Lenz's law

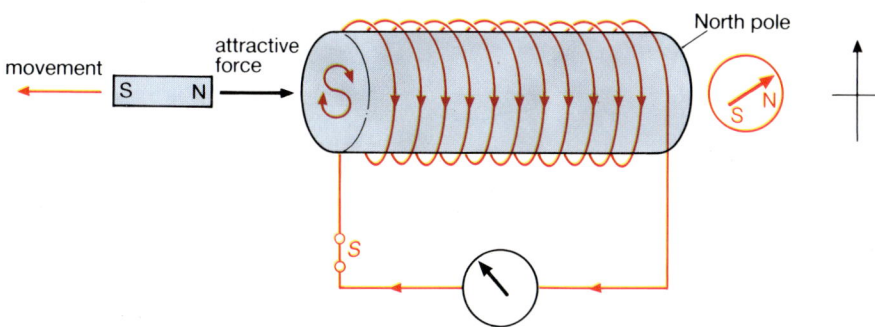

Figure 2(b)

Now the magnet moves away. X is now a south pole so the magnet is attracted to the solenoid. This opposes the motion and so agrees with Lenz's law

An electromagnetic flow meter

Figure 3 shows a way to measure the rate of flow of oil through an oil pipeline. A small turbine is placed in the pipe, so that the oil flow turns the blades round. Some magnets have been placed in the rim of the turbine, so that they move past a solenoid. These moving magnets induce a voltage in the solenoid which can be measured on an oscilloscope (Figure 3(b)). The faster the turbine rotates, the larger is the voltage induced in the solenoid. By measuring this voltage an engineer can tell at what rate the oil is flowing.

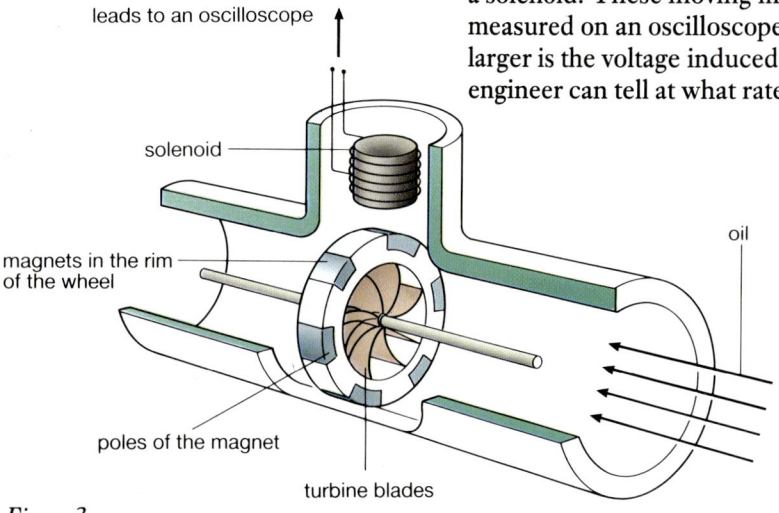

leads to an oscilloscope

solenoid

magnets in the rim of the wheel

oil

poles of the magnet

turbine blades

Figure 3
(a) An electromagnetic flow meter

(b) The oscilloscope trace. The time base is set so that the dot crosses the screen in 0.1 s

Questions

1 (a) In Figure 1 the wire moves up at a speed of 2 m/s. This makes the ammeter kick one division to the right. Say what will happen to the meter in each of the following cases: (i) the wire moves up at 4 m/s, (ii) the wire moves down at 6 m/s, (iii) the wire moves along YY' at 3 m/s, (iv) the magnets are turned round and the wire moved up at 3 m/s, (v) a larger pair of magnets is used so that the length of the wire AB in the magnetic field is doubled. The wire is now moved up at 1 m/s with the direction of the field, as shown in Figure 1.
(b) In the diagram (right), explain which arrangement will produce a larger deflection on the meter.

2 This question refers to the flow meter shown in Figure 3.
(a) What does the oscilloscope trace tell you about the way the magnets are arranged in the rim of the turbine? Are they all arranged with their north poles facing outwards or what?

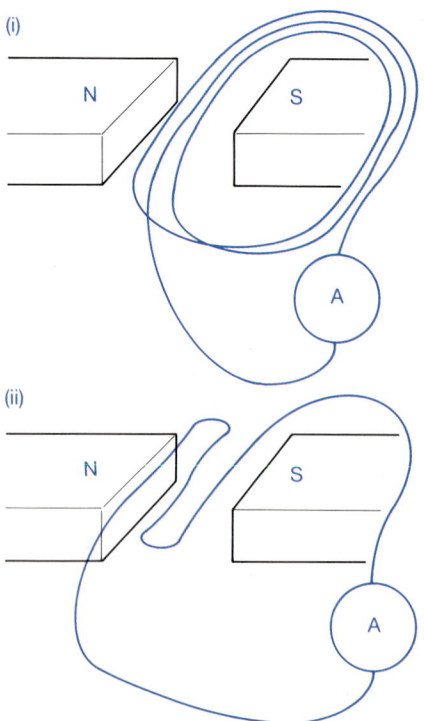

(i)

(ii)

(b) Sketch the trace on the oscilloscope for the following (separate) changes:
(i) the number of turns on the solenoid is made 3 times larger, (ii) the flow of oil is increased, so that the turbine rotates twice as quickly.
(c) Use the information in Figures 3(a) and 3(b) to calculate how many times per second the turbine is rotating.
(d) Gita and Sarah are two apprentice engineers who are using the flowmeter for the first time. This is part of a conversation they have about the meter.
Sarah: Because a current is induced there must be a force that opposes the motion. This means that the turbine will slow down the oil flow.
Gita: The resistance of the oscilloscope is very large so only a small current can flow. So the slowing down effect can be ignored.
Comment on their conversation.

8 Generators

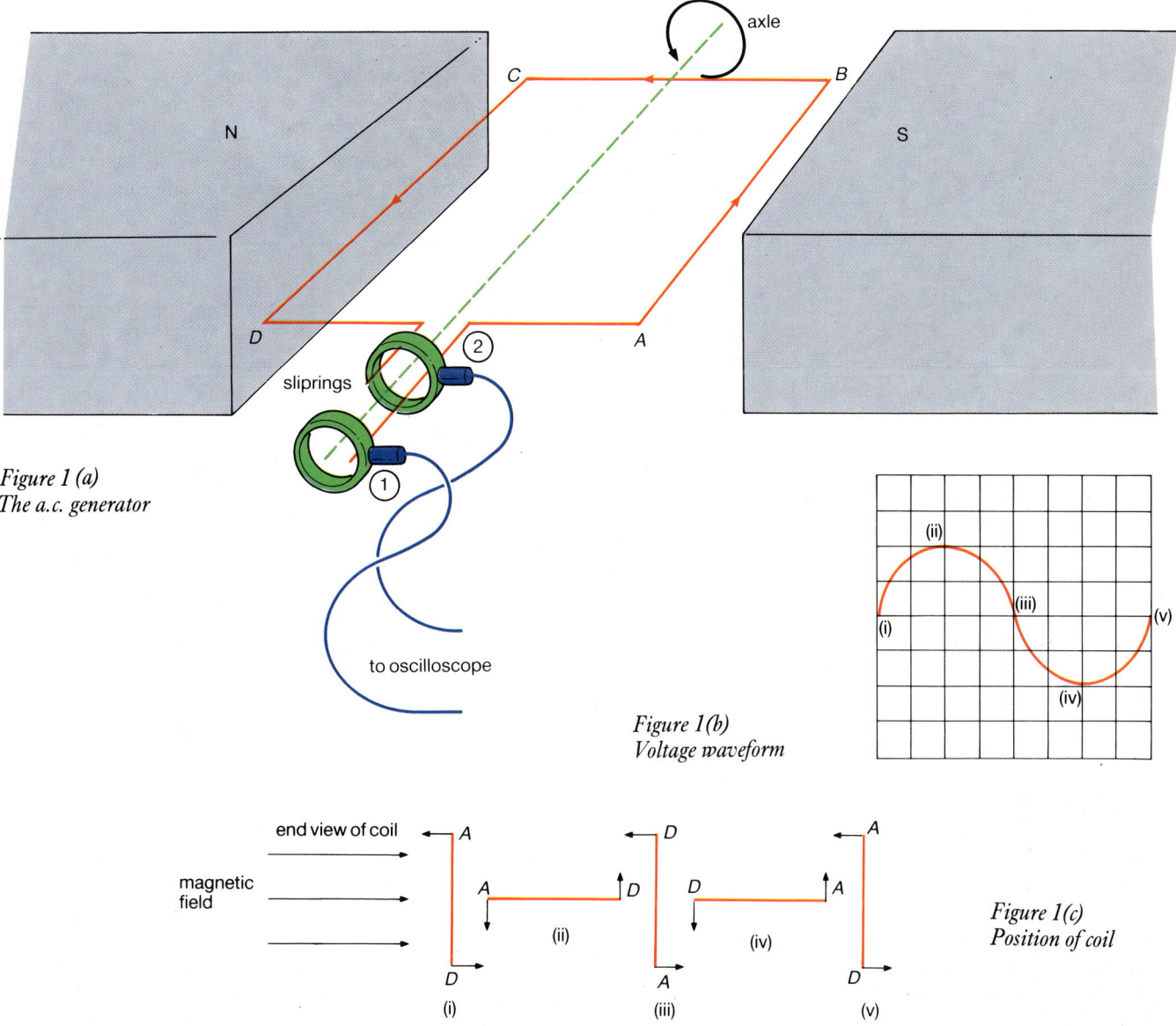

Figure 1 (a)
The a.c. generator

Figure 1(b)
Voltage waveform

Figure 1(c)
Position of coil

The a.c. generator (alternator)

Figure 1 shows the design of a very simple **alternating current** (a.c.) generator.
By turning the axle you can make a coil of wire move through a magnetic field.
This causes a voltage to be induced between the ends of the coil.

You can see how the voltage waveform, produced by this generator, looks on an
oscilloscope screen. In position (i) the coil is vertical with AB above CD. In this
position the sides CD and AB are moving parallel to the magnetic field. No
voltage is generated since the wires are not cutting across the magnetic field lines.

When the coil has been rotated through a ¼ turn to position (ii), the coil
produces its greatest voltage. Now the sides CD and AB are cutting through the
magnetic field at the greatest rate.

In position (iii), the coil is again vertical and no voltage is produced. In position
(iv) a maximum voltage is produced, but in the opposite direction. Side AB is now
moving upwards and side CD downwards.

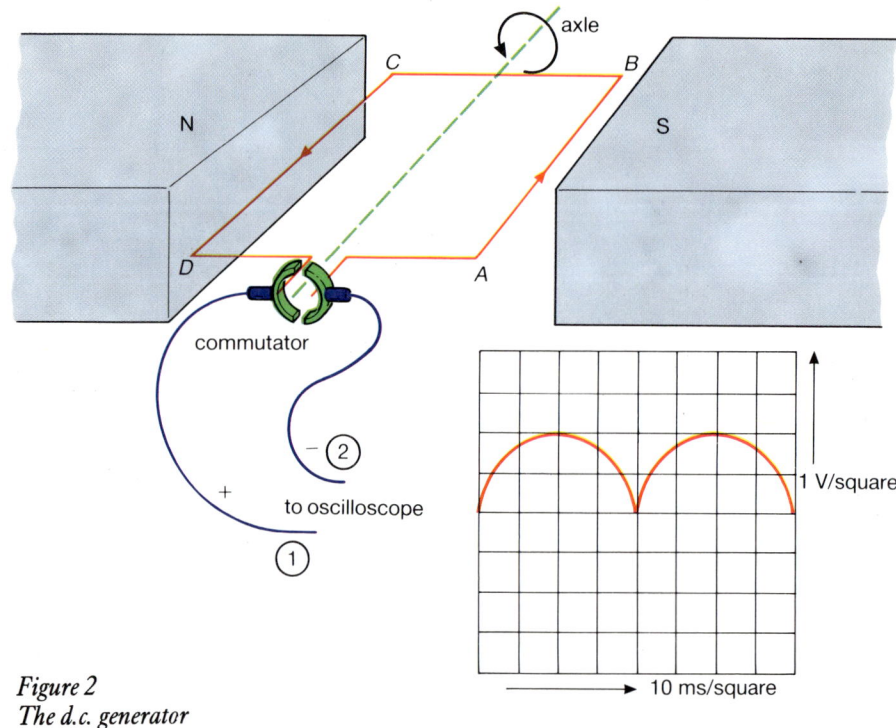

Figure 2
The d.c. generator

The d.c. generator (dynamo)

Figure 2 shows how **direct current** (d.c.) can be generated. The design of the dynamo is very similar to that of the alternator in Figure 1. The difference is that the ends of the coil are now fixed to a split-ring commutator rather than two separate slip rings. Now it does not matter which side, *AB* or *CD*, moves upwards, current will always flow out of one side of the commutator. So direct current is produced.

Producing power on a large scale

The electricity that you use in your home is produced by very large generators in power stations. These generators work in a slightly different way from the ones you have seen so far.

Instead of having a rotating coil and a stationary magnet, large generators have rotating magnets and stationary coils (see Figure 3). The advantage of this set-up is that no moving parts are needed to collect the large electrical current that is produced.

The important steps in the generation of electricity in a power station (Figure 4) are these:
(1) Coal is burnt to boil water.
(2) High pressure steam from the boiler is used to turn a turbine.
(3) The drive shaft from the turbine is connected to the generator magnets, which rotate near to the stationary coils. The output from the coils has a voltage of about 25 000 V.
(4) The turbine's drive shaft also powers the **exciter**. The exciter is a direct current generator which produces current for the rotating magnets, which are in fact electromagnets.

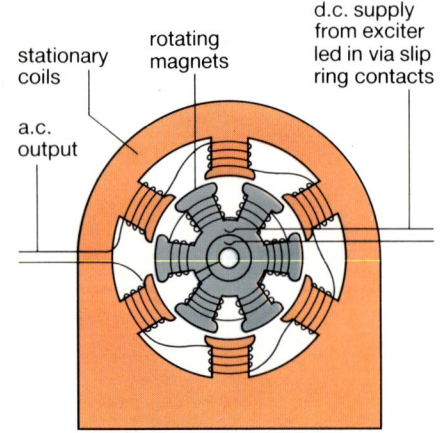

Figure 3
A section through a large generator

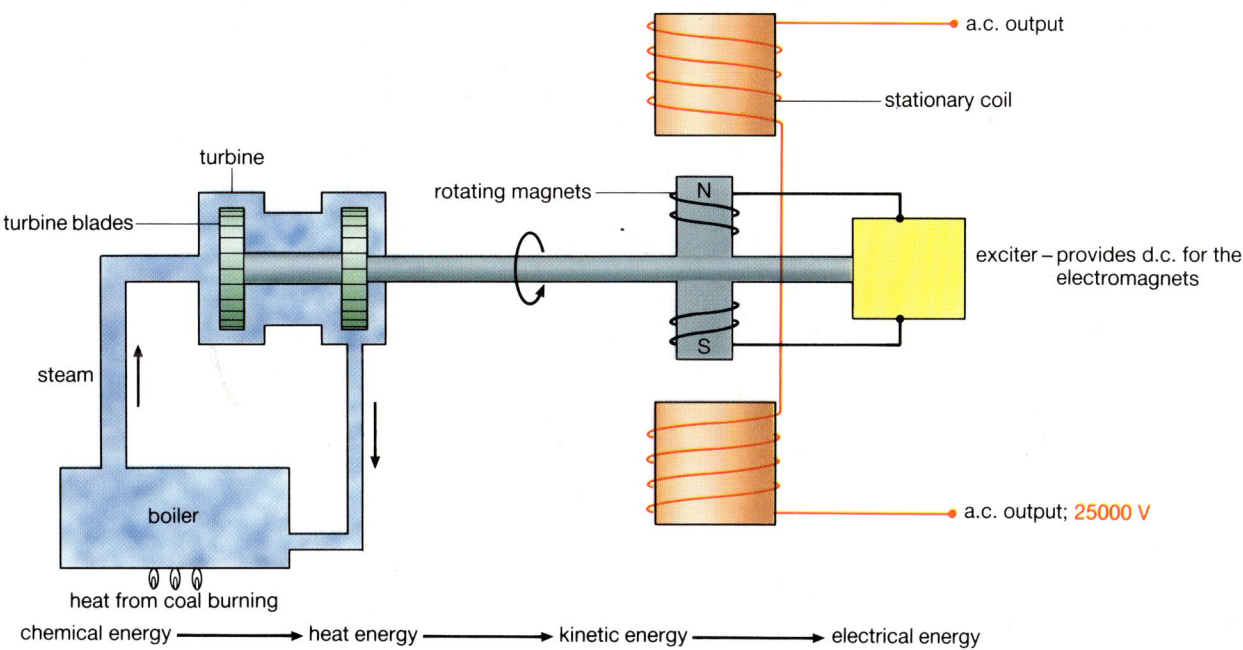

Figure 4
The layout of a power station

Questions

1 (a) Copy Figure 2(b) and mark on the graph the places where the coil is: (i) horizontal, (ii) vertical.

(b) On the same graph show how the voltage appears when the coil is rotated in the opposite direction: (i) at the same speed, (ii) at twice the speed.

2 In the diagram (right) the dynamo is connected to a flywheel by a drive belt. The dynamo is operated by winding the handle. When Surindra turns the handle as fast as she can, the dynamo produces a voltage of 6 V. The graph shows how long the dynamo takes to stop after Surindra stops winding the handle. The time taken to stop depends on how many bulbs are connected to the dynamo.

(a) Explain the energy changes that occur after Surindra stops winding the handle: (i) when the switches S_1, S_2 and S_3 are open (as shown), (ii) when the switches are closed.

(b) Why does the flywheel take longer to stop when no light bulbs are being lit by the dynamo?

(c) Make a copy of the graph. Use your graph to predict how long the flywheel turns when you make it light 4 bulbs.

(d) The experiment is repeated using the bulbs that use a smaller current. They use 0.15 A rather than 0.3 A. Using the same axes, sketch a graph to show how the running time of the dynamo depends on the number of bulbs, in this second case.

9 Introduction to Transformers

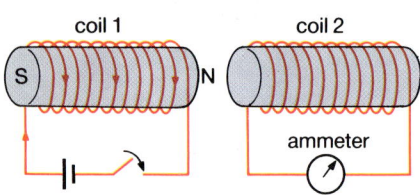

Figure 1

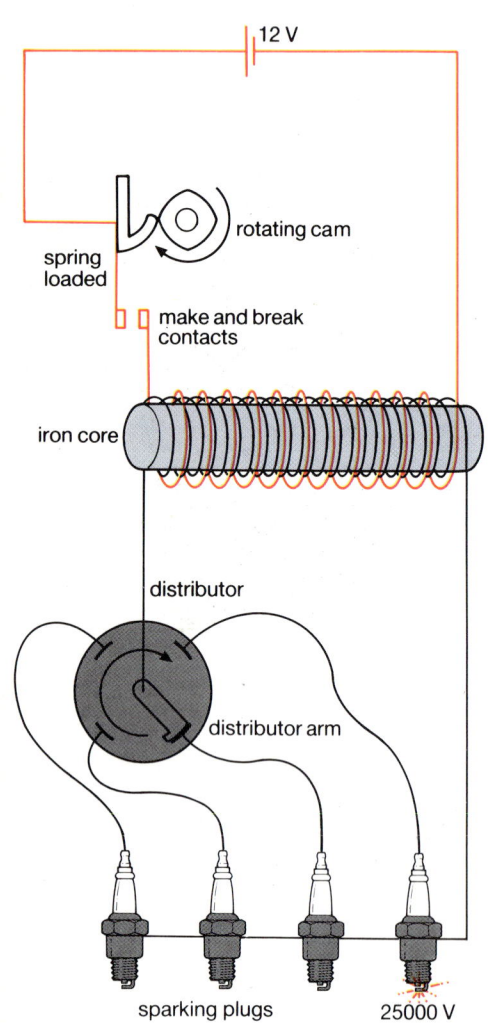

Figure 2
An induction coil in a car

This transformer delivers power to the national grid

Mutual inductance

In Figure 1 (above left) when the switch is closed in the first circuit, the ammeter in the second circuit kicks to the right. For a moment a current flows through coil 2. When the switch is opened again the ammeter kicks to the left.

Closing the switch makes the current through coil 1 grow quickly. This makes the coil's magnetic field grow quickly. For coil 2 this is like pushing the north pole of a magnet towards it, so a current is induced in coil 2. When the switch is opened the magnetic field near coil 1 falls rapidly. This is like pulling a north pole away from coil 2. Now the induced current flows the other way.

The ammeter reads zero when there is a constant current through coil 1. A current is only induced in the second coil by a changing magnetic field. This happens when the switch is opened and closed.

The induction coil

Inside the cylinders of a car's engine, the mixture of air and petrol explodes when it is sparked off by a sparking plug. To make sparks large voltages, about 25 000V, are needed. Figure 2 shows how this is done.

Two solenoids have been wound round an iron core. The red solenoid is connected to the car battery through a pair of make-and-break contacts. When the circuit is complete there is a current. The iron core becomes magnetised and a strong magnetic field is produced. As the cam rotates the circuit is broken. This causes the magnetic field to fall very rapidly. A very large voltage is now induced in the second (black) solenoid, which has a large number of turns. The rotating distributor arm feeds the high voltage to each sparking plug in turn.

Transformers

A **transformer** is made by putting two coils of wire onto a soft iron core as shown in Figure 3. The primary coil is connected to a 2 V alternating current supply. The alternating current in the primary coil makes a magnetic field that rises and then falls again. The soft iron core carries this changing magnetic field to the secondary coil. Now a changing voltage is induced in the secondary coil. In this way, energy can be transferred continuously from the primary circuit to the secondary circuit.

Transformers are useful because they allow you to change the voltage of a supply. For example, model railways have transformers that decrease the mains supply from 240 V to a safe 12 V. These are **step-down transformers**. The transformer in Figure 3 steps *up* the voltage from 2 V to 12 V.

To make a step-up transformer the secondary coil must have more turns of wire in it than the primary. In a step-down transformer the secondary has fewer turns of wire than the primary coil.

The rule for calculating voltages in a transformer is:

$$\frac{V_s}{V_p} = \frac{N_s}{N_p}$$

V_p = primary voltage; V_s = secondary voltage; N_p = number of turns on the primary coil; N_s = number of turns on the secondary.

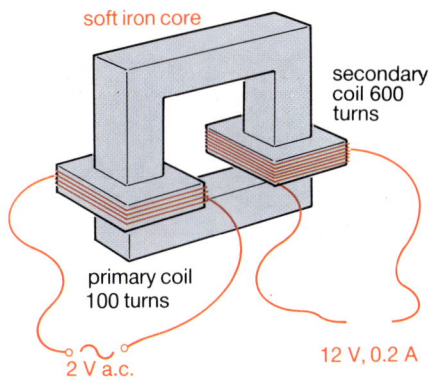

(a) *A step-up transformer*

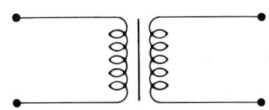

(b) *The circuit symbol for a transformer*

Figure 3

Questions

1 (a) In Figure 1 when the switch is opened the ammeter kicks to the left. Describe what happens to the ammeter during each of the following (i) the switch is closed and left closed so that a current flows through coil 1, (ii) the coils are now pushed towards each other, (iii) the coils are left close together, (iv) the coils are pulled apart.
(b) The battery is replaced by an a.c. voltage supply which has a frequency of 2 Hz. What will the ammeter show when S is closed?
2 (a) Explain why the black solenoid in Figure 2 must have such a large number of turns.
(b) Will the car still work if the black circuit is connected to the battery, and the red solenoid to the distributor?
3 Explain why a transformer does not work when you plug in the primary to a battery.
4 The question refers to Figure 3.
(a) What power is used in the secondary circuit?
(b) Explain why the smallest current that can be flowing in the primary circuit is 0.6 A.
(c) Why is the primary current likely to be a little larger than 0.6 A?
5 The table below gives some data about 4 transformers. Copy the table and fill the gaps.

A small transformer for use in a laboratory

Primary turns	Secondary turns	Primary voltage (V)	Secondary voltage (V)	Step-up or step-down
100	20		3	
400	10 000	10		
	50	240	12	
	5 000	33 000	11 000	

10 More about Transformers

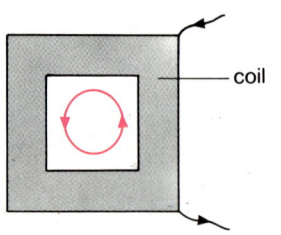

Figure 1
(a) Eddy currents flow in an unlaminated core

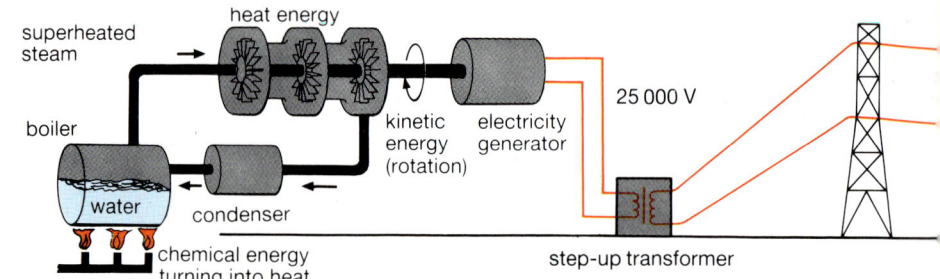

Figure 3
How the power gets to your home. The diagram shows how electricity is generated at a power station, and how it is distributed around the country through the national grid

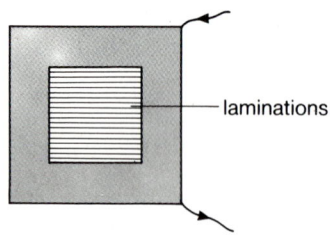

(b) *Laminations help to stop eddy currents flowing*

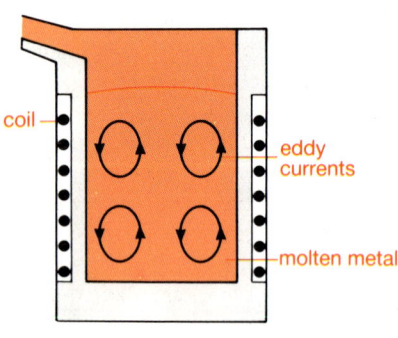

Figure 2
An induction furnace, which puts eddy currents to good use. In this furnace, scrap metal is melted down for re-use. The heating is due to eddy currents that are induced in the metal itself. These are induced by an alternating current flowing through a coil on the outside of the furnace

Power in transformers

We use transformers to transfer electrical power from the primary circuit to the secondary circuit. Many transformers do this very efficiently and there is little loss of power in the transformer itself. For a transformer that is 100% efficient we can write:

> Power supplied by primary circuit = Power used in the secondary circuit
>
> $$V_p \times I_p = V_s \times I_s$$

In practice transformers are not 100% efficient. These are the most important reasons for transformers losing energy:

- The windings on the coils have a small resistance. So when a current flows through them they heat up a little.
- The iron core conducts electricity. Therefore the changing magnetic field from the coils can induce currents in the iron. These are called **eddy currents**. To reduce energy lost in this way the core is laminated. This means the core is made out of thin slices of iron with sheets of insulating material between each slice (Figure 1).
- Energy is lost if some of the magnetic field from the primary coil does not pass through the secondary coil.
- Energy is also used to switch the direction of the domains in the iron core, 50 times each second. This can be done easily in the iron, so energy losses are small.

The national grid

You may have seen a sign at the bottom of an electricity pylon saying 'Danger high voltage'. Power is transmitted around the country at voltages as high as 400 000 V. There is a very good reason for this – it saves a lot of energy. The following calculations explain why.

Figure 4 on the next page suggests two ways of transmitting 25 MW of power from a Yorkshire power station to the Midlands:
(a) The 25 000 V supply from the power station could be used to send 1000 A down the power cables.
(b) The voltage could be stepped up to 250 000 V and 100 A could be sent along the cables.

How much power would be wasted in heating the cables in each case, given that 200 km of cable has a resistance of 10 Ω?

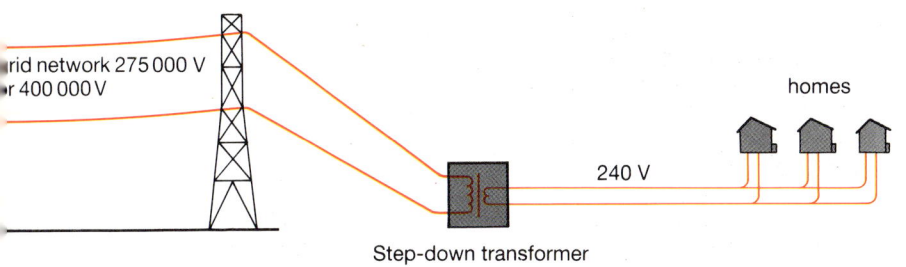

The transformers in this substation are used to change the voltage of the electricity being transmitted through the national grid

(a) Power lost = voltage drop along cable × current

$$= IR \times I$$

$$= I^2R$$

$$= (1000)^2 \times 10$$

$$= 10\,000\,000\,\text{W or }10\,\text{MW}$$

(b) Power lost $= I^2R$

$$= (100)^2 \times 10$$

$$= 100\,000\,\text{W or }0.1\,\text{MW}$$

We waste a lot less power in the second case. The power loss is proportional to the square of the current. Transmitting power at high voltages allows smaller currents to flow along our overhead power lines.

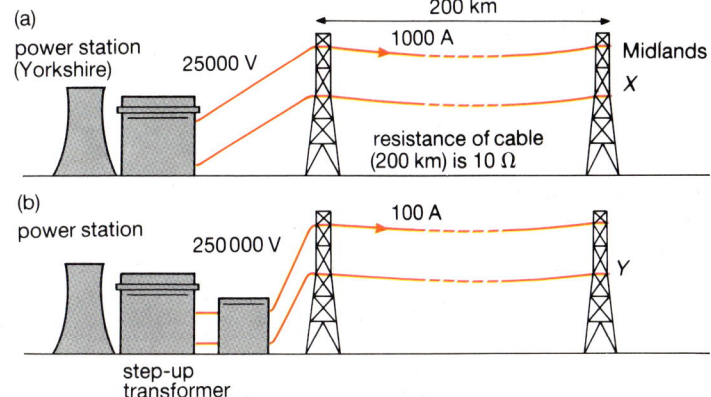

Figure 4

Questions

1 This question is about how you can use a transformer to melt a nail. You will need to use the data provided.
(a) Calculate the voltage across the nail.

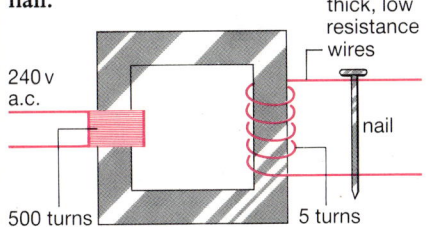

- The nail has a resistance of 0.02 Ω.
- The melting point of the nail is 1540°C.
- The nail needs 10 J to warm it through 1°C.

(b) Calculate the current flowing in:
(i) the secondary circuit,
(ii) the primary circuit.
(c) Calculate the rate at which power is used in the secondary circuit to heat the nail.
(d) Estimate roughly how long it will take the nail to melt. Mention any assumptions or approximations that you make in this calculation.
(e) Explain how a transformer can be used to produce high currents for welding.

2 Explain why the electricity supply in your home is a.c. rather than d.c.
3 Use the data in Figure 4 to calculate the voltage at each of the points X and Y.
4 The data in the table below was obtained using samples of copper, aluminium and steel wires. Each wire was 100 m long and had a diameter of 2 mm. Use this data to explain why our overhead power cables are made out of aluminium with a steel core.

Material	Resistance (Ω)	Force needed to break wire	Density in kg/m³	Cost of wire
Copper	2.2 Ω	320 N	8900	£5.60
Aluminium	3.2 Ω	160 N	2700	£1.60
Steel	127 Ω	1600 N	9000	£0.14

SECTION J: STUDY QUESTIONS

1 Two students, Rajeeb and Emma have designed a new ammeter shown below. They have plotted a calibration graph to show the extension of the spring for a particular current.

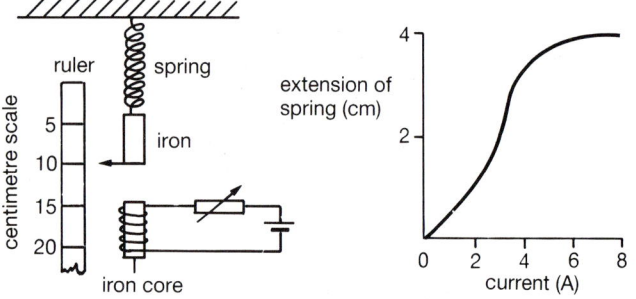

(a) Explain why their device can be used as an ammeter.

(b) Explain the shape of the graph in as much detail as you can.

(c) Here are some remarks that the students made about their ammeter. Evaluate their comments.

(i) *Rajeeb:* Our ammeter is not really any good for measuring currents above 4 A.

Emma: It would be better at measuring large currents if we used fewer turns on the solenoid.

(ii) *Rajeeb:* It does not matter which way the current flows the spring still gets pulled down.

Emma: That's useful, we can use our ammeter to measure a.c.

2 The diagram below represents a bar magnet with five plotting compasses placed near it. The purpose of the plotting compasses is to show the directions of lines of force (magnetic flux lines).

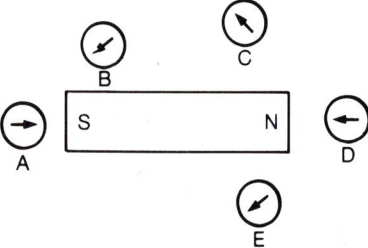

Which compass needle is pointing in the wrong direction?

LEAG

3 The diagram below represents a current carrying conductor placed between the pole-pieces of a U-shaped magnet.

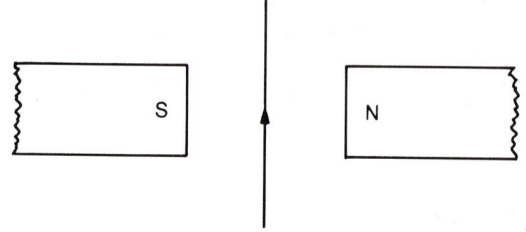

If the current flows in the direction of the arrow, in which direction will the conductor tend to move?

A toward the north-seeking pole of the magnet.

B into the page (and perpendicular to it).

C in the same direction as the current.

D out of the page (and perpendicular to it).

E toward the south-seeking pole of the magnet.

LEAG

4 The diagram represents a step-down transformer.

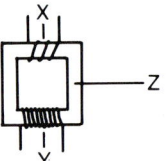

Which of the following lines is correct?

	X	Y	Z
A	secondary winding	primary winding	steel core
B	primary winding	secondary winding	steel core
C	secondary winding	primary winding	copper core
D	primary winding	secondary winding	soft iron cor
E	secondary winding	primary winding	soft iron cor

NEA

5 You can use magnetic tapes to store information. These tapes can be used as computer memories or for recording music. The tape is made of plastic and is coated with very small magnetic particles. The direction of magnetisation of these particles can be changed by applying a strong magnetic field.

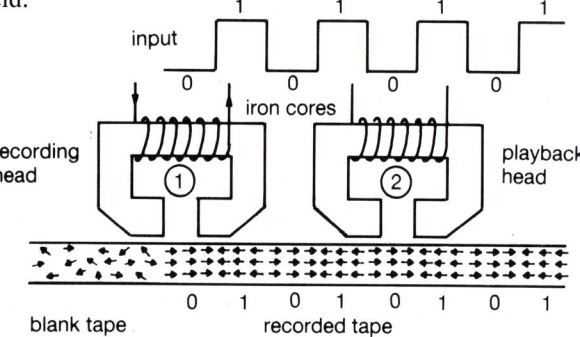

(a) In the diagram a series of pulses are sent into the recording head of a tape recorder. Explain how these manage to produce the pattern of magnetism shown in the tape.

(b) Explain how the same tape could be used as a computer memory or for recording music.

(c) (i) When the tape is passed by the playback head, a voltage is induced in the coil. Explain why.

(ii) Draw a sketch to show the form of this induced voltage. Assume the tape moves at the same speed as it did when it was recording.

(d) To erase the tape an erase head is used. The erase head is supplied with a very high frequency voltage (about 50 kHz). Explain why this high frequency is chosen.

MEG

6 The diagram below shows a coil connected to a sensitive meter. The meter is a 'centre zero' type. (This means that when no current is flowing the needle points to the centre of the scale.)

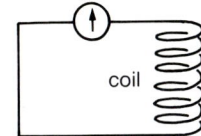

A girl performs two experiments.
First experiment
She pushes a magnet into the coil. A moment later, she removes the magnet. The meter needle deflects (moves away from zero) only when the magnet is moving.

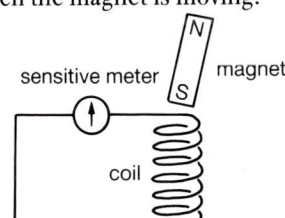

Second experiment
She brings a second coil close. She switches a current, supplied by a battery, on and off in this second coil. The needle deflects only for a very short time each time she switches on or off.

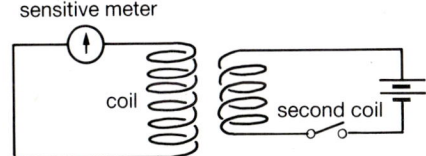

(a) In the first experiment,
(i) what is happening in the coil while the magnet is moving?
(ii) how does the deflection of the needle as the magnet is pushed towards the coil compare with the deflection of the needle as the magnet is pulled away?
(iii) without changing any of the apparatus, how could she make the needle deflection as large as possible?
(b) In both experiments, what is happening in the space around the coil at the times when the needle is deflected?

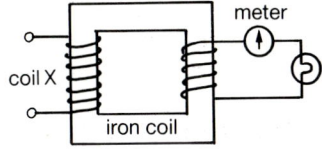

(c) Two coils are wound around an iron core as shown. What, if anything, will you see on the meter if
(i) a constant direct current is passed through coil X?
(ii) the constant direct current in coil X is switched on and then off again?
(iii) an alternating current of frequency 50 Hz is passed through coil X?
(iv) the device in the diagram is a transformer. Name a household device which includes a transformer.

MEG

7 Diagrams 1 and 2 show a current in a wire. Copy *both* diagrams and show on them the magnetic field round the wire.

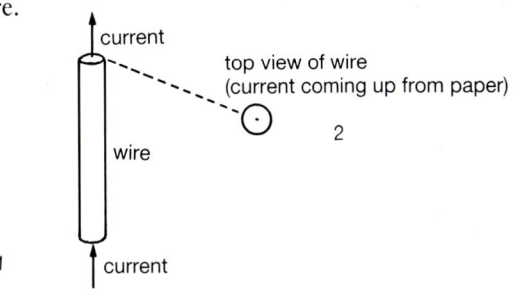

(b) Diagram 3 shows the rotor of a simple d.c. motor.

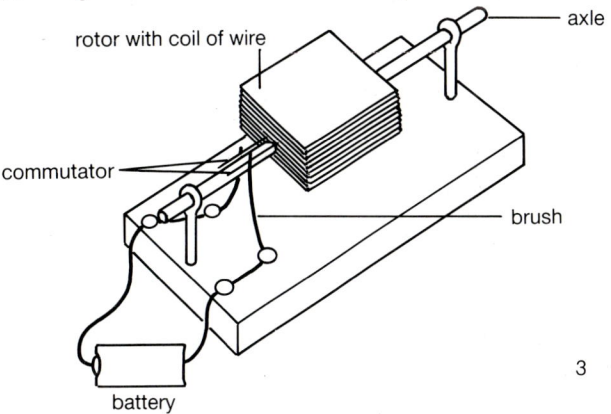

(i) What else is needed to make the rotor turn?
(ii) Explain why the commutator arrangement is needed.

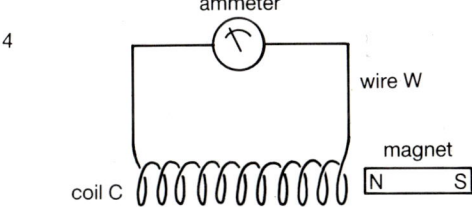

(c) What happens in wire W when the magnet shown in diagram 4 is moved into coil C?
(d) British Telecom is replacing telephone copper wires by optical fibre cables. These are made of glass. Light is transmitted instead of electricity. The reason is that telephone reception is more satisfactory.
With copper wires there is sometimes a 'bad line'. Crackling noises can be heard. With optical fibres this should not happen. Explain what causes crackling on a 'bad line'.

LEAG

8 Power losses in the grid system are reduced by using
A thin wires.
B high wires.
C underground cables.
D high voltages.
E d.c. instead of a.c.

NEA

9 A scientist goes to a deserted region to study the plants and animals there. In order to generate electricity for the home, the scientist decides to build a windmill. The best site for the windmill is several hundred metres from the house. The scientist sets up power lines to carry the electricity over this distance as shown below.

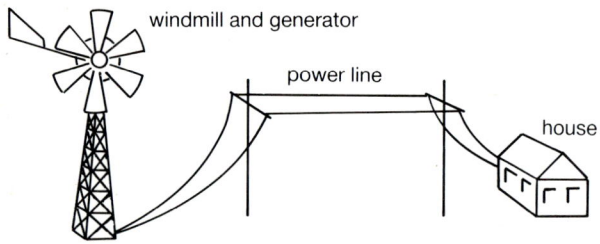
windmill and generator
power line
house

(a) Two generators are available, one providing a.c. supply and the other d.c.. Explain what is meant by (i) a.c. supply (ii) d.c. supply.
(b) The scientist connects a cathode ray oscilloscope (CRO) to the output of each generator in turn.
(i) Name one quantity which can be measured by the CRO.
(ii) Sketch the trace seen on the oscilloscope screen when the a.c. generator is used.

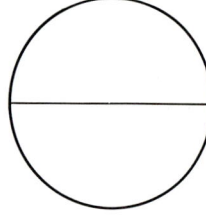
C.R.O. screen
d.c.

(iii) Sketch the trace seen on the oscilloscope screen when the d.c. generator is used.

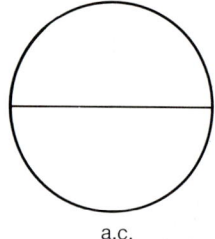
a.c.

(c) The scientist set up the arrangement shown in the diagram below in order to investigate the conditions for the most efficient transmission of power.

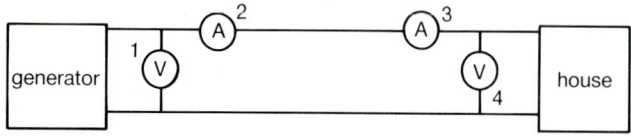
generator   house

(i) Which **two** meters would read the same, regardless of the efficiency?
(ii) How is electrical energy 'lost' from the system between the generator and the house?

(d) The scientist found that the efficiency increased with higher voltages from the generator and decided to use a step-up transformer, shown in the diagram below, at the windmill end of the power line.

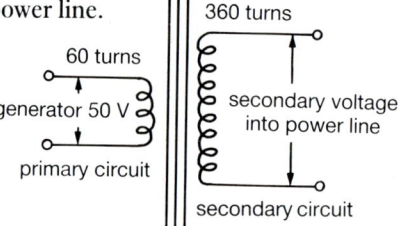
360 turns
60 turns
generator 50 V
secondary voltage into power line
primary circuit
secondary circuit
step-up transformer

(i) State which type of generator must be used with the transformer.
(ii) What material is used to make the core of the transformer?
(iii) Calculate the voltage across the power line, assuming no energy loss.
(e) Several members of the public have protested in the past against the presence of power lines and pylons and have proposed the use of underground cables. Discuss reasons why underground cables have not become popular.

LEAG

10 A small coil is connected to a galvanometer. When the magnet is moved closer to the coil as shown in diagram 1 the galvanometer pointer moves to the right of the zero position.

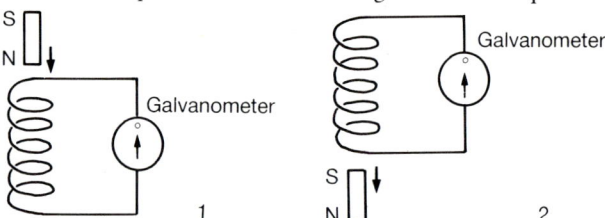
S
N
Galvanometer
Galvanometer
S
N
1   2

What does the galvanometer pointer do when the magnet is moved as in diagram 2?
A It gives a steady reading to the left.
B It gives a kick to the left.
C It does not move.
D It gives a steady reading to the right.
E It gives a kick to the right.

MEG

11 Which one of the following statements about a moving coil loudspeaker is correct?
A The coil should be firmly fastened to the magnet.
B The gap between the poles of the magnet should be as large as possible.
C The loudspeaker will not work if the polarity of the magnet is reversed.
D The coil should have many turns of wire and a low resistance.
E The strength of the magnet is unimportant.

NEA

SECTION K

Electronics

A false colour scanning electron micrograph of the surface of a compact disc. The surface of the disc, which is plastic coated, has been cracked in the shape of a rectangle, to show the music layer beneath. The music is coded in the series of tiny notches which are just visible. Light from a laser is bounced off these notches, and decoded by the player as music

1 Electronic Devices

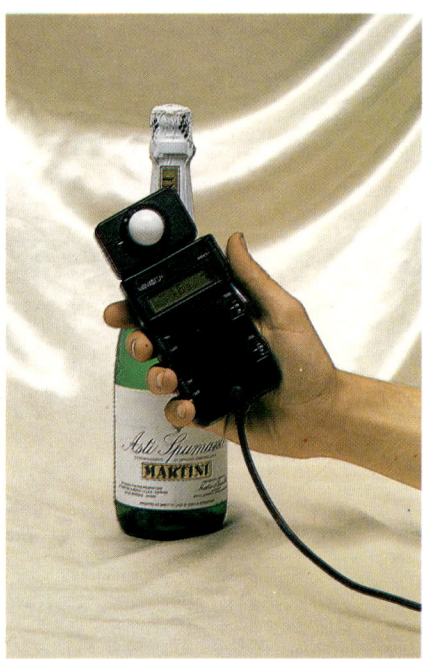

A photographer's light meter uses light dependent resistors. The reading from the light meter enables the photographer to adjust his camera to take a properly exposed photograph

Light dependent resistor (LDR)

The electrical resistance of some materials depends on the brightness of the light around us. When it is dark the resistance of a piece of cadmium sulphide can be a few megohms. In bright sunlight, the resistance drops to a few thousand ohms. Cadmium sulphide can be used to make a **light dependent resistor** (LDR).

Figure 1 shows how you can make a simple light meter using an LDR and an ammeter. When it is bright the resistance of the LDR is low and the reading on the ammeter is high. When it is dark the resistance of the LDR is high, so the current is low. Such a light meter can help a cricket umpire decide when it is too dark to carry on playing safely.

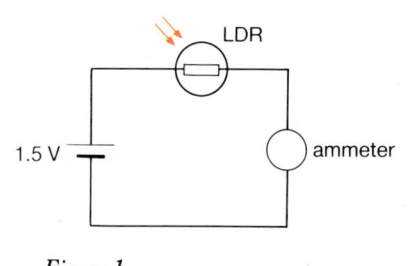

Figure 1
A simple light meter

Silicon diodes

A silicon diode allows current to pass through it in one direction only. Figure 2(a) shows the circuit symbol for the diode. Current flows easily from anode to cathode. However, current cannot flow the other way. When the anode is positive and the cathode negative, we say the diode is **forward biased**. The diode is **reverse biased** when the cathode is positive and the anode negative. Figure 2(b) shows how the current flowing through a diode varies with the applied voltage. Notice that once the voltage across the diode is about 0.7 V, the current increases very rapidly.

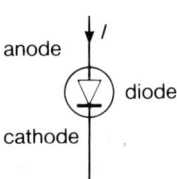

Figure 2
(a)

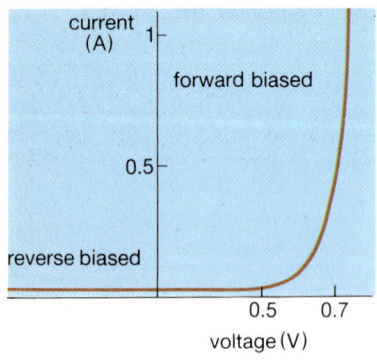

(b)

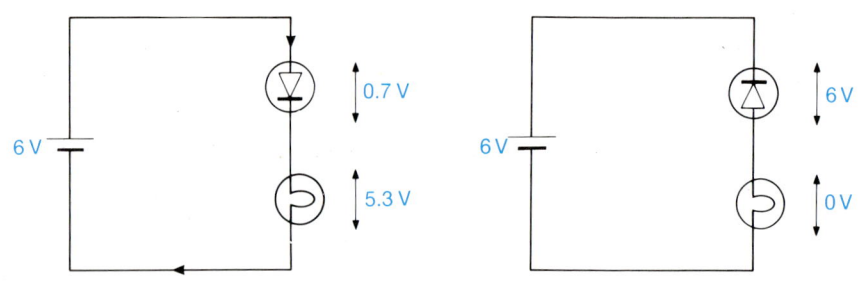

Figure 3
(a) Forward biased   *(b) Reverse biased*

Figure 3 shows a diode in action. When it is forward biased the bulb is lit. There is about 0.7 V across the diode. This means that the voltage across the bulb is 6 V − 0.7 V = 5.3 V. When the diode is reverse biased no current flows. There is no voltage across the light bulb and so 6 V across the diode.

Light emitting diode (LED)

An LED is a diode that gives out (emits) light when it is forward biased. It can emit green, red or yellow light.

When an LED is working, it usually has a voltage of about 2 V across it and a current of 10 mA flowing through it. An LED can easily be damaged if too large a current (50 mA) flows through it. So we protect an LED by putting a resistance in series with it.

Example. What value of resistor do you need for the circuit in Figure 4? The voltage across R must be 4 V (6 V − 2 V). The current flowing through R is 10 mA, or $\frac{10\text{ A}}{1000} = 0.01$ A.

$$R = \frac{V}{I}$$
$$= \frac{4\text{ V}}{0.01\text{ A}}$$
$$= 400\ \Omega.$$

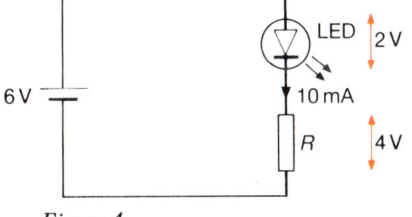

Figure 4

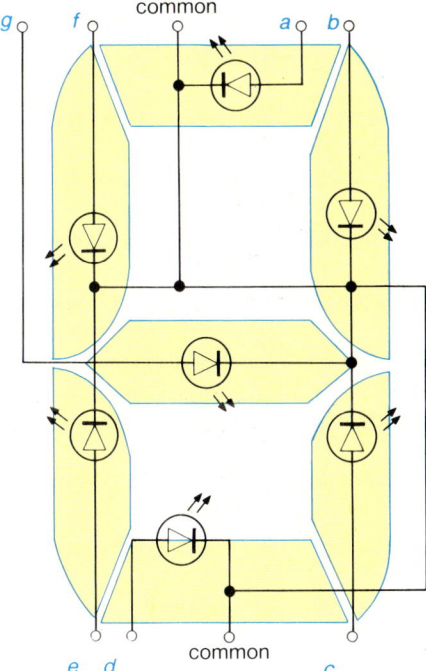

Figure 5
LEDs can be used to display numbers. This is the circuit diagram for a seven segment display. The coloured bars show how the numbers are made up

Displaying numbers

A common use of LEDs is to make a **7 segment display**. Here 7 LEDs are arranged to make a figure-of-eight pattern (Figure 5). Each LED can be switched on separately. Depending on which bars are lit, the display shows a different number. It is possible to display all the numbers from 0 to 9. You often see 7 segment displays on calculators and clocks.

Questions

1 Look carefully at the light meter in Figure 1.
(a) What is the resistance of the LDR when it is just too dark to play cricket safely?
(b) Sketch a graph to show how the resistance of the LDR depends on the brightness of the light falling on it.
2 (a) Which bulbs light in these circuits?

(b) Which bulbs light if you turn the batteries round?

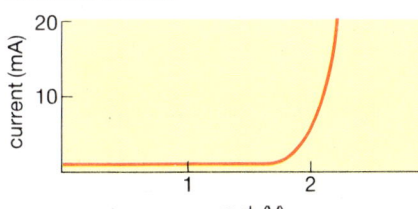

3 The graph shows how the current through an LED varies with the applied voltage. The LED works normally when there is a current of 20 mA through it.

(a) What is the voltage across the diode when the current is 20 mA?
(b) For each of the circuits below choose a resistor that will make the diode(s) work correctly. You may only choose from the resistors given.

• Values of available resistors (Ω)
33, 47, 56, 82, 100, 120, 150

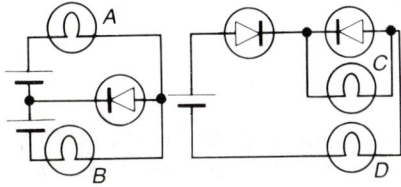

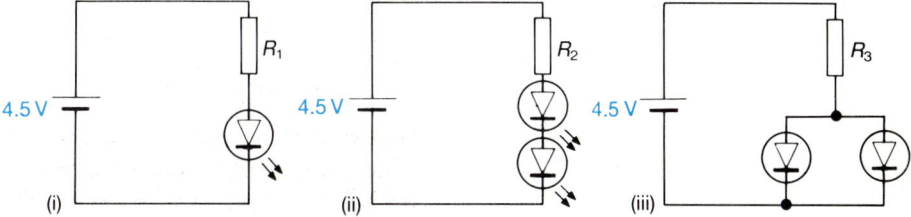

2 Rectification and Smoothing

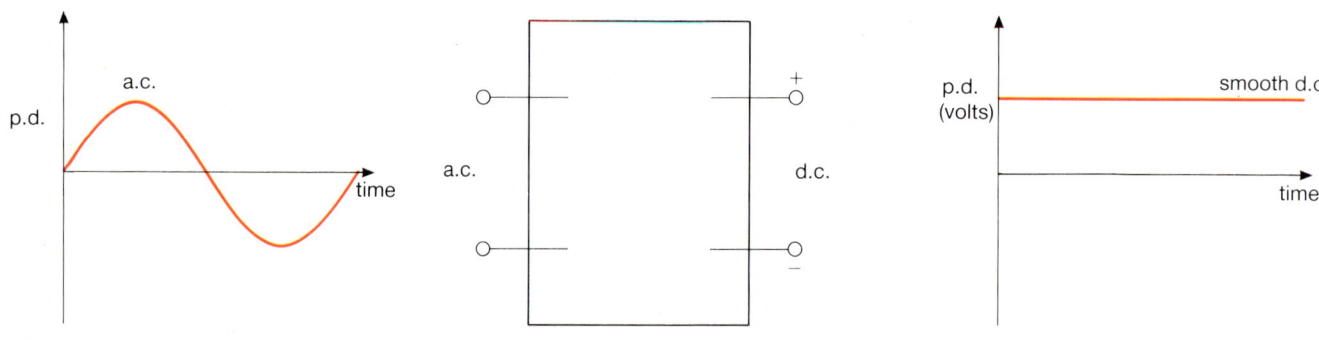

Figure 1
Rectification
(a) The voltage produced by the mains

(b) The rectifier changes a.c. to d.c.

(c) The smoothed d.c. voltage

The mains electricity supply produces alternating current (a.c.). Sometimes we want to turn this supply into a direct current (d.c.) supply. This process is called **rectification**. We also want to have a 'smooth' d.c. supply. This means that the d.c. supply must be constant (Figure 1).

Half-wave rectifiers

Figure 2 shows the simplest way of rectifying an a.c. supply, to produce d.c. for a resistor, in which a single diode is used. This is called **half-wave rectification**.

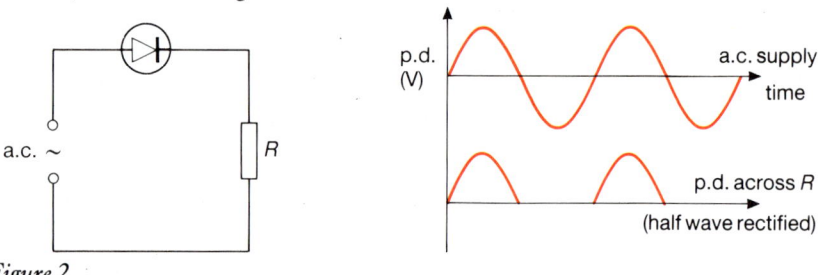

Figure 2
Half-wave rectification

The a.c. supply tries to drive current first one way, then the other, through the resistor. However, the diode only allows current to flow one way. Half of the time, current flows through the resistor. The other half of the time there is no current, because the diode is reverse biased. To make a smooth d.c. supply, we need to use a **capacitor**.

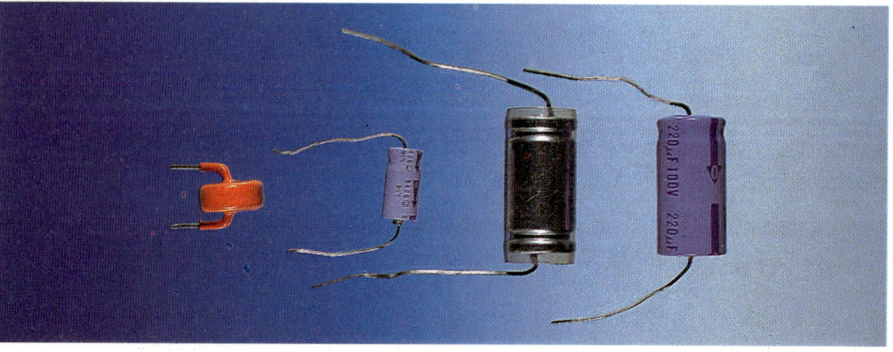

Some practical capacitors

Capacitors

A capacitor is like a rechargeable battery. It stores a small amount of charge and energy. This energy can be used to make a current flow, for a short time, through a resistance or a light bulb. The construction of capacitors is explained in Figure 3.

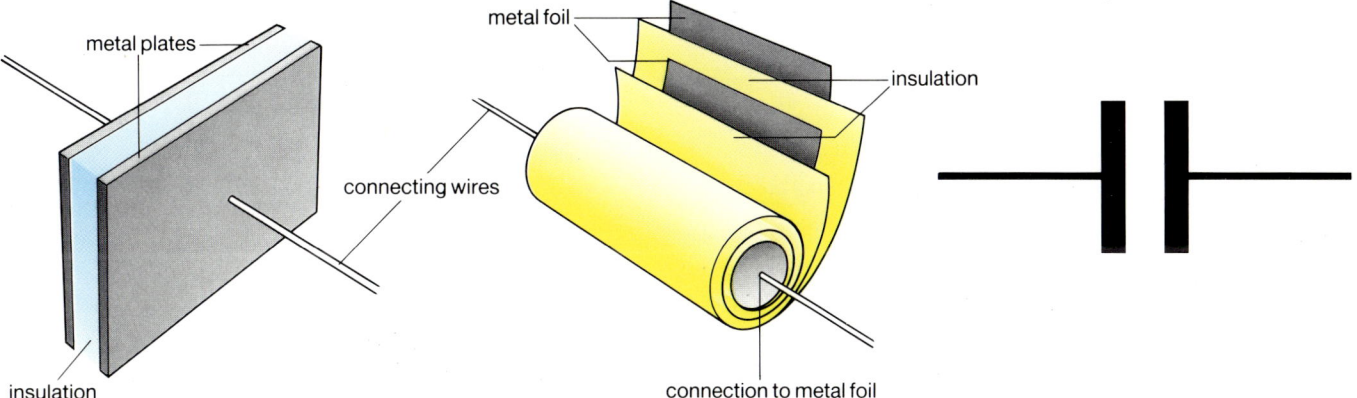

Figure 3
(a) *The simplest form of capacitor is made from two metal plates, with insulation between them*

(b) *Practical capacitors are made like a swiss roll. Metal foil, separated by layers of insulation, is rolled up*

(c) *Circuit symbol for a capacitor*

The size of a capacitor is measured in units called **farads**. A farad is a very large unit; so we usually use microfarads or μF to measure the size of a capacitor. 1 000 000 μF = 1 farad.

In Figure 4(a) the capacitor is charged by connecting it to the battery. When the switch is closed there is a quick pulse of current. This leaves one side of the capacitor with positive charge, and the other side with negative charge. Once the capacitor is charged, no more current passes through it.

In Figure 4(b) the capacitor is discharged by connecting it to a resistor and LED. Charge flows, until the positive charge has neutralised the negative charge. A large capacitor stores more charge than a smaller capacitor. This means that a large capacitor keeps the current going for a longer time. So the LED lights for longer. A large resistance also makes the current last longer. When R is big, I is small. So the capacitor loses its charge more slowly.

For a slow discharge, $R \times C$ must be large

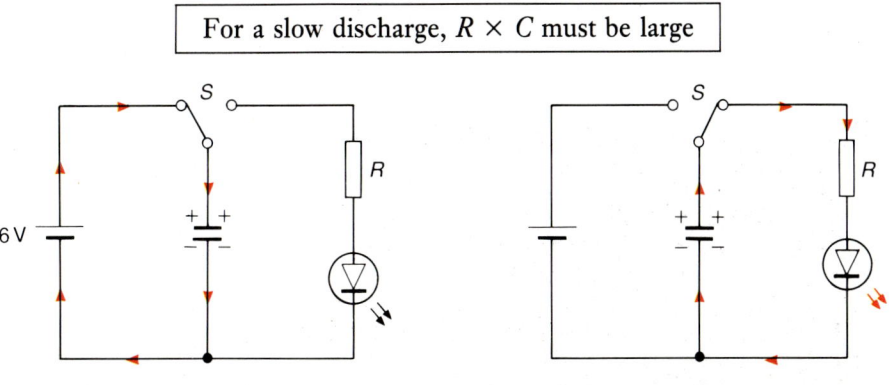

Figure 4
(a) *Charging the capacitor*

(b) *Discharging the capacitor*

241

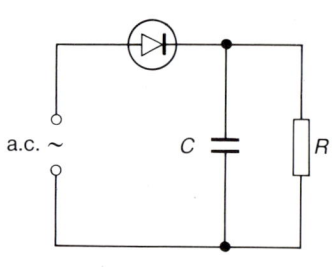

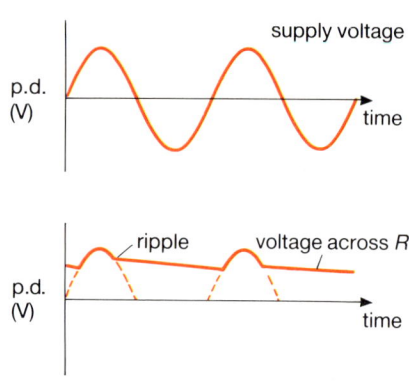

Figure 5
Smoothing the voltage from an a.c. supply

Smoothing

Figure 5 shows how to smooth the voltage from an a.c. supply, using a large capacitor. Each time the a.c. supply reaches its maximum voltage, the capacitor gets charged up to this voltage. When the a.c. voltage drops, the voltage across the resistor stays roughly constant. This is because the capacitor acts like a battery to provide the extra current. Now the d.c. voltage shows only a small 'ripple'.

Increasing the current drawn from the supply makes the ripple worse. This is because the capacitor cannot supply enough charge. You can make the ripple less by using a large capacitor.

Questions

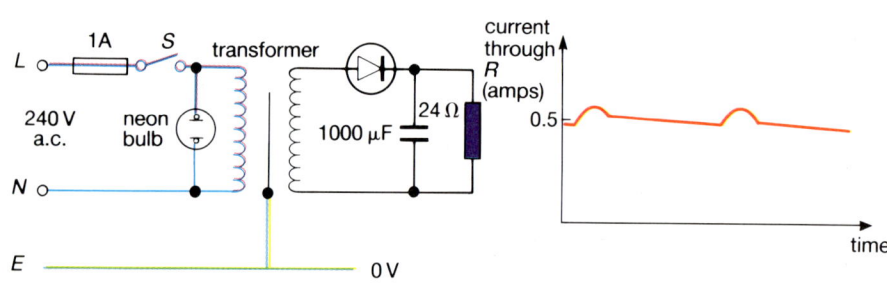

1 The circuit above shows a d.c. power supply which plugs into the mains.
(a) What does the transformer do?
(b) Why is the core of the transformer earthed?
(c) Why does the live wire have: (i) a switch, (ii) a fuse?
(d) Why is there a neon bulb?
(e) The graph shows how the current through R changes with time. Copy the graph. Add to it two further graphs to show the current when these two separate changes are made:
(i) $C = 2000\ \mu F$, (ii) $R = 12\ \Omega$

2 This diagram (right) shows a bridge rectifier. It produces a voltage which is 'full-wave rectified'.
(a) The arrows in the diagram show the current path when A is positive. Copy the diagram and show the current path when B is positive and A negative.
(b) Now explain the shape of the graph showing the voltage across R. Why is it called 'full-wave rectified'?
(c) Explain how a capacitor can be used to smooth this rectified voltage. Show on your diagram how the capacitor should be connected to the circuit.

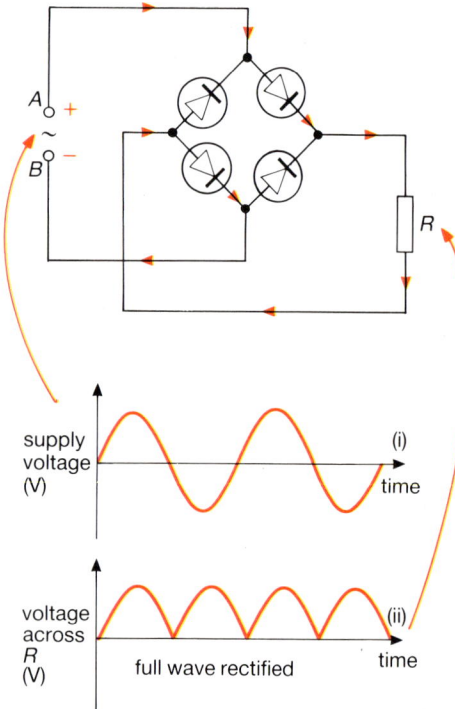

3 Transistors

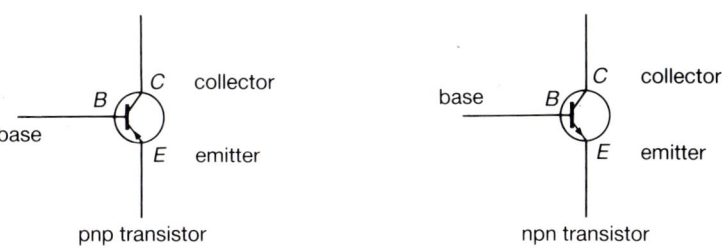

V_{BE} (V)	I_B (mA)	I_C (mA)	I_E (mA)
0	0	0	0
0.3	0	0	0
0.6	0	0	0
0.7	1	100	101

Table 1

Figure 1
These are the circuit symbols for two types of transistor. The arrows show the direction of the current through the emitter.

The outside of a transistor is nothing special to look at. But the transistor is the key to all modern electronic circuits. It looks like a small piece of plastic or metal, with three long legs sticking out. These legs are connecting wires to the transistor's three terminals. These terminals are called the **base**, the **emitter** and the **collector**. Transistors are made out of specially manufactured germanium or silicon. Figure 1 shows the circuit symbols for two types of transistor.

Figure 2 shows a circuit that will help you to understand the action of a transistor. A 6 V battery has been connected across the collector and the emitter. The power supply connected between the base and the emitter can vary the voltage between 0 and 1 V. Table 1 shows how the voltage between the base and the emitter, V_{BE}, affects the currents going into and out of the transistor. The currents are called the base, collector and emitter currents, (I_B, I_C and I_E).

Turning a transistor on and off

When there is a voltage of about 0.7 V between base and emitter, the transistor is on. This means a current goes into the base and a current flows from collector to emitter. If the voltage (V_{BE}) is less than 0.7 V, the transistor is off. No current flows into or out of it.

The transistor is so useful because it can be switched on and off by a small change in base voltage. It is an electronic switch with no moving parts, so it does not wear out. In computers, transistors are switched on and off millions of times a second! That is why computers can calculate so quickly.

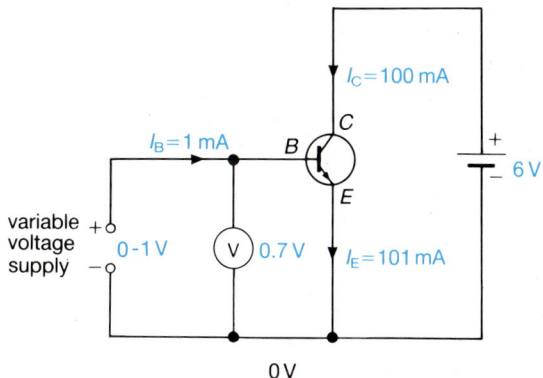

Figure 2
A circuit showing how a transistor works

Controlling the transistor

Figure 3, below, shows one way to control the base current of a transistor. A variable resistor is connected between the base and the positive side of the battery. When the resistance is very high (1 MΩ) no current flows into the base. The lamp is off. When the resistance is made lower (1 Ω) current flows into the base. Then the lamp is on.

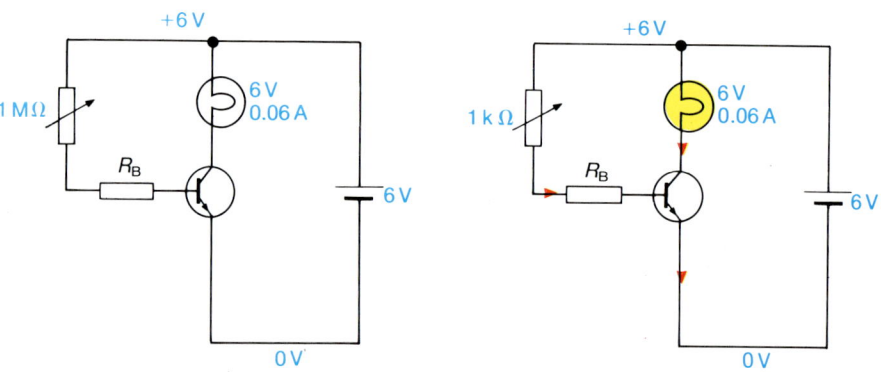

Figure 3
Controlling the base current of a transistor
(a) High resistance; current cannot flow
into the base, so the lamp is off

(b) Low resistance; a small current flows
into the base, so the lamp is on

The extra resistance, R_B, is there to protect the transistor. If the resistance of the variable resistor is made zero, a large current would flow into the base and cause damage.

A potential divider (Figure 4) can also be used to control a transistor. The two resistors are in series, so the same current flows through them. However, resistor X is five times bigger than resistor Y. So the voltage across X is five times bigger than the voltage across Y. There is a voltage of 5 V across AB and a voltage of 1 V across BC. The voltage at A is 6 V, at B, 1 V, and at C it is 0 V.

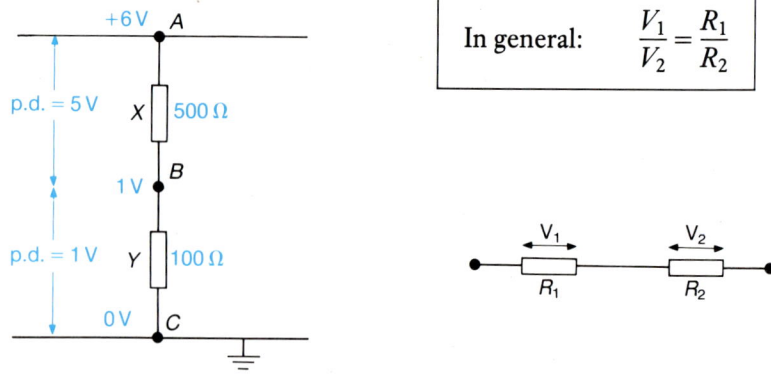

In general: $\dfrac{V_1}{V_2} = \dfrac{R_1}{R_2}$

Figure 4
A potential divider (a)

(b)

Example. Which light bulb will be on in Figure 5?

For the light bulb to be on, the transistor must be on. This happens when the voltage at B is 0.7 V or more.

In each circuit the resistance between A and C is 1200 Ω.

In (a) $\frac{1}{12}$ of the resistance is between BC. So the voltage across BC is $\frac{1}{12}$ of 6 V = 0.5 V. The voltage at B is 0.5 V and the bulb is **off**.

In (b) $\frac{1}{6}$ of the resistance is across BC. So the voltage across BC is $\frac{1}{6}$ of 6 V = 1 V. The voltage at B is 1 V and the bulb is **on**.

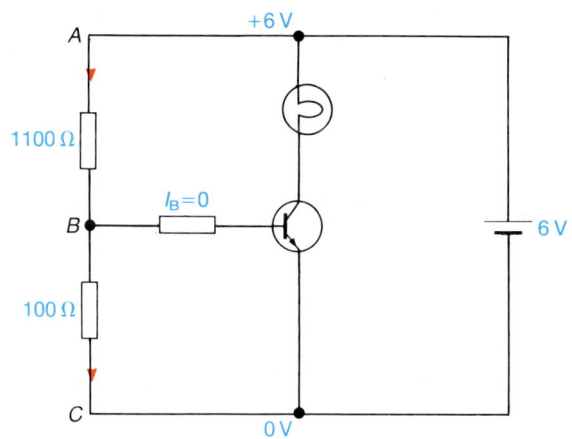

Figure 5 (a)

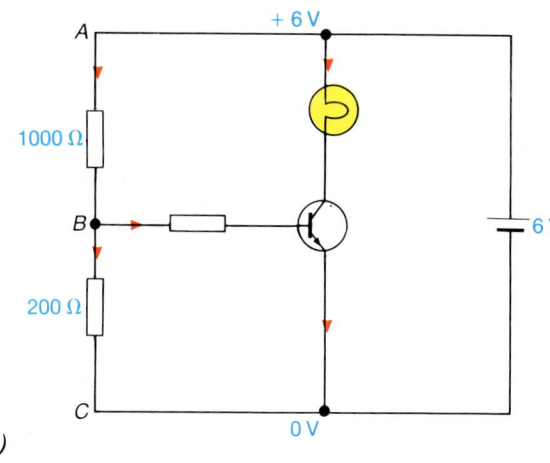

(b)

Questions

1 In this question you may assume that the voltmeters V_1 and V_2 have a very high resistance, so you can ignore any current that flows through them.

(a) What do the voltmeters read in the circuit below when: (i) $R_1 = 5$ kΩ; $R_2 = 10$ kΩ, (ii) $R_1 = 2$ kΩ; $R_2 = 2$ kΩ, (iii) $R_1 = 300$ Ω; $R_2 = 1500$ Ω?

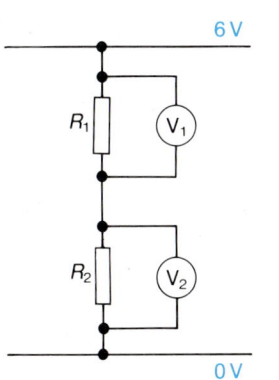

(b) (i) what does V_1 read in the circuit below? (ii) What does V_2 now read? (iii) What is the value of R_3?

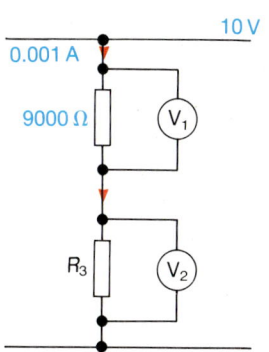

2 In the diagrams shown opposite the transistors are similar to the transistor shown in Figure 3. The light bulbs

need a current of about 100 mA (0.1 A) to make them light. Explain which of the bulbs will light in these diagrams.

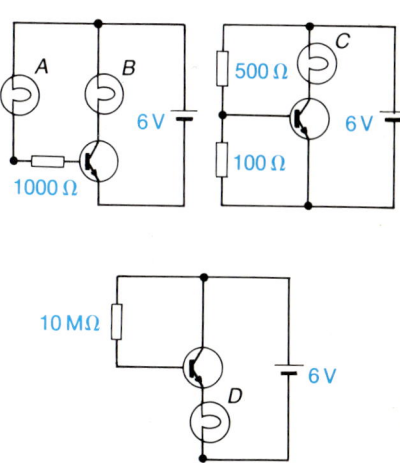

4 Using Transistors

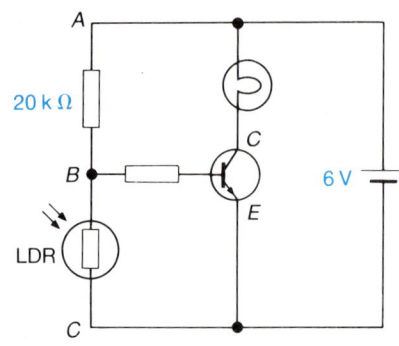

Figure 1
Making a light come on in the dark

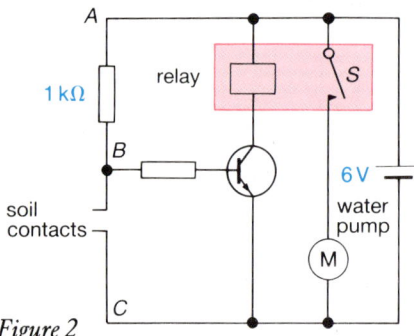

Figure 2
A tomato-watering system

Here are some circuits which use transistors to switch things on and off.

Turning on a light in the dark

In daylight an LDR has a resistance of about 500 Ω. So when it is light the voltage across BC is very small (Figure 1). This is because the resistance of an LDR is much smaller than the 20 kΩ resistor. Now the voltage at B is close to zero. The transistor is switched off and the bulb is out.

When it is dark the resistance of the LDR becomes very high, about 1 MΩ (1 000 000 Ω). The voltage across the 20 kΩ resistor is small compared with the voltage across the LDR. Now the voltage at B is nearly 6 V. The transistor is switched on and the lamp lights.

An automatic tomato waterer

The circuit in Figure 2 makes sure your tomatoes get watered when you are away on holiday. The soil contacts are placed into the soil around the tomatoes. When the soil is wet it conducts electricity well. This means that there is a short circuit between the base and the emitter of the transistor. So the transistor is switched off. When the soil dries out the resistance between B and C becomes large. This makes the voltage at B rise and the transistor is switched on. Now the current flowing through the transistor energises the relay coil. The magnetic field from the coil makes the switch S close. This switches the pump on. We need a relay to switch on the pump because it uses a current of 2 A. This is a far larger current than can flow through the transistor.

Transistors make it easy to keep these tomatoes watered while you are away on holiday

Controlling the temperature

This circuit might help you grow tomatoes in winter. The idea is to turn on a heater when your greenhouse gets too cold. The resistance of the thermistor changes as shown in the graph (Figure 3). The thermistor's resistance is highest when it is cold.

We can calculate what value the variable resistor must be set to, so that the heater is switched on when the temperature drops below freezing point (0°C) (Figure 3). From the graph you can see that the resistance of the thermistor is 700 Ω at 0°C. For the transistor to be switched on, the voltage drop across the thermistor needs to be 0.7 V. This means there must be 5.3 V (6.0 V − 0.7 V) across the variable resistor (Figure 4b).

The same current goes through the resistor and the thermistor. (We ignore any small base current going into the transistor.)

$$\text{So } I = \frac{V}{R} = \frac{\text{voltage across resistor}}{\text{resistance of } R} = \frac{\text{voltage across thermistor}}{\text{resistance of thermistor}}$$

$$\text{or} \qquad \frac{5.3\text{ V}}{R} = \frac{0.7\text{ V}}{700}$$

$$\text{So} \qquad R = 700\ \Omega \times \frac{5.3}{0.7}$$

$$= 5300\ \Omega$$

Adjusting R allows you to control the temperature of your greenhouse.

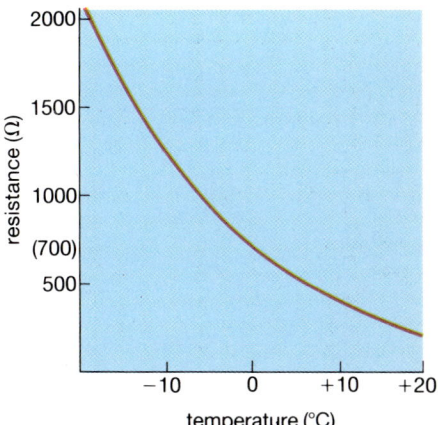

Figure 3
The thermistor resistance changes with temperature

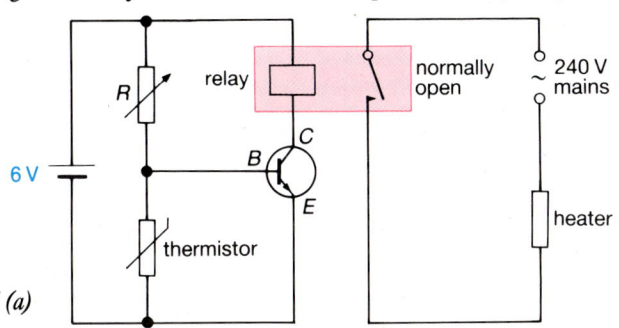

Figure 4 (a)

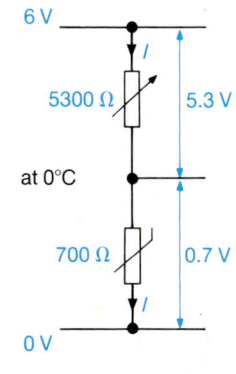

(b)

Questions

1 Jamal has designed the circuit below, so that his light bulb will come on in the dark. Has he got his circuit right? Explain your answer.

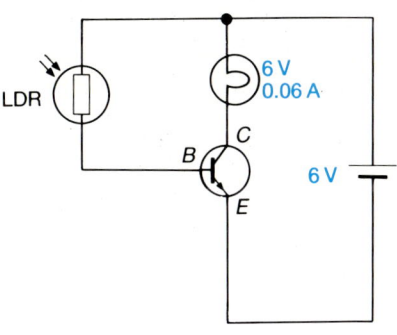

2 Look at the circuit in Figure 4(a). You have decided that you want the heater on the greenhouse to switch on when the temperature is 10°C.

(a) What is the resistance of the thermistor at this temperature?
(b) What is the voltage across the thermistor when the transistor is switched on?
(c) Therefore, what is the voltage across the variable resistor when the transistor is on?
(d) Now work out what value R must be set to, so that the heater comes on at a temperature of 10°C.

3 This question is about making a time control switch. When the switch is pressed and released, the capacitor gets charged up. For a while the voltage at B is high. The transistor is on and the bulb lights. However the capacitor loses its charge through R. After a while the voltage at B drops. This makes the bulb go out.

(a) Explain where you might use this idea in your house.
(b) What changes could you make to the circuit to make the bulb light for longer?

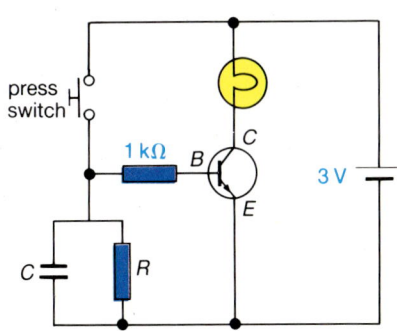

5 Transistor Amplifiers

Current gain

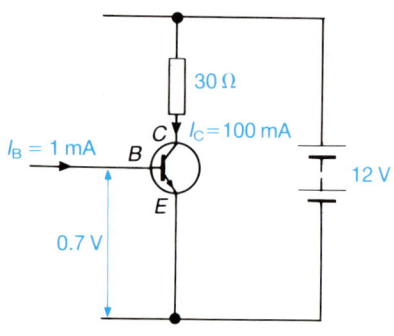

Figure 1

In Figure 1 the voltage between the base and the emitter is 0.7 V. The current going into the base is 1 mA and the current going into the collector is 100 mA. So a small base current controls a much larger collector current. The **current gain** of this transistor is 100.

$$\text{current gain} = \frac{I_C}{I_B}$$

The gain varies for different transistors; it can be anywhere between 10 and 1000.

The transistor is very sensitive to changes in its base current. Figure 2 shows that I_C is proportional to I_B. So if we double the base current to 2 mA the collector current increases to 200 mA.

This behaviour of the transistor is very important. It allows us to build **amplifiers**. When a small alternating current goes into the base, a much larger alternating current flows into the collector (Figure 2).

Instead of the resistor in Figure 1, you can put in a loudspeaker or earphone. The amplifier in your stereo system uses lots of transistors. When you play a tape, a small current goes into the base of a transistor. A much larger current flows through the loudspeaker. Now you can hear the music.

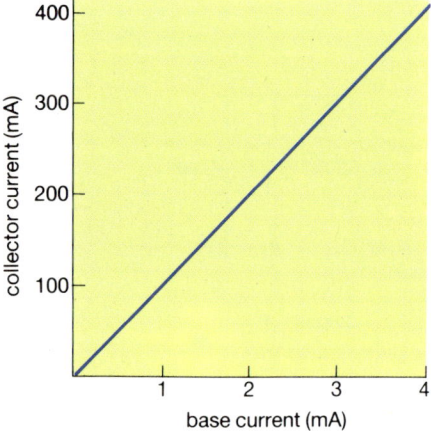

Figure 2
The collector current is proportional to the base current

(not drawn to scale)

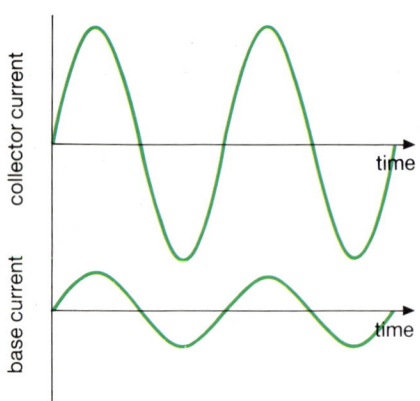

Figure 3

Some amplifiers are for personal use

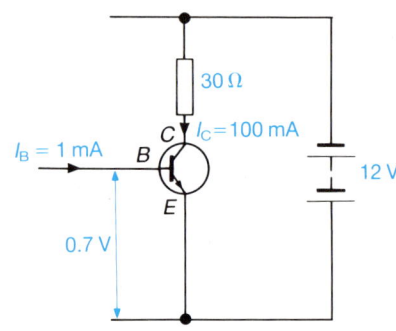

Figure 4
A circuit for a simple amplifier

Making an amplifier

Figure 4 shows how to make a simple amplifier. A microphone is used to pick up a speaker's voice. A loudspeaker relays what is said to an audience:
- The base resistor R_B is there to allow a small current to go into the base. This switches the transistor on. We call this biasing the transistor.
- When someone speaks into the microphone, a small changing voltage is produced across it. This signal passes through the capacitor, and makes changes to the base current.

Big amplifiers are needed for concerts in places as large as Wembley Stadium. This photograph shows the Live Aid concert in 1985

- The small changes to the base current are amplified by the transistor. The current is now large enough to drive the loudspeaker.
- It is important to have a capacitor in the circuit. Capacitors allow alternating currents through, but not direct currents. This means that the signal can get through to the transistor. Without the capacitor the biasing of the transistor would be upset. This is because a direct current would go from the base through the microphone. Now the base would be at the wrong voltage, and the transistor would not work properly.

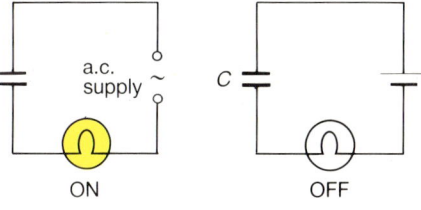

Figure 5
A capacitor passes a.c. but not d.c.

Questions

1 Below you can see a transistor circuit driving a loudspeaker. An alternating current is going into the base (graph 1). Graph 2 shows how the base current controls the collector current.

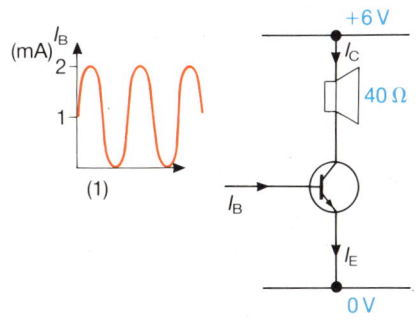

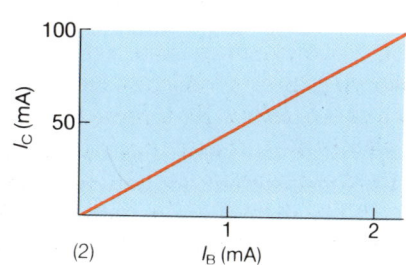

(a) What is the current gain of the transistor?
(b) What are the maximum and minimum values of: (i) I_B, (ii) I_C?
(c) Sketch a graph to show how I_C varies with time.
(d) When I_B is 1 mA, how big is the current I_E?

2 Look again at the diagram in question 1.
(a) When the collector current I_C is 50 mA, what is the voltage across the loudspeaker?
(b) Now explain why the largest collector current that can flow through this circuit is 150 mA.
(c) A very large base current is now fed into the transistor (graph 3). The music now sounds odd. Can you explain why? This is called distortion.

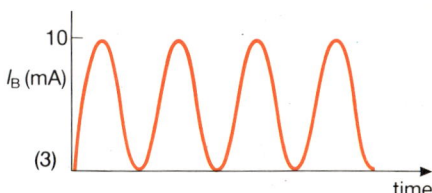

6 Operational Amplifiers

Here you can see a photograph of two operational amplifiers. Operational amplifiers like these are used in many electrical devices

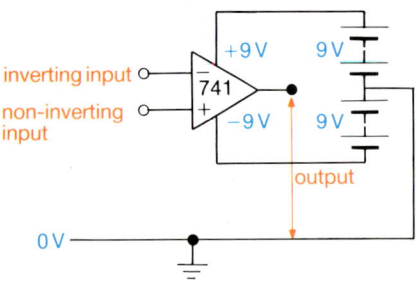

Figure 1
The 741 op-amp

Input	Output
V+ > V−	+ 9 V
V− > V+	− 9 V

Table 1

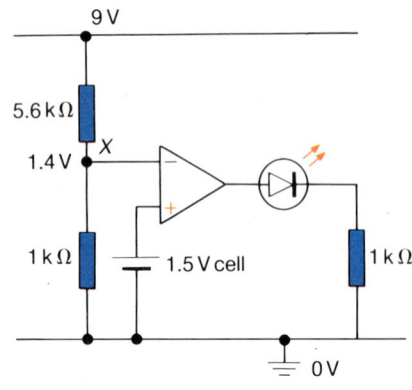

Figure 2
A cell tester

Operational amplifiers are used for comparing and amplifying voltages. One of the most widely used operational amplifiers (or **op-amps**) is the 741. This op-amp contains about 20 transistors, a capacitor and several resistors. All these components are on a single piece of silicon. This is called a **silicon chip**. The 741 op-amp is an example of an **integrated circuit**.

Figure 1 shows the various connections to the 741. It is powered by two 9 V batteries. It has two inputs and an output. When the non-inverting input voltage, V+, is greater than the inverting input voltage, V−, the output voltage is +9 V. When V− is greater than V+ the output voltage is −9 V (table 1). (The 741 must always be connected to two batteries. We do not usually show the connections on circuit diagrams. This makes the diagrams easier to follow.)

A battery tester

When you are making an electronic circuit, it is really annoying to find that you have a flat battery. A 741 can be used as a battery (or cell) tester. Figure 2 shows the idea. A potential divider fixes the non-inverting input at about 1.4 V. The cells to be tested should produce a voltage of about 1.5 V. When the voltage is greater than 1.4 V the output is +9 V. Then the LED is forward biased and it lights. If the battery is flat (voltage less than 1.4 V) the output is −9 V. Then the LED is reverse biased and it does not light.

Making an amplifier

The 741 can also be used as a voltage amplifier. The output of the 741 depends on the voltage difference between the two inputs. The 741 has a very high voltage gain. When the difference between the two inputs is about 0.001 mV, the output is about 1 V.

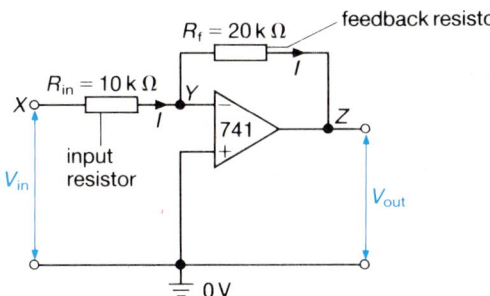

Figure 3
An inverting amplifier. (The connections of the op-amp to the power supply are not shown)

To make a stable amplifier we apply some **negative feedback**. The output is connected to the inverting input through a feedback resistor. When the op-amp has negative feedback we can say:

- The two inputs have very nearly the same voltage, V+ = V−
- No current flows into the op-amp itself. (In fact a current of about 0.000 000 000 01 A goes into the op-amp, but it is so small we will ignore it!)

We will now apply these two rules to the amplifier shown in Figure 3. Since V+ = V−, the point Y must be at earth voltage (0 V). Also, since no current goes into the op-amp, the same current I goes through the input and feedback resistors.

$$I = \frac{V_{XY}}{R_{in}} = \frac{V_{YZ}}{R_f}$$

V_{XY} = voltage between X and Y
V_{YZ} = voltage between Y and Z

So

$$\frac{V_{in} - 0}{R_{in}} = \frac{0 - V_{out}}{R_f}$$

or

$$\frac{V_{out}}{V_{in}} = -\frac{R_f}{R_{in}}$$

In this example $\frac{R_f}{R_{in}} = 2$. So the amplifier multiplies this voltage by −2.

The minus sign means that the phase of the voltage has been changed (Figure 4). The output voltage of an op-amp can be no bigger than the supply voltage. If the supply voltage is from two 9 V batteries, you can get no more than 9 V out of the amplifier. This means that if the input voltage is too big, the output gets distorted (Figure 5).

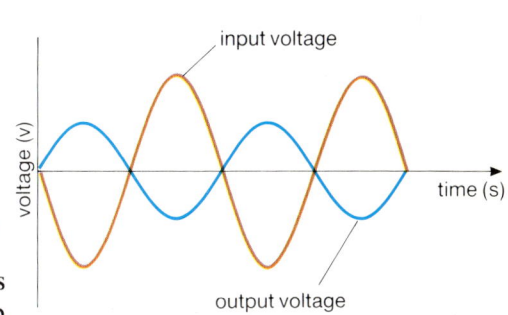

Figure 4
The input and output voltages are out of phase

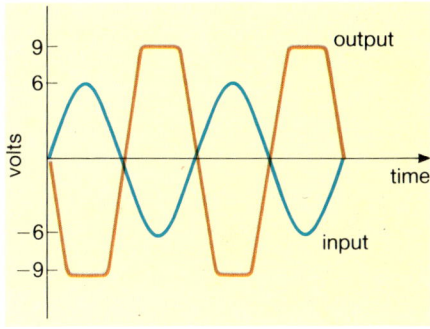

Figure 5
This amplifier tries to multiply the voltage by 2, but it cannot produce more than 9 V

Questions

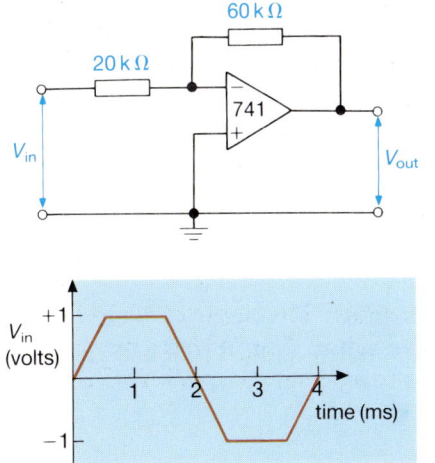

1 On the left you can see a circuit diagram for an operational amplifier. The graph shows how the input voltage varies with time. Copy the graph. Add to it a second graph to show how the output voltage varies with time. Give as much detail as possible.

2 This question is about using an op-amp to add voltages (see circuit on right).
(a) What is the voltage at C?
(b) Calculate the current I_1 and I_2.
(c) Now calculate the current I_3.

(Remember no current goes into the op-amp).
(d) What is the voltage at D?
(e) Explain how this circuit could be useful in an audio-mixer.

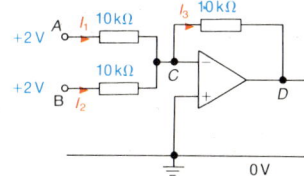

3 Explain why the voltage at point X in Figure 2 is about 1.4 V.

7 Logic Gates

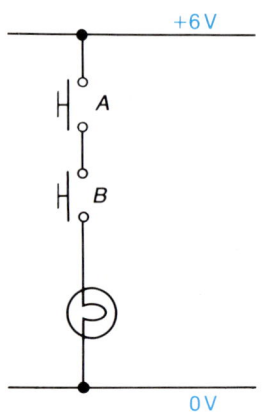

Figure 1
A two-state system: the bulb is on or off

Table 1

State of switch		State of bulb
A	**B**	
open	open	OFF
closed	open	OFF
open	closed	OFF
closed	closed	ON

Two states: ON and OFF

Logic gates are widely used in computers and other electronic systems. They collect and display information.

A very simple example of a logic gate is shown in Figure 1. To turn the light bulb on, you have to press both switches. It is no good just pressing one of them. The behaviour of the circuit is summarised in table 1. This simple circuit can collect information about your fingers! When two fingers press A and B the bulb lights.

Figure 2 shows a similar circuit that works with relays. This time the circuit collects information about voltages. The relay switches in the circuit are normally open. However, if the voltage at A is high (for example, above 3 V), the top switch is closed. So when A and B are both above 3 V the lamp switches on.

Truth tables

Truth tables show the behaviour of an electronic system in shorthand. A high voltage (above 3 V in our example) is defined as **logic state '1'**. A low voltage (below 3 V) is defined as **logic state '0'**.

You can describe the behaviour of the circuit in Figure 2 like this. When both A AND B are 1, Q is 1; otherwise Q is 0. Table 2 summarises this. We call this sort of table a **truth table**.

The circuit you have just been looking at makes an **AND-gate**. This gate was made with switches and relays. Logic gates are also made easily with transistors in integrated circuits. The advantages of integrated circuit logic gates are: they are cheap, small and can be switched millions of times each second. The circuit symbol for an AND-gate is shown above Table 2.

All integrated circuits need batteries to power them. We usually leave out connections to batteries in the circuit diagrams though. This helps to make the circuits easy to follow.

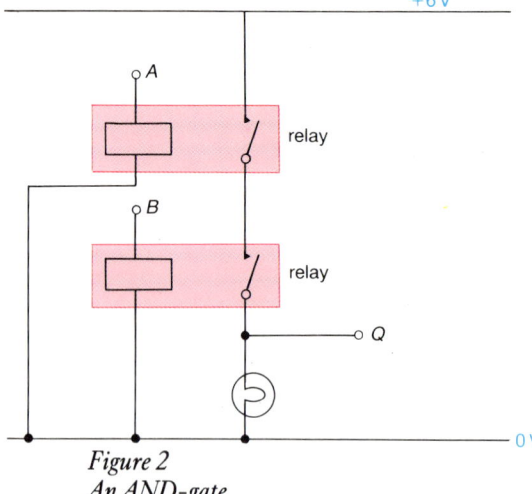

Figure 2
An AND-gate

State of inputs		State of outputs Q
A	**B**	
0	0	0
0	1	0
1	0	0
1	1	1

Table 2 Truth table for an AND-gate, with its circuit symbol.

More logic gates

- An **OR-gate** is shown on the next page. When A OR B (OR both) is 1 then Q is 1 (Table 3).
- A **NOT-gate** is shown on the next page. When A is NOT 1, Q is 1 (Table 4).

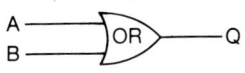

State of inputs		State of outputs Q
A	B	
0	0	0
0	1	1
1	0	1
1	1	1

Table 3 Circuit symbol and truth table for an OR-gate.

State of input A	State of output, Q
0	1
1	0

Table 4 Circuit symbol and truth table for a NOT-gate.

Input A	Input B	Q
0	0	1
0	1	0
1	0	0
1	1	0

Table 5. Circuit symbol and truth table for a NOR-gate.

Input A	Input B	Q
0	0	1
0	1	1
1	0	1
1	1	0

Table 6. Circuit symbol and truth table for a NAND-gate.

- A **NOR-gate** symbol and truth table are shown on the right. When neither A NOR B is 1, Q is 1 (Table 5).
- A **NAND-gate** symbol and truth table are shown below right. When both A AND B are NOT 1, Q is 1 (Table 6).

Questions

1 This question is about working out logic states. For example when A is 1, B is 0. What are the states of the points C and D?

2 The circuit below shows a security system for a car.

(a) Which switches have to be closed to turn on the starter motor?

(b) What happens if any other switch is closed?

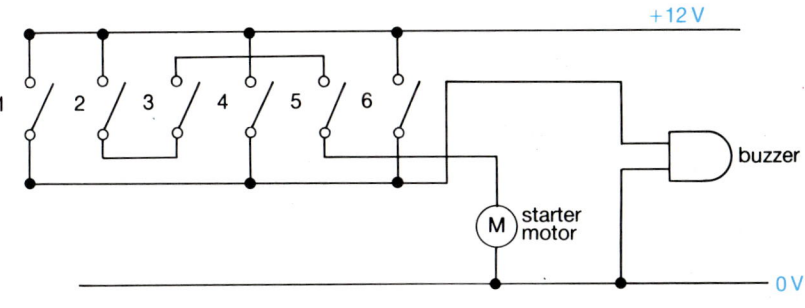

3 What single gate can be made from the AND-gate and the NOT-gate? Explain your answer.

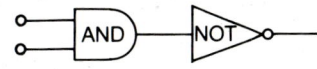

8 Using some Logic

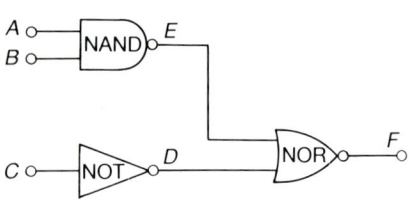

Figure 1

A	B	C	D	E	F
0	0	0	1	1	0
0	0	1	0	1	0
0	1	0	1	1	0
1	0	0	1	1	0
0	1	1	0	1	0
1	0	1	0	1	0
1	1	0	1	0	0
1	1	1	0	0	1

Table 1

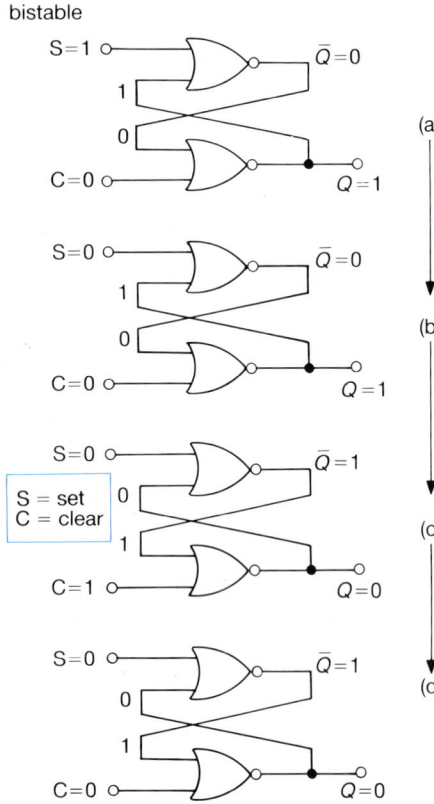

Figure 2
This shows how a bistable changes from one state to another

A BBC microcomputer contains about half a million NOR gates

Analysing logic systems

More complicated logic systems can be made by using a combination of gates. In Figure 1, how does the output F depend on the states of A, B and C? You can work out this problem by constructing a truth table (Table 1).

First you write out the columns A, B and C, working out all the possible combinations of states. The state of E can be worked out from the truth table of a NAND-gate. E is only 0 when A and B are both 1. D is 0 when C is 1; D is 1 when C is 0. You can now fill in the columns D and E. D and E are the inputs to the NOR-gate; F is the output. F is 1 when D and E are 0.

So this circuit makes a 3-input AND gate. When A AND B AND C are 1, F is 1. Otherwise F is 0.

The bistable

The **bistable** or **flip-flop** is the basis of electronic computer memories. A bistable can be made from two NOR gates, as shown in Figure 2. The sequence of diagrams shows how the bistable works. You should remember that the output (Q or $\bar{Q}$) of a NOR gate is only 1 when both inputs are 0. If either input (or both) is 1, the output is 0.

In diagram (a) the bistable is set with S = 1 and C = 0. Since S = 1, $\bar{Q}$ = 0. Now both inputs to the lower NOR gate are 0, so Q = 1.

In diagram (b), S has changed to 0. This does not affect Q or $\bar{Q}$. So Q = 1 still.

However, in diagram (c) S = 0 and C = 1. This means that Q = 0 and $\bar{Q}$ = 1. The bistable has been cleared.

In diagram (d) C goes back to 0. This does not affect Q or $\bar{Q}$.

Notice that the bistable remembers which of S or C was *last* in the logic state 1. The output, Q, of the bistable has two states. Q is either 0 or 1. We say that the bistable remembers one **bit of information**. The word 'bit' is short for binary digit. Computers calculate only in terms of binary numbers. All binary numbers can be expressed in terms of noughts and ones.

The electronic memory for a home computer has about 250 000 bistables inside it. These can be accommodated on a few small silicon chips. When you turn your computer off, it forgets what is stored in its electronic memory. So to store information permanently, you must transfer it to a magnetic disc or tape.

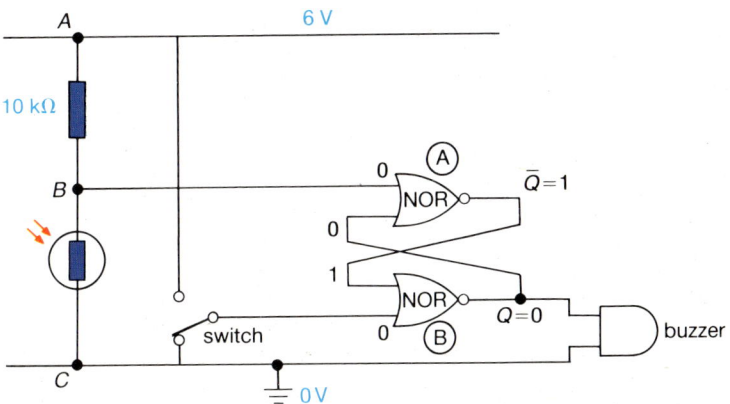

Figure 3
A latched burglar alarm

An electronic latch

Bistables are often used in alarm systems. This is because they work as **electronic latches**. When the alarm has been triggered, the latch keeps the alarm on.

Figure 3 shows a circuit for a 'latched' burglar alarm. An infra-red beam of light falls onto a LDR. This keeps the LDR's resistance low. B is at a low voltage. Both inputs of NOR gate A are low. This makes $\bar{Q}$ high, and Q low. The buzzer is off.

If the burglar now walks across the infra-red beam, the resistance of the LDR goes high for a moment. Now B goes high and $\bar{Q}$ goes low. Both inputs to NOR gate B are low and Q goes high. The buzzer sounds the alarm. But the alarm stays on because the bistable 'remembers' that the burglar was there.

Questions

1 (a) Why is an infra-red beam of light used in the burglar alarm?
(b) Explain how you can turn the burglar alarm off.
(c) Think of another way that you could use an electronic latch.

3 The circuit below shows part of a security system for a car ferry. S_1 closes when the bow doors are shut. S_2 opens when the passenger gangway is lifted. S_3 is an ignition switch on the bridge. Explain how the system works.

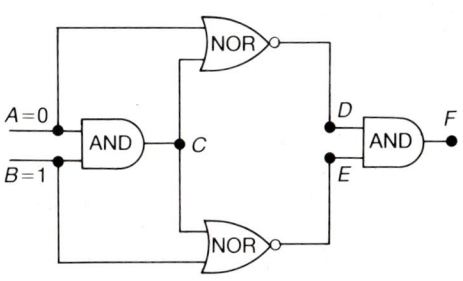

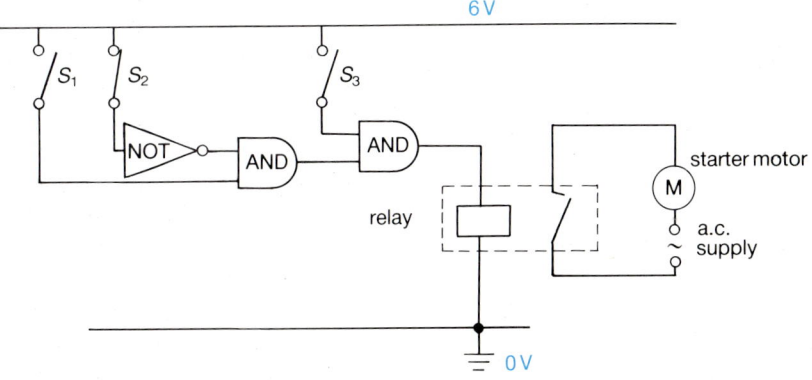

2 (a) In the diagram above, work out the logic state of the points C, D, E, F.
(b) When F is in logic state 1, what are the states of A and B?

SECTION K: *STUDY QUESTIONS*

The following are five components which can be used in electronic circuits:

A capacitor
B diode
C resistor
D thermistor
E transistor

1 Which component allows current to flow in one direction only?
2 Which component can store energy?
3 Which component has three connections?
4 Which component can act as a switch?

LEAG

5 In which circuit will the lamp come on when switch S is closed and stay on even when switch S is then opened?

MEG

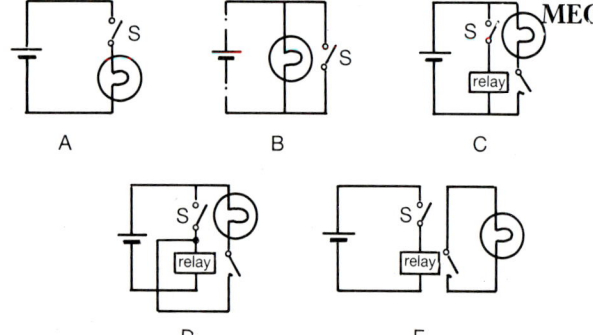

6 In the circuits shown each ammeter reads 100 mA.

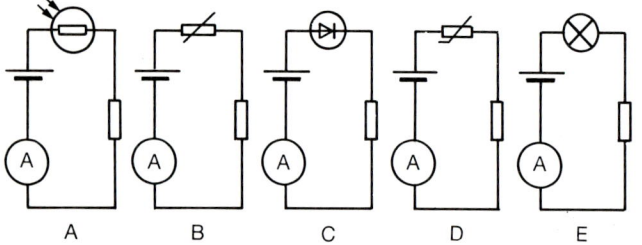

The cell in each circuit is then reversed. In which circuit does the ammeter now read 0 mA?

MEG

7 In the circuit shown, the lamp is on when the switch is closed.

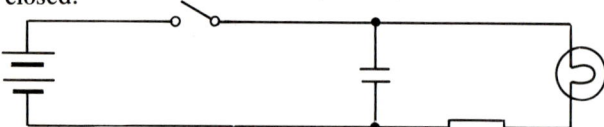

What happens to the lamp when the switch is opened?

A It goes off immediately.
B It goes off after a delay.
C It stays on.
D It goes off and then goes on after a delay.
E It flashes on and off many times.

MEG

8 Which circuit could be used to measure light intensity?

MEG

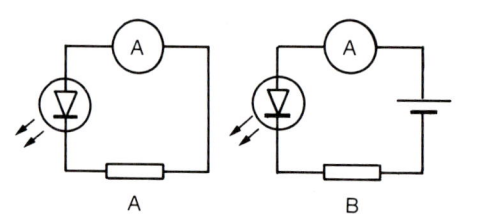

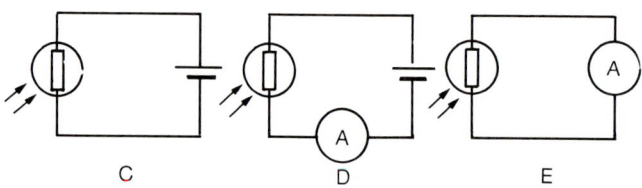

9 These are the truth table and symbol for the AND gate.

Inputs		Outputs
0	0	0
0	1	0
1	0	0
1	1	1

In which one of the circuits will the light-emitting diode be lit?

MEG

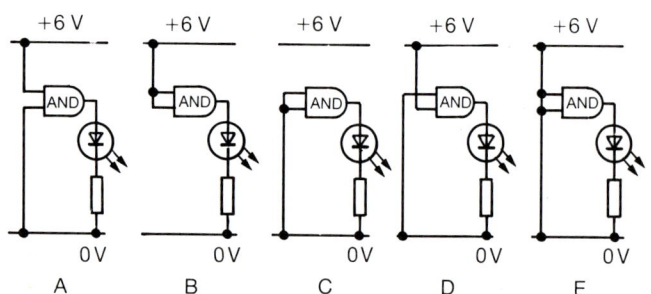

10 The circuit below controls the air conditioning in a factory. When it becomes too hot in the factory the fan switches on. The graphs shows how the resistance of the thermistor changes with temperature. Explain why the fan switches on when the temperature rises to 25°C.

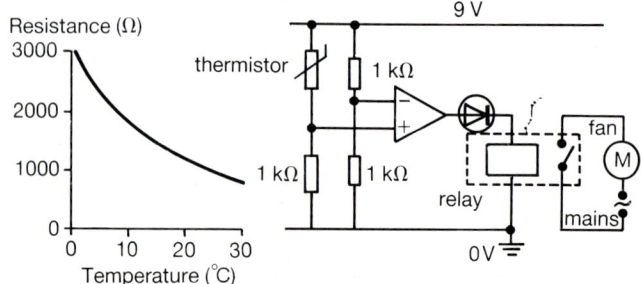

256

11 What is the output voltage in these circuits?

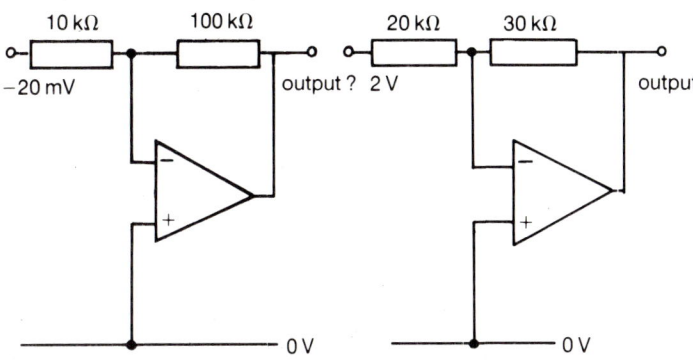

12 This diagram shows how an op-amp can be used as a non-inverting amplifier.

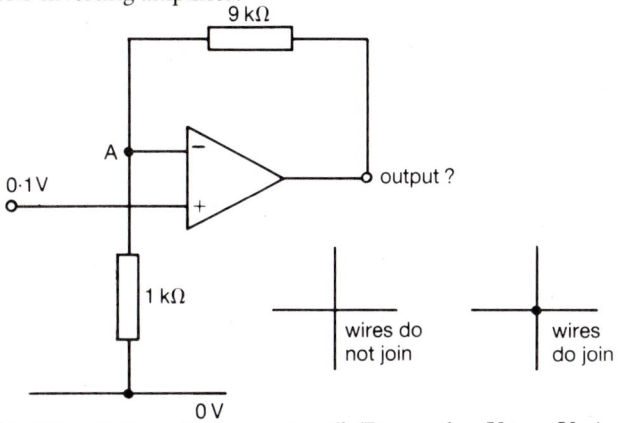

(a) What is the voltage at point A? (Remember $V+ = V-$).
(b) Calculate the current flowing through the 1 KΩ resistor.
(c) What current goes through the 9 KΩ resistor?
(Remember no current goes into the op-amp).
(d) Now calculate the output voltage.
(e) What is the voltage gain of this amplifier?

13 The circuit in the diagram below controls a motor.

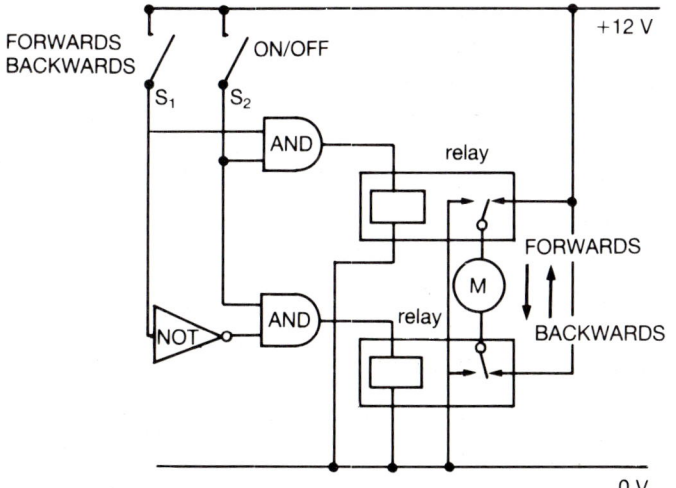

14

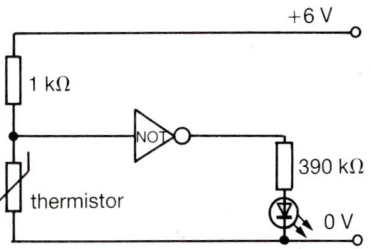

(a) For the circuit drawn above, calculate the input voltage to the NOT gate (the voltage across the thermistor) when the thermistor has a resistance of 500 Ω.
This graph shows the input/output voltage characteristics of the NOT gate used in this circuit.

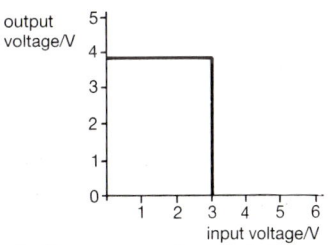

(b) What is the output voltage of the NOT gate when the thermistor has a resistance of 500 Ω?
(c) When the light emitting diode is on there is a voltage of 1.8 V across it. Calculate the current through the 390 Ω resistor.
(d) Write down what the answers to parts (a) and (c) will be if the thermistor and the 1 kΩ resistor are interchanged.

MEG

15 Mr. Corrigan owns an amusement arcade. He is having some trouble with Tommy who is very rough with the pinball machines. So Mr. Corrigan decides to fit anti-tilt devices to his pinball machines. When Tommy tips the machine, the power supply turns off. The circuit is shown below.

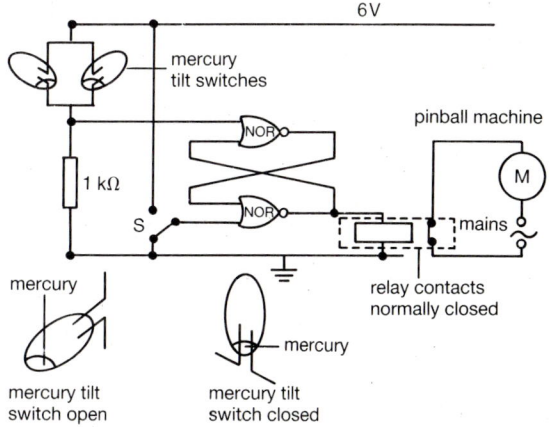

(a) Explain how the mercury tilt switches work.
(b) Why are there two mercury tilt switches?
(c) What do the two NOR gates do?
(d) Why are the relay contacts normally closed?
(e) Explain how the machine can be started again by putting another coin in.

STUDY QUESTIONS

16 (a) A NOT logic gate can be added to the output of a two-input OR logic gate to produce a two-input NOR (NOT OR) logic gate. The completed truth tables for a NOT and an OR gate are given below.

NOT

A	P
0	1
1	0

OR

A	B	P
0	0	0
0	1	1
1	0	1
1	1	1

A and B are inputs. P is the output. Copy and complete the truth table for a two input NOR gate.

A	B	P
0	0	
0	1	
1	0	
1	1	

(b) The circuit symbol for a two-input NAND (NOT AND) gate is shown below.

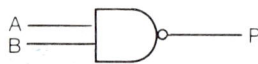

Also given is the truth table for a two-input NAND gate.

A	B	P
0	0	1
0	1	1
1	0	1
1	1	0

In the design of logic circuits, the NAND gate is a basic 'building block' as in the following circuit.

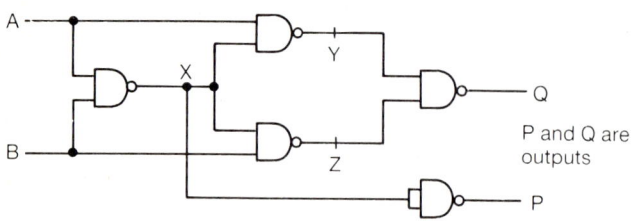

(i) Copy and complete the truth table for this circuit.

A	B	X	Y	Z	P	Q
0	0	1				
0	1	1				
1	0	1				
1	1	0				

(ii) What arithmetic operation does this circuit perform?

(c) The output of a logic circuit can be displayed using an LED and associated series resistor, as in the circuit shown below.

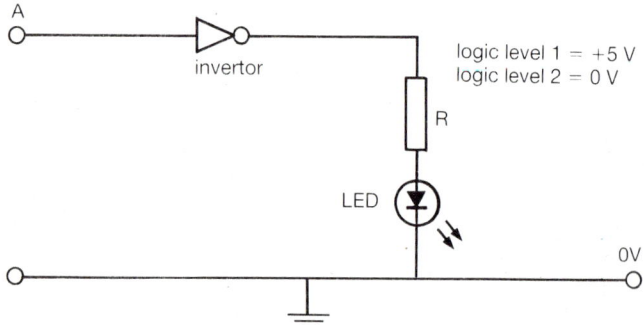

(i) When the LED is lit, what is the logic state at A?
(ii) The LED has a potential difference drop of 2 V across it and a current of 10 mA flowing through it when it is lit. What is the potential difference across resistor R when the LED is lit?
(iii) Calculate the resistance of R.
(iv) Why is the resistor R needed?
(d) In a factory, a particular piece of machinery has an alarm system to warn the operator of a fault. The system has a blue lamp (alight when the machine is operating normally), a red warning lamp and a buzzer. There is a fault detection system and an 'operator acknowledge' button. If a fault is detected, the blue lamp goes out and the red lamp comes on and the buzzer sounds. When the operator acknowledges the fault by pressing the button, the buzzer stops but the red light stays on.
Complete the output columns of the truth table below to show which states are wanted.

INPUT		OUTPUT		
FAULT	OPERATOR ACKNOWLEDGE	RED	BUZZER	BLUE
0	0			
0	1			
1	0			
1	1			

SECTION L
Radioactivity

This is the Conqueror, a nuclear powered submarine. The small nuclear reactor on board means that the Conqueror can remain submerged for many months without any need to refuel. It only surfaces to replenish food and other supplies, or for maintenance

1 Introducing Nuclear Physics

At the Moscow Summit in 1988, the superpowers (the USA and USSR) agreed to reduce their stores of nuclear missiles. The picture shows Ronald Reagan and Mikail Gorbachev shortly after signing the agreement

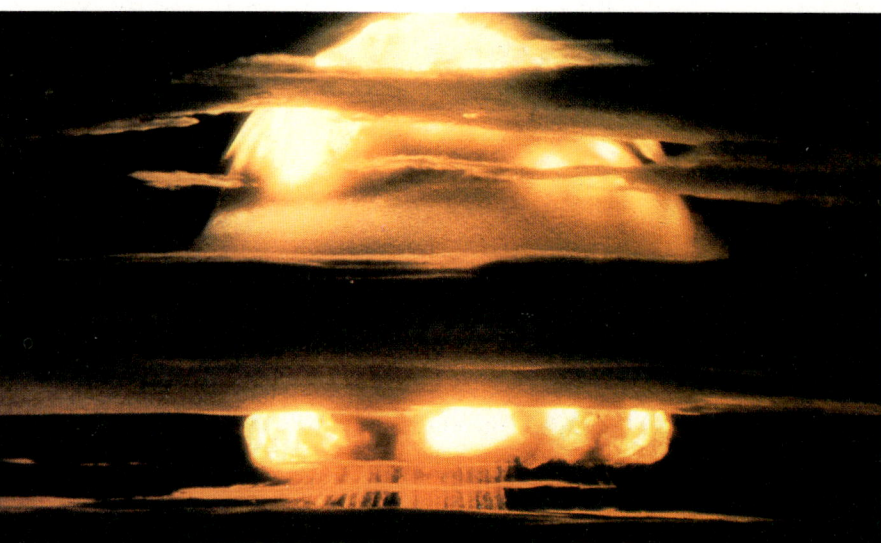

A nuclear war could completely destroy the world we live in. Many people near the explosion would be killed, and many more would die from the radiation emitted. In addition, the exploding bombs would drive smoke and dust high into the atmosphere, which could block out the Sun for long afterwards, resulting in yet more people dying of cold in the following years. This is called a nuclear winter

A lot of people feel strongly that we should get rid of our nuclear weapons completely

The debate about the use of nuclear technology is one of the most widely discussed topic of our lives today. The Americans and Soviets meet often in an attempt to limit their use of nuclear weapons. Although some progress has been made recently, they find it difficult to agree how this should be done.

In the national news we frequently hear about the use of nuclear energy to produce electrical energy. On television we often see people arguing angrily about this issue. On one side we will hear scientists claim that the use of nuclear power is quite safe, and yet we will hear others claim that the use of nuclear power has caused more people than usual to die from cancer.

Which side is right? This is a difficult question to answer, but it is your job as a responsible citizen to read about these matters and to decide for yourself. You should understand why radioactive materials can be harmful to us, and also why they are useful to us.

However, before you can understand these issues you need to learn about nuclear physics and radioactivity.

The nuclear model of the atom

Earlier in the book you read about atoms and how looking at Brownian motion helped us to discover them. We thought of atoms as hard bouncy balls that exerted a pressure by hitting the walls of their container. You now need to know something about the insides of atoms.

In 1909 Geiger and Marsden discovered a way of exploring the insides of atoms. They directed a beam of **alpha particles** at a thin sheet of gold foil. Alpha particles were known to be positively-charged helium ions, He^{2+}, which were travelling very quickly. They had expected all of these energetic particles to pass straight through the thin foil. Much to their surprise they discovered that a very small number of them bounced back, although most of them travelled through the foil without any noticeable change of direction.

Rutherford produced a theory to explain these results; this is illustrated in Figure 1. He suggested that the atom is made up of a very small positively-charged **nucleus**, which is surrounded by **electrons** which are negatively-charged. His idea was that the electrons orbit around the nucleus in the same way that planets orbit around the sun. The gap between the nucleus and electrons is large, in fact the diameter of the atom is about 100 000 times larger than the diameter of the nucleus itself. Because so much of the atom is empty space most of the alpha particles could pass through it without getting close to the nucleus. Some particles passed close to the nucleus and so the positive charges of the alpha particle and the nucleus repelled each other causing a small deflection. A small number of particles met the nucleus head on, these were turned back the way they came. The fact that only a very tiny fraction of the alpha particles bounced backwards tell us that the nucleus is very small indeed. Rutherford proposed that the positive charge in the nucleus was carried by **protons**. Hydrogen has one proton, and this positive charge is balanced by the negative charge of one electron; helium has two protons whose charge is balanced by two electrons which orbit the nucleus.

Nuclear fuels are used to generate electricity at Dungeness B power station in Kent

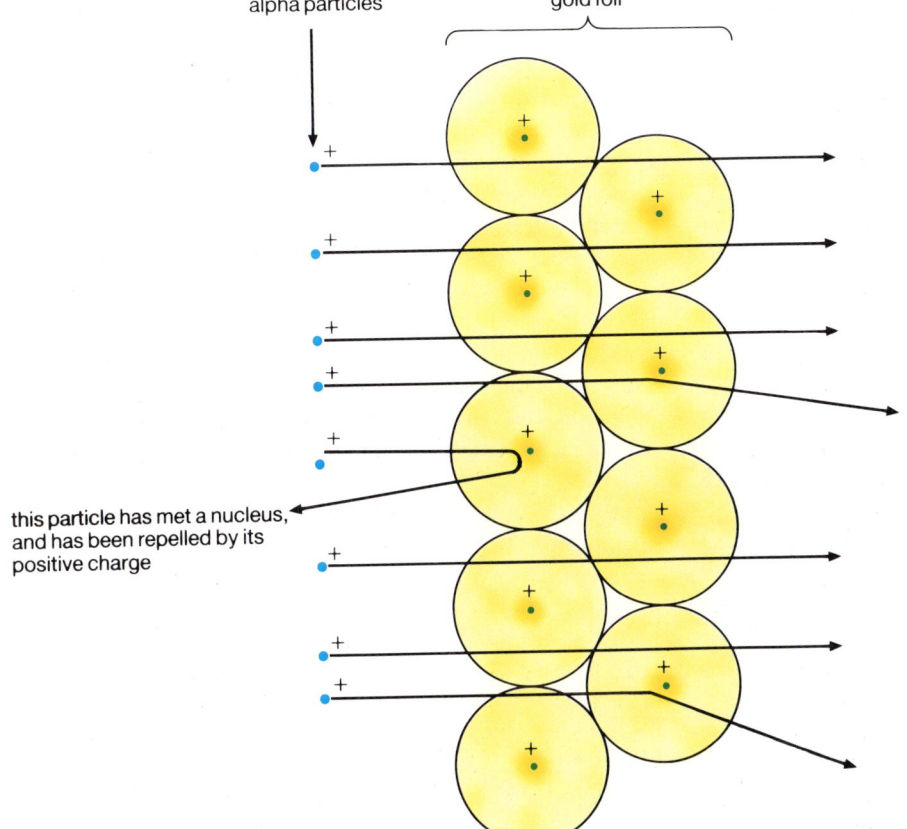

alpha particles

gold foil

this particle has met a nucleus, and has been repelled by its positive charge

Figure 1
Most of the alpha particles pass straight through the gold foil or are deflected slightly, but a very small number bounce back

Questions

1 Explain carefully why Geiger and Marsden's experiment led us to believe that an atom has a nucleus.
2 In Geiger and Marsden's original experiment about one out of every 10 000 alpha particles 'bounced back'. What effect would the following changes have had on that number?
(a) using a thicker gold foil.
(b) using alpha particles that travelled more slowly.
(c) using a copper foil of the same thickness.

2 Atomic Structure

Part of the CERN super proton synchrotron, near Geneva. Here, protons hurtle around a 7 kilometre long tube. They travel so fast that it takes them only 0.00002 seconds to go round once. The synchrotron helps nuclear physicists to understand more about the structure of the nucleus

Particle	Mass*	Charge*
proton	1	1
neutron	1	0
electron	$\dfrac{1}{1840}$	-1

*by comparison with a proton's
Table 1

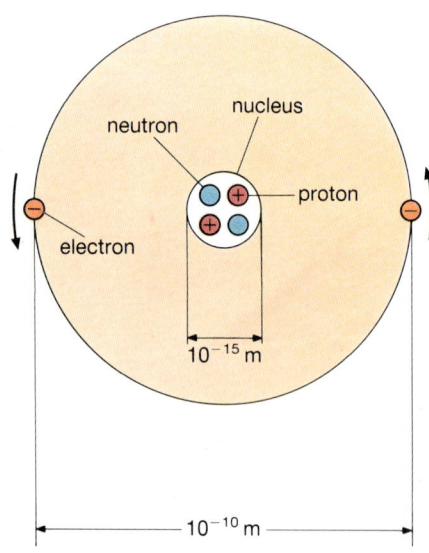

Figure 1
The helium atom; this is not drawn to scale – the diameter of the nucleus is about 100 000 times smaller than that of the atom itself

Neutrons, protons and electrons

Rutherford's model of the atom that you met in the last section is essentially the one that we accept today, except that it was not quite as simple as he thought. There is another particle in the nucleus called the **neutron**. The neutron was discovered after people realised that the nucleus of most atoms contained some extra mass. For example, the mass of a helium atom is four times that of a hydrogen atom, but it has only two protons in the nucleus. The neutron has no charge but it has the same mass as a proton. In comparison with the proton or neutron, the electron has virtually no mass but it carries a negative charge the same size as a proton's charge (see Table 1).

A hydrogen atom has 1 proton and 1 electron; it is electrically neutral because the charges of the electron and proton cancel each other. A helium atom has 2 protons and 2 neutrons in its nucleus, and 2 electrons outside that. The helium atom is also neutral because it has the same number of electrons as it has protons; it has 4 times the mass of a hydrogen atom because it has 4 particles in the nucleus (see Figure 1 and Table 2). You should remember that nearly all of an atom's mass is in the nucleus.

element	hydrogen, H	helium, He	lithium, Li
number of electrons	1	2	3
number of protons	1	2	3
number of neutrons	0	2	4
number of particles in nucleus	1	4	7
mass relative to hydrogen	1	4	7

Table 2

Ions

Atoms are electrically neutral since the number of protons balances exactly the number of electrons. However, it is possible either to add extra electrons to an atom, or to take them away. When an electron is added to an atom a **negative ion** is formed; when an electron is removed a **positive ion** is formed. Some examples are given in Table 3. The name ion is also used to describe charged molecules.

element	number of protons	number of electrons	total charge	ion
helium, He	2	1	+1	He^+
magnesium, Mg	12	10	+2	Mg^{2+}
chlorine, Cl	17	18	−1	Cl^-

Table 3

Proton and nucleon numbers

The number of protons in the nucleus of an atom determines what element it is. Hydrogen atoms have 1 proton, helium atoms 2 protons, uranium atoms 92 protons. The number of protons in the nucleus decides the number of electrons that are to be found surrounding the nucleus. The number of electrons determines the chemical properties of an atom. The number of protons in the nucleus is called the **proton** (or **atomic**) **number of the atom** (symbol **Z**). So the proton number of hydrogen is one, $Z = 1$.

The mass of an atom is decided by the number of neutrons and protons added together. Scientists call this number the **nucleon** (or **mass**) **number of an atom**. The name nucleon refers to either a proton or a neutron.

> Proton number = number of protons
>
> Nucleon number = number of protons and neutrons

For example, an atom of carbon has 6 protons and 6 neutrons. So its proton number is 6, and its nucleon number is 12. To save time in describing carbon we can write it as $^{12}_{6}C$; the nucleon number appears on the left and above the symbol C, for carbon, and the proton number on the left and below.

Nucleon number 12
Proton number 6 $\quad$ **C**

A carbon atom which has a single positive charge on it would be written as $^{12}_{6}C^+$.

Isotopes

Not all the atoms of a particular element have the same mass. For example, two carbon atoms might have nucleon numbers of 12 and 14. The nucleus of each atom has the same number of protons, 6, but one atom has 6 neutrons and the other 8 neutrons. Atoms of the same element (carbon in this case) which have different masses are called **isotopes**. These two isotopes of carbon can be written as carbon-12, $^{12}_{6}C$ and carbon-14, $^{14}_{6}C$.

Questions

1 (a) An oxygen atom has 8 protons, 8 neutrons, and 8 electrons. What is its: (i) its proton number? (ii) nucleon number?
(b) Why is the oxygen atom electrically neutral?
2 How many protons, neutrons and electrons are there in each of the following atoms?
(a) $^{17}_{8}O$ $\qquad$ (c) $^{238}_{92}U$
(b) $^{235}_{92}U$ $\qquad$ (d) $^{17}_{8}O^{2-}$
3 Write an essay to describe the structure of an atom. In your essay, you ought to mention terms such as proton number, nucleon number, ions and isotopes.

3 Radioactivity

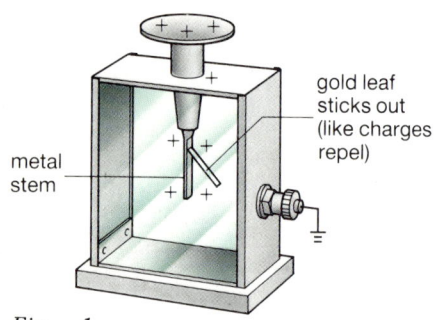

Figure 1
(a) Positively-charged electroscope

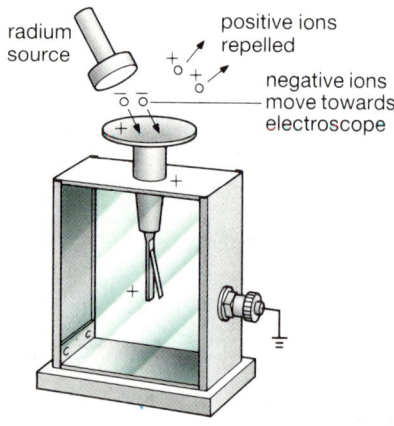

(b) Negative ions neutralise the electroscope

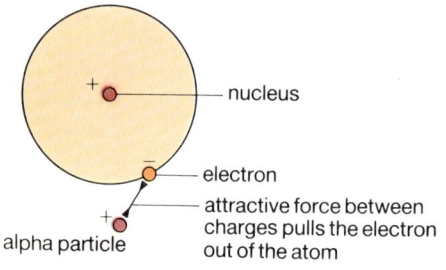

Figure 2

The nucleus of an atom is usually very stable; the atoms that we are made of have been around for thousands of millions of years. Atoms may lose or gain a few electrons during chemical reactions, but the nucleus does not change during such processes.

However, there are some atoms which have unstable nuclei which throw out particles to make the nucleus more stable. The first element discovered which emitted these particles was radium, and the name **radioactivity** was given to this process. There are three types of particle which can be released:

● **Alpha particles** are the nuclei of helium atoms, so they have a nucleon number of 4 and a proton number of 2. They have 2 positive charges since they are helium atoms stripped of their 2 electrons. When an alpha particle is emitted from a nucleus it causes it to change into another nucleus with a nucleon number 4 less and a proton number 2 less than the original one. It is usually only very heavy elements that emit alpha particles, for example:

$$\underset{\substack{\text{uranium}\\\text{nucleus}}}{^{238}_{92}\text{U}} \rightarrow \underset{\substack{\text{thorium}\\\text{nucleus}}}{^{234}_{90}\text{Th}} + \underset{\substack{\text{alpha particle}\\\text{(helium nucleus)}}}{^{4}_{2}\text{He}}$$

This is called **alpha decay**

● **Beta particles** are electrons. In a nucleus there are only protons and neutrons but a beta particle can be created and thrown out of a nucleus when a neutron turns into a proton and an electron. Since an electron has a very small mass, when it leaves a nucleus it does not alter the nucleon number of that nucleus. However, the electron carries away a negative charge so the removal of an electron increases the proton number of a nucleus by 1. For example, carbon-14 decays into nitrogen by emitting a beta particle.

$$\underset{\substack{\text{carbon}\\\text{nucleus}}}{^{14}_{6}\text{C}} \rightarrow \underset{\substack{\text{nitrogen}\\\text{nucleus}}}{^{14}_{7}\text{N}} + \underset{\substack{\text{beta}\\\text{particle}}}{^{0}_{-1}\text{e}}$$

This is called **beta decay**

● When some nuclei decay, by sending out an alpha or beta particle, they also give out a **gamma ray**. Gamma rays are electromagnetic waves, like radio waves or light. They carry away from the nucleus a lot of energy, so that the nucleus is left in a more stable state. Gamma rays have no mass or charge, so when one is emitted, there is no change to the nucleon or proton number of a nucleus (see Table 1).

	Particle lost from nucleus	Change in nucleon number	Change in proton number
alpha decay α	helium nucleus $^{4}_{2}\text{He}$	−4	−2
beta decay β	electron $^{0}_{-1}\text{e}$	0	+1
gamma decay γ	electromagnetic waves	0	0

Table 1

Ionization

All three types of radiation (alpha, beta and gamma) cause **ionization** and this is why we must be careful when we handle radioactive materials. The radiation makes ions in our bodies and these ions can then damage our body tissues (see page 272).

We can demonstrate the ionizing effect of a radium source by holding it close to a charged gold leaf electroscope, (Figure 1). The electroscope is initially charged positively so that the gold leaf is repelled from the metal stem. When a radium source is brought close to the electroscope, the leaf falls, showing that the electroscope has been discharged. The reason for this is that the alpha particles from the radium create ions in the air above the electroscope as the charges on these particles pull some electrons out of air molecules (Figure 2). Both negative and positive ions are made; the positive ones are repelled from the electroscope, but the negative ones are attracted so that the charge on the electroscope is neutralised. It is important that you understand that it is not the charge of the alpha particles that discharges the electroscope, but the ions that they produce.

Background radiation

There are a lot of rocks in the Earth that contain radioactive uranium, thorium and potassium, and so we are always exposed to some ionizing particles. In addition the Sun emits lots of protons which can also create ions in our atmosphere. These two sources make up **background radiation**. Fortunately the level of background radiation is quite low and in most places it does not cause a serious health risk.

In some jobs, people are at a greater risk. X-rays used in hospitals also cause ionization. Radiographers make sure that their exposure to X-rays is as small as possible. In nuclear power stations neutrons are produced in **nuclear reactors**. The damage caused by neutrons is a source of danger for workers in that industry.

The background radiation in Britain is highest in Cornwall. This is due to radioactive elements in the granite rocks

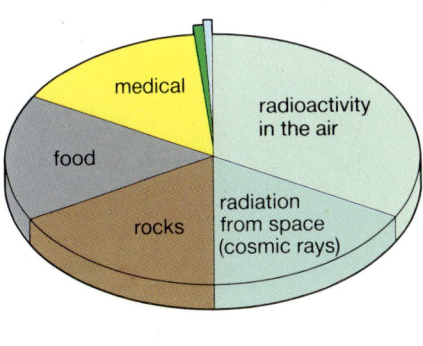

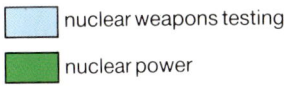

nuclear weapons testing

nuclear power

Figure 3
Sources of radiation in Britain

Questions

1 Fill in the gaps in the following radioactive decay equations:

(a) $^3H \rightarrow {}_2He + {}^0e$

(b) $^{229}_{90}Th \rightarrow Ra + {}^4_2He$

(c) $^{14}_6C \rightarrow ? + {}^0_{-1}e$

(d) $^{209}_{82}Pb \rightarrow {}_{83}Bi + ?$

(e) $^{225}_{89}Ac \rightarrow {}_{87}Fr + ?$

2 $^{238}_{92}U$ decays by emitting an alpha particle and two beta particles; what element is produced after those three decays?

3 Explain what effect losing a gamma ray has on a nucleus.

4 Explain carefully how a radioactive source that is emitting only alpha particles can discharge a negatively charged electroscope.

5 (a) What is background radiation and where does it come from?

(b) Use the pie chart (Figure 3) to discuss whether the nuclear power industry in the UK, is likely to cause a serious health hazard.

4 More on α, β and γ Radiation

Detecting particles

We make use of the ionizing properties of radiations to detect them. This is done using a **Geiger-Muller (GM) tube**. Figure 1 shows how such a tube works. A metal tube is filled with argon under low pressure; inside the tube there is a thin wire anode. A potential difference of about 450 V is applied between the inside and outside of the tube. When alpha, beta or gamma radiation enters the tube the argon atoms inside are ionized. These ions are then attracted to the electrodes in the tube so a small current flows. This current is then amplified and a counter can be used to count the number of particles entering the tube.

A false colour photograph of cloud chamber tracks. The green tracks are caused by alpha particles. One of them (yellow) collides with a proton (red) in the hydrogen gas that fills the chamber

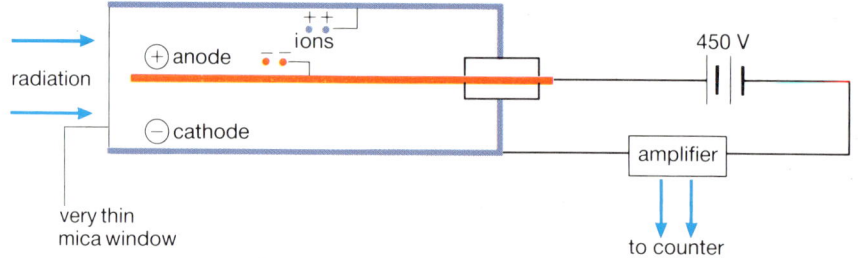

Figure 1
A Geiger-Muller tube

Cloud chambers

Another way of detecting radiation is to use a **cloud chamber** (Figure 2). The bottom of the cloud chamber is kept cold by placing some solid carbon dioxide ('dry ice') underneath the metal base plate. The inside of the chamber is filled with alcohol vapour. When a radioactive source is placed inside the cloud chamber, tracks are formed in the dense alcohol vapour. These white tracks can be seen clearly against the black bottom of the chamber. The alcohol molecules condense to leave a vapour trail in the region where ions have been produced by the passage of a particle. The tracks left by alpha particles are straight and thick; these show us that alpha particles are very strongly ionizing. Beta particles do not ionize air so strongly, so their tracks are much thinner. Gamma rays leave nearly no track at all because they produce few ions in a given distance.

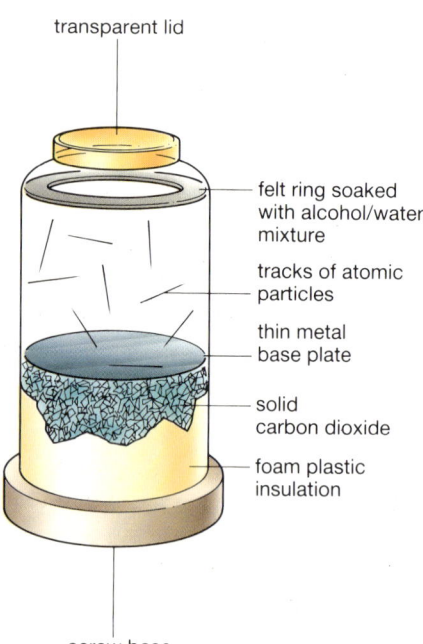

Figure 2
A cloud chamber

Properties of radiation

- Alpha particles will travel about 5 cm through air and they will be stopped by a sheet of paper (Figure 3). They ionize air very strongly. Alpha particles travel at speeds of about 10^7 m/s. This is more slowly than beta or gamma rays travel. They can be deflected by a very strong magnetic field, but the deflection is very small indeed because alpha particles are so massive.
- Beta particles can travel several metres through air and they will be stopped by a sheet of aluminium a few millimetres thick (Figure 3). They do not ionize air as strongly as alpha particles. Beta particles travel at speeds just less than the speed of light (3×10^8 m/s). Beta particles can be deflected quite easily by a magnetic field, because they are such light particles (see Figure 4).

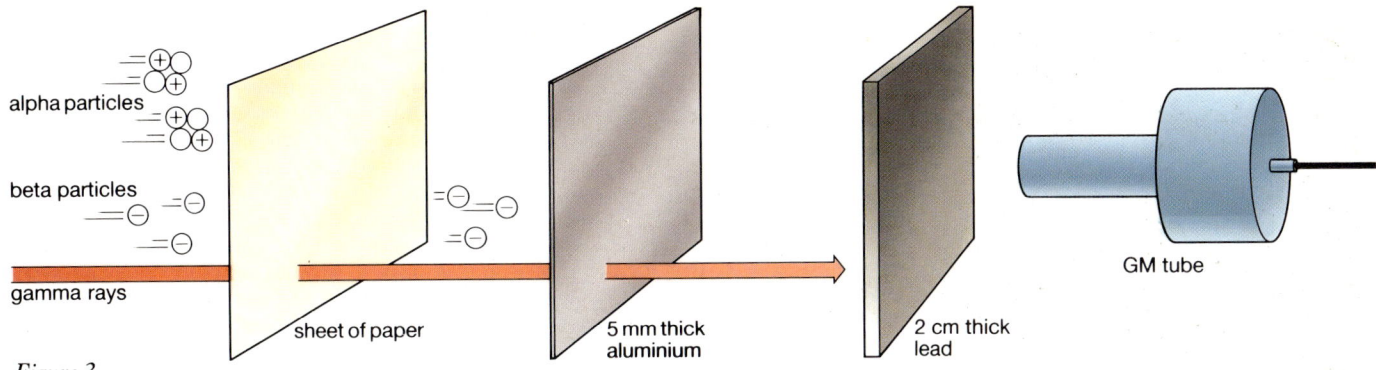

Figure 3

● Gamma rays can only effectively be stopped by a very thick piece of lead (Figure 3). They are electromagnetic waves, so they travel at the speed of light. Gamma rays only ionize air very weakly, and they cannot be deflected by a magnetic field because they carry no charge.

Alpha particles will cause most damage to your bodies if they get inside you; this could happen if you were to breathe in a radioactive gas such as radon. A school alpha source is not dangerous (unless you swallow it!) because the particles cannot penetrate your skin. On the other hand you want to avoid handling any gamma ray sources since these rays can get right into the middle of your body and cause damage there.

Radiation	Nature	Speed	Ionizing power	Penetrating power	Deflection magnetic field
alpha α	helium nucleus	10^7 m/s	very strong	stopped by paper	very small indeed
beta β	electron	just less than 3×10^8 m/s	medium	stopped by aluminium	large
gamma γ	electro-magnetic waves	3×10^8 m/s	weak	stopped by thick lead	none

Table 1. A summary of radiation properties.

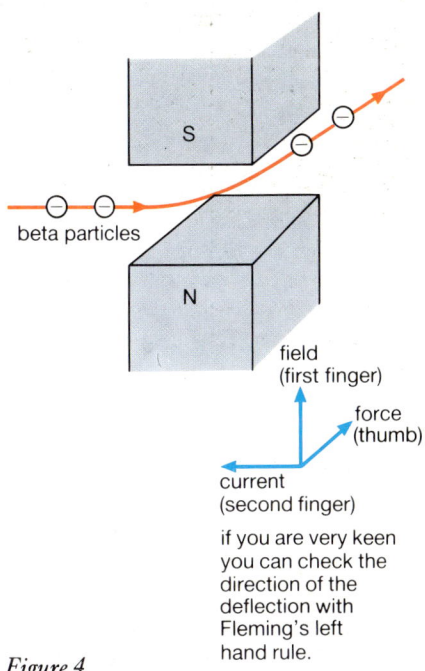

if you are very keen you can check the direction of the deflection with Fleming's left hand rule.

Figure 4
The deflection of beta particles by a magnetic field. Note that the deflection is not towards the poles of the magnet, but at right angles to them

Questions

1 Explain carefully how you could use a Geiger-Muller tube and some pieces of paper, aluminium and lead to show that radium emits alpha, beta and gamma radiations.

2 Which type of radiation is most dangerous to us?

3 A beta particle can be deflected by a magnetic field. Why can an alpha particle only be deflected a little and a gamma ray not at all?

4 Why do gamma rays leave only very faint tracks in a cloud chamber?

5 This question is about testing the thickness of a metal sheet. The metal sheet moves past a β-source and Geiger counter at a speed of 0.2 m/s.

(a) Plot a graph from the data below. Explain why the count rate changes over the first 50 s.

(b) Explain why the count rate drops and then rises again, after 50 s.

Count rate (s⁻¹)	75	80	77	73	76	75	63	57	50	55	67	75	77
Time (s)	0	10	20	30	40	50	60	70	80	90	100	110	120

5 Radioactive Decay

Throw	Number of coins left
0	1000
1	500
2	250
3	125
4	62
5	31
6	16
7	8
8	4
9	2
10	1

Table 1 Coin tossing experiment

Time (hour)	Number of nuclei left
0	1 000 000
1	500 000
2	250 000
3	125 000
4	62 500
5	31 250
6	15 620
7	7 810
8	3 900
9	1 950
10	980

Table 2 The number of nuclei left in a sample: half life 1 hour

We know that the atoms of some radioactive materials decay by emitting alpha or beta particles from their nuclei. But it is not possible to predict when the nucleus of one particular atom will decay. It could be in the next second, or sometime next week, or not for a million years.

The radioactive decay of an atom is rather like tossing a coin. You cannot say with certainty that the next time you toss a coin, that it will fall head up. However, if you throw a lot of coins you can start to predict how many of them will fall heads up. You can use this idea to help you understand how radioactive decay happens. You start off with a thousand coins; if any coin turns up head up then it has 'decayed' and you must take it out of the game. Table 1 shows the likely result (on average). Every time you throw a lot of coins about half of them will turn up heads.

Radioactive materials decay in the same way. If we start off with a million atoms then we find that after a period of time (say one hour), half of them have decayed. In the next hour we find that half of the remaining atoms have decayed, leaving us with a quarter of the original number (Table 2). The period of time taken for half the number of atoms to decay in a radioactive sample is called the **half life**, and it is given the symbol $t_{1/2}$. You can see that after 10 tosses of the coins, nearly all the coins have fallen heads up, and that after 10 half lives, nearly all of the nuclei have decayed.

Measurement of half life

If you look at Table 2 you can see that the number of nuclei that decayed in the first hour was 500 000, then in the next hour 250 000 and in the third hour 125 000. So as time passes not only does the number of nuclei left get smaller but so does the rate at which the nuclei decay. So by measuring the decay rate of a radioactive sample we can determine its half life.

Figure 1 shows how we can measure the half life of radon, which is a gas. (You are not allowed to use radioactive materials until you are over 16, so you cannot do this experiment yourself.) The gas is produced in a plastic bottle; we can give the bottle a squeeze and force some gas into a chamber which is fixed on to the end of a GM tube. The GM tube is attached to a ratemeter, which tells us the rate at which radon is decaying in the chamber. We measure the rate of decay on the ratemeter every 10 seconds and plot a graph of the count rate against time, Figure 2. We can see that the count rate halves every 50 seconds, so that is the half life of radon.

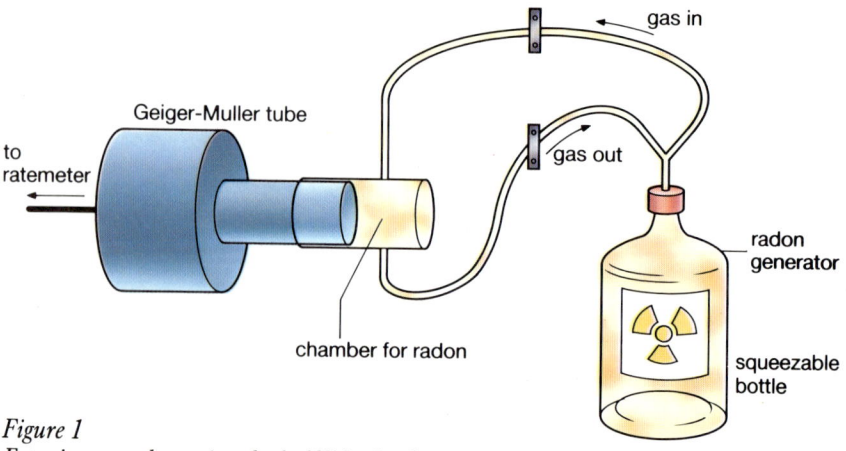

Figure 1
Experiment to determine the half life of radon

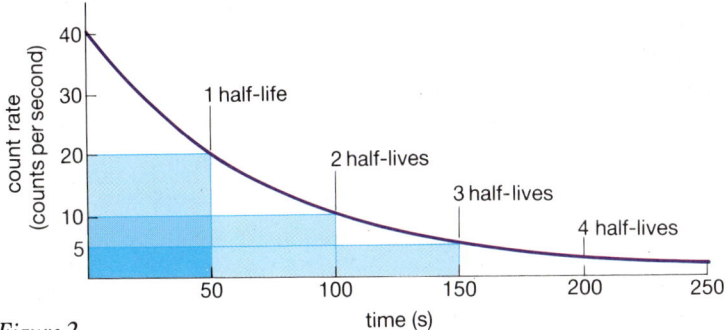

Figure 2

Dating archaeological remains

Carbon-14, $^{14}_6C$, is a radioactive isotope; it decays to nitrogen with a half life of about 5500 years. All living things (including you) have a lot of carbon in them, and a small fraction of this will be carbon-14. When a tree dies, for example, the radioactive carbon will begin to decay, and after 5500 years the fraction of carbon-14 in the dead tree will be half as much as you would find in a living tree. So by measuring the amount of carbon-14 in ancient relics, scientists can calculate their age. With this technique it was possible to date the Turin Shroud, which was thought to be the burial shroud of Christ.

Disposal of radioactive waste

Nuclear power stations produce radioactive waste materials, some of which have half lives of hundreds of years. These waste products are packaged up in concrete and steel containers and are buried deep underground or are dropped to the bottom of the sea. This is a controversial issue; some scientists tell us that radioactive wastes produce only a very low level of radiation, and that the storage containers will remain intact for a very long time. Others worry that these products will contaminate our environment and believe it is wrong to leave radioactive materials that could harm future generations.

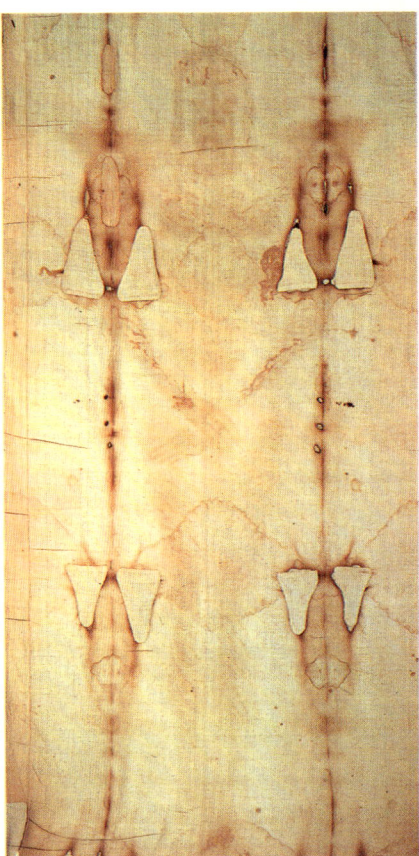

The Turin Shroud. Radiocarbon dating has shown that the shroud was made in the 13th century

Questions

1 A GM tube is placed near to a radioactive source with a long half life. In three 10 second periods the following number of counts were recorded: 150, 157, 145. Why were the three counts different?

2 A radioactive material has a half life of 2 minutes. What does that mean? How much of the material will be left after 8 minutes?

3 The following results for the count rate of a radioactive source were recorded every minute. Plot a graph of the count rate (*y*-axis) against time (*x*-axis), and use the graph to work out the half life of the source.

Counts per second	time (minute)
100	0
59	1
34	2
20	3
12	4
7	5

4 Why does radioactive waste worry some people?

5 When doing an experiment to measure the half life of radon you will also detect some background radiation. How can you correct for this?

Containers for radioactive waste materials must be strong enough to prevent leakage. Here, a technician uses a geiger counter to check a drum containing radioactive waste at Hinkley Point nuclear power station

6 Nuclear Fission

Fission

You read earlier (page 264) that the nuclei of some large atoms were unstable and that to become more stable they lost an alpha or a beta particle. Some heavy nuclei, ^{235}U for example, may also increase their stability by **fission**. Figure 1 shows how this works. Unlike alpha or beta decay, which happens at random, the fission of a nucleus is usually caused by a neutron hitting it. The uranium nucleus absorbs this neutron and turns into a ^{236}U nucleus, which is so very unstable that it splits into two smaller nuclei. The nuclei that are left are not always identical; two or three energetic neutrons are also emitted in the process. The remaining nuclei are usually radioactive and will decay by the emission of beta particles to form more stable nuclei.

The fission process releases a tremendous amount of energy. The fission of a nucleus provides about 40 times more energy than the release of an alpha particle from a nucleus. Fission is important because we can control the rate at which it happens, so that we can use the energy released to create electrical energy.

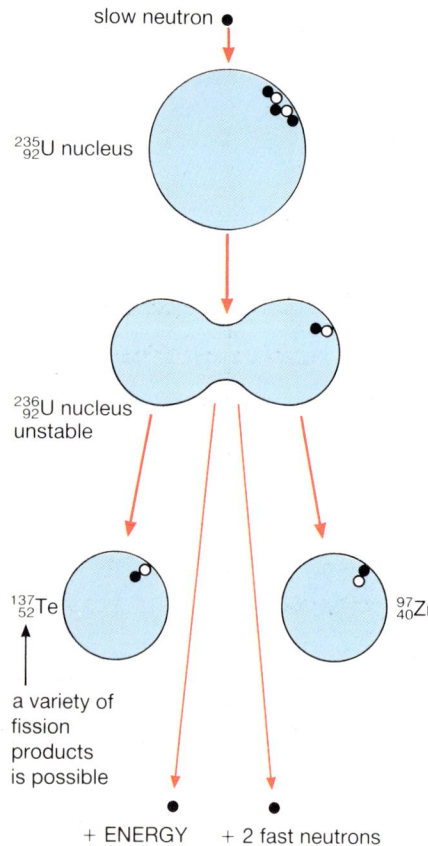

slow neutron

$^{235}_{92}U$ nucleus

$^{236}_{92}U$ nucleus
unstable

$^{137}_{52}Te$ $^{97}_{40}Zr$

a variety of
fission
products
is possible

+ ENERGY   + 2 fast neutrons

Figure 1
The fission of a uranium-235 nucleus

Chain reaction

Once a nucleus has divided by fission, the neutrons that are emitted can strike other neighbouring nuclei and cause them to split as well. This chain reaction is shown in Figure 2. Depending on how we control this chain reaction we have two completely different uses for it. In a controlled chain reaction, on average only one neutron from each fission will strike another nucleus and cause it to divide. This is what we want to happen in a power station. In an uncontrolled chain reaction all the neutrons from each fission strike other nuclei. This is how nuclear bombs are made. It is a frightening thought that a piece of pure uranium-235 the size of a tennis ball has enough stored energy to flatten a town.

Nuclear power stations

Figure 3 shows the essential points of a gas-cooled **nuclear reactor**. The energy released by the fission processes in the uranium fuel rods produces a lot of heat. This heat is carried away by carbon dioxide gas which is pumped around the reactor. The hot gas then boils water to produce steam, which can be used to work the electrical generators.

- The **fuel rods** are made of uranium-238, 'enriched' with about 3% uranium-235. ^{238}U is the most common isotope of uranium, but it is only ^{235}U that will produce energy by fission.
- The fuel rods are embedded in graphite, which is called a **moderator**. The purpose of a moderator is to slow down neutrons that are produced in fission. A nucleus is split more easily by a slow-moving neutron. The fuel rods are long and thin so that neutrons can escape. Neutrons leave one rod and cause another nucleus to split in a neighbouring rod.
- The rate of production of energy in the reactor is carefully regulated by the **boron control rods**. Boron absorbs neutrons very well, so by lowering them the reaction can be slowed down. In the event of an emergency they are pushed right into the core of the reactor and the chain reaction stops completely.

n → $^{235}_{92}U$

$^{235}_{92}U$ → n, n, n

$^{235}_{92}U$ → n, n, n

$^{235}_{92}U$ → n, n, n

Figure 2
A chain reaction in uranium-235

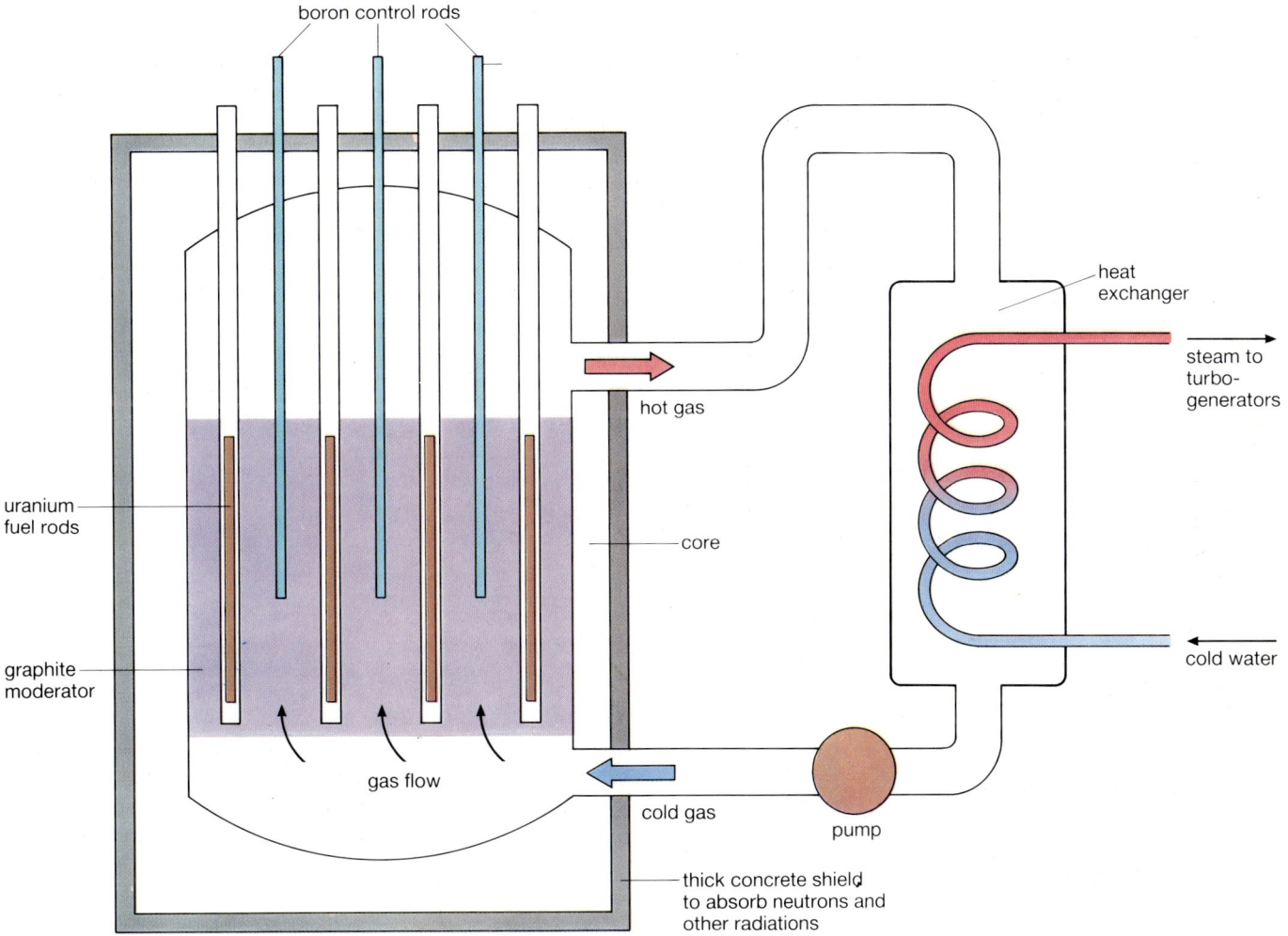

Figure 3
A gas-cooled nuclear reactor

Questions

1 Explain what is meant by nuclear fission. In what way is fission (i) similar to (ii) different from radioactive decay?

2 What is a chain reaction? Explain how the chain reaction works in a nuclear bomb and in a nuclear power station.

3 The following questions are about the nuclear reactor shown in Figure 3.

(a) What is the purpose of the concrete shield surrounding the reactor?

(b) Why is carbon dioxide gas pumped through the reactor?

(c) Which isotope of uranium produces the energy in the fuel rods?

(d) Will the fuel rods last for ever?

(e) What is the purpose of the graphite moderator?

(f) What would you do if the reactor core suddenly got too hot?

4 This question is about producing energy inside the nuclear reactor core, shown in Figure 3. This core has 1700 uranium fuel rods, each 1 m long with a diameter of 30 mm. Use the data provided to answer the following questions.

(a) How much ^{235}U is there in the core?

(b) What is the total amount of heat energy that this amount of ^{235}U can release?

(c) How long will this amount of nuclear fuel last for, if the core produces power continuously?

- Mass of one fuel rod is 14 kg
- 3% of the fuel is ^{235}U
- Power produced in the reactor core is 2400 MW
- 1 kg of ^{235}U produces 10^{14} J of heat energy
- There are 3×10^7 s in 1 year

7 The Hazards of Radiation

Measuring radiation

When scientists try to work out the effect on our bodies of a dose of radiation, they need to know how much energy each part of the body has absorbed. After all, the damage done to us will depend on the amount of energy that each kilogram of body tissue absorbs. The unit used to measure a **radiation dose** is the **gray**, (symbol Gy). A dose of 1 Gy means that each kilogram of flesh absorbs 1 joule of energy.

$$1\,\text{Gy} = 1\,\text{J/kg}$$

Some radiations are more damaging than others, so scientists prefer to talk in terms of a **dose equivalent**, which is measured in **sieverts** (symbol Sv).

$$\text{dose equivalent (Sv)} = Q \times \text{dose (Gy)}$$

Q is a number that depends on the radiation, as shown in Table 1. Alpha particles are very strongly ionizing and cause far more damage than a dose of beta or gamma radiation that carries the same energy. Usually the amounts of radiation that we are exposed to are very small. Most people receive about 1/1000 sievert each year, this is a **millisievert** (symbol 1 mSv).

Q	Type of radiation
1	beta particles/gamma rays
10	protons and neutrons
20	alpha particles

Table 1

The Chernobyl nuclear reactor near Kiev in Russia. During an unauthorised experiment in 1986, the core of the number 4 reactor unit melted, causing an explosion which released a large amount of radioactive material. This will be a very serious health hazard for many years to those living close to the reactor. Winds carried some of the radioactive material to many countries in Europe

Risk estimates

On 26 April 1986 there was an explosion in the Russian nuclear reactor at Chernobyl, causing a large leakage of radiation. During May the **background count** in Britain increased causing us all to be exposed (on average) to an extra dose equivalent to 0.1 mSv. This is a small dose and no worse than going on holiday in Cornwall, where granite rock areas produce low amounts of radiation. However, estimates have been made to suggest that over the next 30 years, extra people will get cancer as a result of the Chernobyl disaster.

Research suggests that for a population of 1000, about 12 fatal cancers will be caused by a dose equivalent of 1 Sv. In Britain, the population is about 50 million, so the number of deaths expected by a dose for all of us of 1 Sv would be:

$$\frac{12}{1000} \times 50\,000\,000 = 600\,000$$

However, the dose from Chernobyl was only 0.0001 Sv, so the estimated number of deaths from the Chernobyl disaster, in Britain over the next 30 years, is about $600\,000 \times 0.0001 = 60$.

How dangerous is radiation?

Radiation affects materials by ionizing atoms and molecules. When an atom is ionized, electrons are removed or added to it. This means a chemical change has occurred. In our bodies such a chemical change could cause the production of a strong acid which will attack and destroy cells.

- **High doses** of radiation will kill you. There is only a 50 per cent chance of surviving a dose equivalent to 4 Sv. A 10 Sv dose equivalent would give you no chance of survival. Such high doses kill too many cells in the gut and bone marrow for your body to be able to work normally. You could be exposed to such doses in a nuclear war, and people certainly died in Hiroshima and Nagasaki as a result of such doses.

- **Moderate doses** of radiation below 1 Sv will not kill you. Damage will be done to cells in your body, but not enough to be fatal. The body will be able to replace the dead cells and the chances are that you would then recover totally. However, a study of the survivors from Hiroshima and Nagasaki shows that there is an increased chance of dying from cancer some years after the radiation dose. Even so, you would only have a chance of about 1 in 100 of getting cancer from such levels of radiation.

- **Low doses** of radiation, below 10 mSv, are thought to have little effect on us. However, some people think that any exposure to radiation will increase your chances of getting cancer.

There can be no doubt that radiation doses can cause cancer or leukaemia (see Table 2). Uranium miners are exposed to radon gas, and girls who painted luminous watch dials were exposed to radium.

Workers in the nuclear power industry are exposed to more radiation than the rest of the population. Special monitoring and remote control are used to keep this extra radiation exposure as small as possible.

Questions

1 In Table 1, the value of Q for alpha radiation is 20. Why is it so high?
2 What does a sievert measure?
3 Summarise the effects of high, moderate and low doses of radiation on us.
4 Many modern watches do not have luminous dials. Instead they have small lights that turn on at the press of a switch. Explain why lights are safer.
5 Use the data in Table 2 to show that exposure to alpha radiation is more likely to cause cancer than exposure to gamma radiation or X-rays.
6 In a nuclear reactor disaster about 200 workers are exposed to a radiation dose equivalent to 2 Sv. Use the data in the text to estimate the number of them likely to die from cancer some time after the accident.

Source of radiation	Type of radiation	Number of people studied	Extra number of cancer deaths caused by radiation
uranium miners	alpha	3400	60
radium luminisers	alpha	800	50
medical treatment	alpha	4500	60
medical treatment	X-rays	14000	25
Hiroshima bomb	gamma rays and neutrons	15000	100
Nagasaki	gamma rays	7000	20

Table 2. This table illustrates the connection between radiation and the increased chance of cancer.

8 Nuclear Power: The Future?

The outside (left) and inside (right) of a 'torus' (a hollowed-out doughnut shape) at a laboratory in Oxfordshire. A strong magnetic field traps ionised hydrogen inside the torus. Researchers hope to generate a sustained fusion reaction in such a vessel

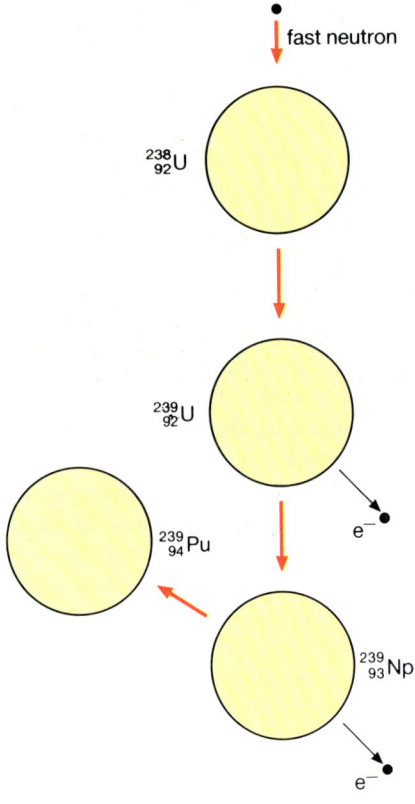

fast neutron

$^{238}_{92}U$

$^{239}_{92}U$

e^-

$^{239}_{94}Pu$

$^{239}_{93}Np$

e^-

Figure 1
Production of plutonium from uranium

At the moment about 15% of our electricity is generated using nuclear power. Nearly all the rest is produced in coal-burning power stations. In 1987 the government announced plans to build more nuclear power stations, and it now seems likely that early next century as much as 70% of our electricity will come from nuclear power.

The reason for the increased use of nuclear fuels is that our fossil fuels (coal and oil) are running out; once they are gone we cannot replace them. In Britain we have saved up about 20 000 tonnes of uranium – how long will that last? At present we do not use our uranium very efficiently. Only 0.7 per cent of uranium is the fissionable ^{235}U, which produces energy in most reactors. However, in 1976 at Dounreay (Scotland) a new type of power station opened. This has a **fast breeder reactor**, which uses a new fuel, **plutonium**.

Plutonium is an element that does not occur naturally, but it is produced from uranium, as shown in Figure 1. Plutonium nuclei will release energy by fission, but the process is triggered by the nucleus absorbing a fast neutron (hence the name 'fast reactor'). By producing plutonium from uranium our stocks of nuclear fuel will last for a few hundred years.

Nuclear fusion

The energy that is produced inside our Sun comes from the fusing together of hydrogen nuclei. **Fusing** means melting together, which is a good description of the process. At the centre of the Sun the temperature is about 10 000 000 K; at these temperatures the nuclei of atoms are stripped of all their surrounding electrons, and they are moving very quickly indeed. Fusion involves two small nuclei colliding and sticking together to form a larger nucleus. As in the fission of a large nucleus, the fusion of two small nuclei releases a lot of energy (Figure 2).

At Culham in Oxfordshire, attempts are being made to get energy from nuclear fusion. This is an extremely difficult project; as you can imagine producing conditions similar to the inside of a star is no easy matter! If this experiment is successful it might solve the problem of producing electricity for a long time to come.

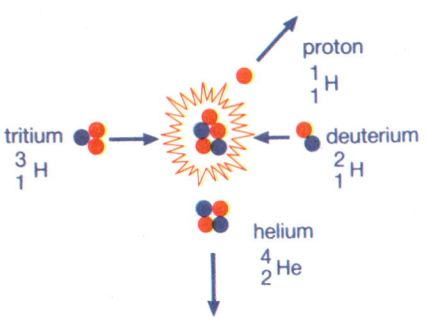

proton
^1_1H

tritium
^3_1H

deuterium
^2_1H

helium
^4_2He

Figure 2
The fusion of deuterium and tritium

The nuclear debate

Is nuclear power the way into the next century, or should we be looking to produce energy from natural sources such as wind and the tides? A lot of people are pressing for more nuclear power stations and others are strongly against them. This is an important issue and you should have your own ideas about it. Below are listed some points that someone in favour of nuclear power might make, and also some points that could be made by someone against it.

Views of someone in favour of nuclear power stations.

- Fossils fuels are running out, so nuclear power provides a convenient way of producing electricity.
- Nuclear power stations produce a very small level of radiation. The extra radiation is very little in comparison with the background radiation, and is not a health hazard.
- Coal-fired power stations put out more radiation into the atmosphere than nuclear power stations, because coal is naturally radioactive. Burning coal also produces acid rain.
- Radioactive waste can be safely stored.
- The chances of a large nuclear accident in this country are very small. Our technology is far better than the Russians, so an accident like Chernobyl could not happen here.
- Accidents happen anyway; nobody seems to worry about the number of deaths caused in road accidents. Is anybody suggesting banning cars?
- People do not understand radiation. That's why they are afraid of it.

The prototype fast reactor at Dounreay power station

A lot of people are worried by the dumping of radioactive waste

Views of someone against nuclear power stations.

- Fossil fuels are running out, so we should be looking to conserve energy. Research should be done to use wind and wave power.
- Nuclear power stations produce dangerous quantities of radioactive waste. The government has ordered Sellafield to stop discharging waste into the Irish Sea. Statistics show that children are more likely to die of leukaemia near Sellafield.
- Coal-fired power stations cause acid rain and produce radiation. They should be closed down as well.
- It is irresponsible to store radioactive wastes with long half-lives; it pollutes the environment for our grandchildren.
- The fact remains that a power station blew up in 1986, belching radioactive stuff all over Europe. We may have escaped lightly, but a lot of people in Russia will die as a result.
- Cars have got nothing to do with it.
- We *do* understand radiation. That's why we're afraid of it.

Questions

1 Why does the production of plutonium allow us to get more energy than we would if we used uranium as a nuclear fuel?
2 Fast reactors do not have moderators. Explain why.
3 What is nuclear fusion? Why is it much more difficult to control nuclear fusion than nuclear fission?
4 You have now read a lot about nuclear energy. Write a short essay to explain whether it is the answer to our energy problems. After thinking about the arguments above, who do you agree with?

9 Uses of Radioactive Materials

Radioactive materials have a great number of uses in medicine, industry and agriculture. People who work with radioactive materials must wear radiation badges which record the amount of radiation to which they are exposed.

Medicine

- Radioactive **tracers** help doctors to examine the insides of our bodies. Iodine-131 is used to see if our thyroid glands are working properly. The thyroid is an important gland in the throat which controls the rate at which our bodies function. The thyroid gland absorbs iodine, so a dose of radioactive iodine (the tracer) is given to a patient. Doctors can then detect the radioactivity of the patient's throat, to see how well the patient's thyroid is working.
- Cobalt-60 emits very energetic gamma rays. These rays can damage our body cells, but they can also kill bacteria. Nowadays nearly all medical equipment such as syringes, dressings and surgeons' instruments is first packed into sealed plastic bags, and then they are exposed to intense gamma radiation. In this way all the bacteria are killed and so the equipment is sterilised.
- The same material, cobalt-60, is used in the treatment of cancers. Doctors direct a strong beam of radiation on to the cancerous tissue to kill the cancer cells. The treatment is very unpleasant and causes the patient to be very ill, but it is often successful in slowing down the growth or completely curing the cancer.

To check the amount of radiation that workers in a nuclear power station are exposed to, they wear special radiation-sensitive badges, like the ones in this photograph. At the end of each month the sensitive film in the badges is developed. It can then be seen how much radiation the wearer has been exposed to

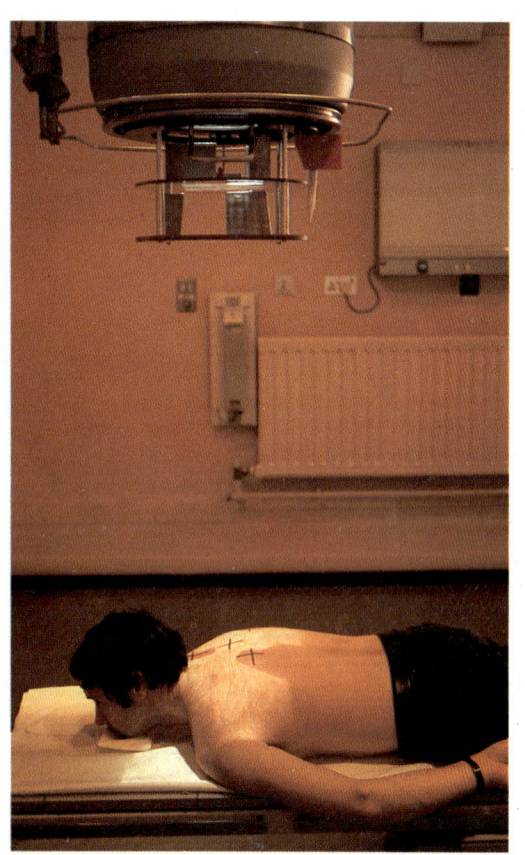

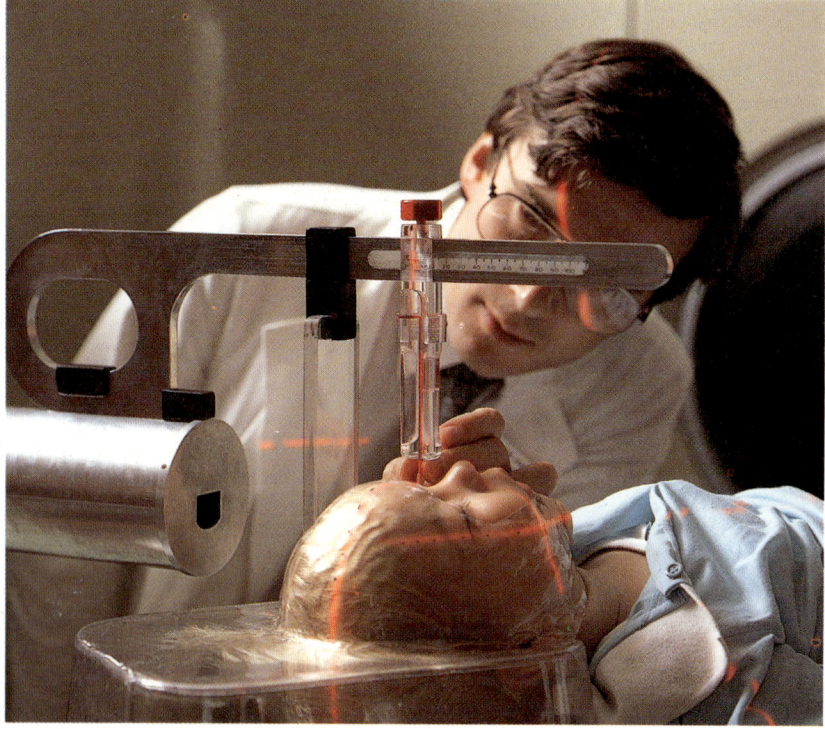

Gamma rays can be used to destroy cancer cells. Here, a child with a cancerous growth in the eye is being treated with gamma rays from cobalt-60. The red light is used to direct the equipment at the cancerous tissue, before the gamma rays are released. The photograph on the left shows a patient with Hodgkins disease (cancer of the lymph nodes) being treated by radiotheraphy

Industry

- Radioactive tracers may be used to detect leaks in underground pipes (Figure 1). The idea is very simple; the radioactive tracer is fed into the pipe and then a GM tube can be used above ground to detect an increase in radiation level and hence the leak. This saves time and money because the whole length of the pipe does not have to be dug up to find the leak.
- A radioisotope of iron is used in industry to estimate the wear on moving parts of machinery. For example, car companies want to know how long their piston rings last for. A piston ring which has radioactive iron in it is put into an engine and run for several days. At the end of the trial, the oil from the engine can be collected, and from the radioactivity of the oil the engineers can calculate how much of the piston ring has worn away.

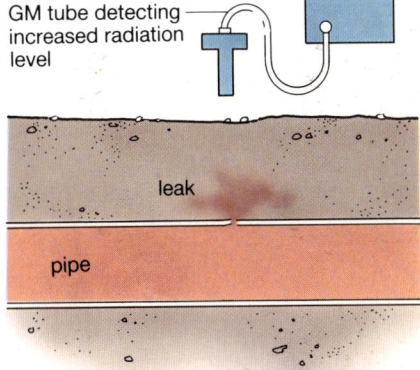

GM tube detecting increased radiation level

leak

pipe

Figure 1
How to find leaks in pipelines without digging

A technician lowers a casket of fruit into a pool of water used to study irradiated fruit. The casket is irradiated by a source of X-rays in the pool. This technique can be used to preserve foods that would otherwise become rotten very quickly

Agriculture

- Tracers are used in agriculture too. Phosphates are vital to the growth of plants and are an important component of fertilisers. Radioactive phosphorus-32 is used as a tracer to show how well plants are absorbing phosphates.
- In some countries gamma radiation is used to prolong the shelf-life of food. Gamma rays are very penetrating so this process can be used on pre-packaged or frozen foods. Gamma rays kill the bacteria in the food and so can eliminate the chance of food poisoning. However, the gamma rays will also kill some cells in the food itself and therefore can alter the taste considerably. This process is not allowed in Britain, although we probably import irradiated food.
- Gamma rays are also used to help produce new types of crops. Large doses of gamma rays will kill cells, but smaller doses can cause mutations to the cells, which will change the nature of the crop. The seeds of crops are exposed to gamma radiation to encourage mutations. The new crops may show desirable qualities, like being stronger or producing a greater yield. These successful mutations can be kept and used in the fields.

Questions

1 Explain carefully why radioactive tracers are useful.
2 Do radioactive tracers show different chemical properties to other isotopes of the same element?
3 What uses does gamma radiation have in medicine and agriculture?
4 In the text you will have read that radioisotopes are useful in industry. Clearly they have important economic results; it saves a lot of money to find a leak with the help of a radioisotope. However, the people who find the leak and the general public will be exposed to radiation in the process and therefore there will be a health hazard. How can we balance the possible loss of human life with the saving of money? What is an acceptable risk?

1 A silver atom has a proton number of 47 and a nucleon number of 107. The atom is neutral.
(a) The atom contains protons and neutrons. How many of each does the atom have? Where are the protons and neutrons?
(b) What other particle does the atom have? How many of these particles are there?
(c) Why is the atom neutral?
(d) Silver-108 is another isotope of silver. Explain, with as much detail as possible, what this means.

2 One of the isotopes of lithium has in its nucleus 3 protons and 3 neutrons.
How many protons and neutrons could be found in the nucleus of another isotope of lithium?
A 2 protons and 3 neutrons
B 2 protons and 4 neutrons
C 3 protons and 4 neutrons
D 4 protons and 2 neutrons
E 4 protons and 4 neutrons

LEAG

3 A cloud chamber contains radioactive gas. The diagrams below show tracks of alpha particles at one minute intervals.

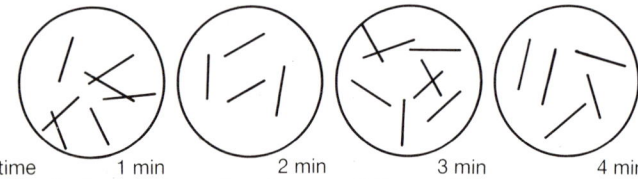

time   1 min   2 min   3 min   4 min

Why is the number of tracks seen different from minute to minute?
A Alpha-particles are emitted in a random fashion.
B The gas has a half-life of 1 minute.
C We cannot count the tracks quickly enough.
D The amount of radioactive gas increases and decreases.
E The cloud chamber does not contain enough vapour to show the tracks properly.

MEG

4 The diagram below represents a source of radiation inside a sealed box. The box is made of sheet lead which is 2 cm in thickness.

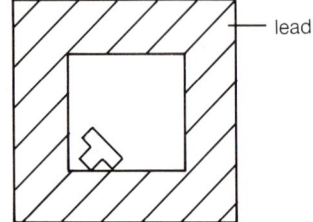
lead

If radiation can be detected from the outside of the box what is the source most likely to be emitting?
A X-rays
B alpha particles
C beta particles
D gamma radiation
E ultra-violet light

LEAG

5 (a) The table below shows some of the isotopes of the element uranium, the types of radiation they emit and their half-lives.

(i)   What will be the number of neutrons in the nucleus which remains when an atom of uranium-235 decays by the emission of an alpha particle?
(ii)   Uranium-235 can also decay by a process called nuclear fission. What happens to the uranium nucleus in this process?
(iii) When the Earth was formed it is thought that there was about 64 times as much uranium-235 present on Earth than is the case today. What does this suggest about how long ago the Earth was formed?
(iv) If you were given equal masses of uranium-238 and uranium-239 which would you expect to be the most radioactive? Explain your answer.
(b) The following passage might have been written by a newspaper reporter:

"There has been a leak of radiation from the nuclear reprocessing works. *The radiation was in the form of a gas.* About 10 g of the gas radon-220, which has a half life of 52 seconds, escaped. The escape is not thought to be dangerous *as the gas becomes harmless after 104 seconds.*"

The phrases in italics are incorrect. State and explain what is wrong with them.
(c) Radioactive sources have many uses in industry and medicine. Here are two examples:
1. The examination of welds in steel pipes.
2. The examination of diseased organs in the human body.
Choose **one** of these examples. State and explain what kind of radiation is suitable and explain how it is used in the example you have chosen.

MEG

Atomic number, Z (number of protons)	Mass number, A (number of protons and neutrons)	Type of radiation	Half life
92	235	alpha	7.1×10^8 years
92	236	alpha	2.4×10^7 years
92	237	beta	6.8 days
92	238	alpha	4.5×10^9 years
92	239	beta	23.5 minutes

6 Energy is released
A in nuclear fusion but not in nuclear fission.
B in nuclear fission but not in nuclear fusion.
C in neither nuclear fission nor nuclear fusion.
D in both nuclear fission and nuclear fusion.
E in chemical reactions only.

NI

7 The brain can suffer from a particularly nasty cancer called a glioblastoma. This cancer penetrates the brain and cannot be cured by surgery. Instead the neurosurgeon gives the patient an injection which contains some boron. The boron is absorbed by the glioblastoma. Then the patient is irradiated with neutrons. The following nuclear reaction occurs:

$$^{10}_{5}B \; + \; ^{1}_{0}n \; \rightarrow \; ^{7}_{3}Li \; + \; ^{x}_{y}He$$

(a) Copy the equation and fill in the missing number x and y.
(b) The boron nucleus splits up to form lithium and helium. What is this process called? Explain why the lithium and helium nucleus move away from each other very quickly.
(c) Explain how this process can kill the glioblastoma.
(d) Why would this process be dangerous for healthy patients?

8 A radioactive substance has a half-life of 6 hours. At 12 noon the count rate from a sample of this substance is 1000 counts per minute. At what time will the count rate be approximately 250 counts per minute?
A 3 p.m.   **B** 6 p.m.   **C** 9 p.m.   **D** 12 midnight
E 12 noon the next day

MEG

9 An experiment to measure the half-life of caesium-140 is done in a region where there is a high background count. The results of the experiment are shown in the table. Determine the half-life of caesium-140 by plotting a suitable graph.

Count rate (counts/s)	Time (s)
68	0
52	30
40	60
32	90
24	120
20	150
16	180
14	210
12	240
8	270
10	300

10 The age of archaeological remains can be found using carbon dating. All living things contain small amounts of carbon-14; this is a radioactive isotope. The concentration of carbon-14 is the same for all living things. But when the creature (or plant) dies the carbon-14 decays. Its half-life is about 5700 years. So after that time the concentration of the carbon-14 has halved.

Cro-Magnon man is one of our ancestors. Five adult skeletons were found near Les Eyzies in France. A sample of charcoal from this site of 1 g produced a radioactive count of 0.5 counts per minute. A modern sample of charcoal of the same mass produces a count rate of 32 counts per minute. Both counts were corrected for background radiation.

How long ago did Cro-magnon man live?

11 A student used a Geiger-Muller tube and electronic counting apparatus to investigate three different radioactive materials, which were labelled source X, source Y, and source Z. The student placed each source 10 mm from the window of the Geiger-Muller tube.

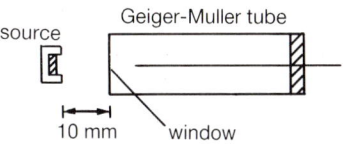

The table shows the results the student obtained with different materials placed between the source and the window of the tube.

Material between source and window	Count rate (number of counts per minute)		
	Source X	Source Y	Source Z
air only	256	132	320
thin paper	268	39	305
0.5 mm of aluminium	254	24	183
5 mm of lead	185	19	21
With no source present at all	20		

(a) What was the effect of the thin paper on the radiation from (i) source X? (ii) source Y?
(b) What was the effect of the 5 mm of lead on the radiation from source Z?
(c) Use *the evidence in the table* to reach conclusions about the behaviour of the radiation from the three sources. (You do not have to name the types of radiation.)
(d) Why is there a count rate when there is no source present?
(e) For this experiment why must the window of the Geiger-Muller tube be very thin?

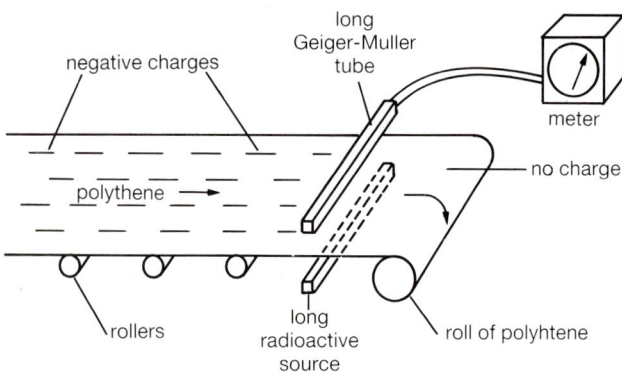

negative charges

long Geiger-Muller tube

meter

polythene →

no charge

rollers

long radioactive source

roll of polyhtene

12 This diagram shows a method which is used in factories to check the thickness of polythene being produced. In this case a long radioactive source is placed below the whole width of the polythene and a long Geiger-Müller tube is placed above it.

A Geiger-Müller tube is used for detecting the presence of radioactive radiation. The output pulses of current from the tube may go into a counter, an amplifier connected to a loud speaker or a meter.

In this application, the reading on the meter can then be used as a measure of thickness of the polythene; the thicker the polythene, the lower the meter reading.

When the polythene passes over the rollers, effects of friction cause it to become negatively charged. The presence of the radioactive source enables the polythene to become discharged.

(a) It is suggested that because radioactive decay is random this method for checking thickness gives better results when the polythene is going through slowly.

(i) What is meant by *random*?

(ii) Why is the result likely to be more reliable when the polythene is going through slowly?

(b) Radiation from the radioactive source ionises the air. This produces many positive and negative ions from atoms in the air. Discuss how these ions are affected by the negative charge on the polythene and hence explain how the polythene becomes discharged.

(c) If sources of similar activity giving either α or β radiation were available, which one would be better for

(i) measuring the thickness of the polythene,

(ii) discharging the roll? Explain your choice in each case.

(d) In view of the presence of the radioactive material state any **two** suitable precautions which should be observed for the safety of workers in the factory.

(e) A buyer of polythene visits the factory and is alarmed by the use of radioactive sources in the method shown. He is concerned that the polythene may become radioactive.

(i) Why has he no real cause for concern?

(ii) Explain briefly an experimental check which you could make to show that there is no cause for concern.

LEAG

13 Plutonium-241 is unstable and it decays by giving out an alpha particle. This is the start of a long decay series. By the emission of more alpha and beta particles, eventually a stable isotope of bismuth is made.

The table shows the decay series. Copy it and fill in the gaps.

Element	Symbol	Radioactive emission
Plutonium	$^{241}_{94}\text{Pu}$	α
Uranium	$^{237}_{92}\text{U}$	β
Neptunium	$^{?}_{?}\text{Np}$	α
Protactinium	$^{?}_{91}\text{Pa}$	β
Uranium	$^{233}_{?}\text{U}$	?
Thorium	$^{229}_{90}\text{Th}$	α
Radium	$^{?}_{?}\text{Ra}$	β
Actinium	$^{?}_{?}\text{Ac}$	?
Francium	$^{221}_{87}\text{Fr}$	?
Astatine	$^{217}_{85}\text{At}$	α
Bismuth	$^{?}_{?}\text{Bi}$	?
Polonium	$^{213}_{84}\text{Po}$	α
Lead	$^{?}_{?}\text{Pb}$	β
Bismuth	$^{209}_{83}\text{Bi}$	stable

14 The equation below shows part of the reaction in a nuclear reactor.

$$^{235}_{92}\text{U} + ^{1}_{0}\text{n} \longrightarrow ^{236}_{92}\text{U}$$

(a) Explain the significance of the numbers 235 and 92.

(b) $^{236}_{92}\text{U}$ atoms are unstable and disintegrate spontaneously into fragments approximately equal in size, together with two or three fast-moving neutrons and a large amount of energy. What is this process called and what is the source of the energy?

(c) $^{235}_{92}\text{U}$ is much more likely to absorb slow-moving (thermal) neutrons than fast-moving neutrons. Describe how neutrons may be slowed down in the reactor core.

(d) Explain what is meant by a chain reaction.

(e) How is the rate of energy production in the reactor core controlled?

(f) The energy is produced in the form of heat in the reactor core. How is the heat removed from the core and how is it converted into electricity?

(g) When the fuel rods are withdrawn from the reactor core they are *radioactive*, containing *isotopes* with long *half-lives*. Explain the terms in italics.

(h) Outline three precautions which must be taken to ensure safe operation of the reactor.

NEA

Acknowledgements

The following companies, institutions and individuals have given permission to reproduce photographs in this book:

Acorn Computers (243, 254); Adams IAL Limited (224, both); Airfotos (138); All-Action Photographic (167, middle); Allsport (16, both; 26, bottom; 44; 46; 51, middle left; 58, top; 78, top left; 127); Alton Towers (70); Associated Press (144; 260, top left); Barnaby's Picture Library (67; 81; 260, top right; 275; 260); Black and Decker (221); Bob Thomas Sports Photography (42, top left; 58, bottom; 59, middle right); Channel 4 (204, top right); Chris Bonington (96); British Airways (10); British Antarctic Survey (110, bottom); British-Israeli Foreign Affairs Committee (35); British Rail (75); British Steel Strip Products; Bruce Coleman Limited (8, bottom; 169); Canon (165); Central Electricity Generating Board (80; 81; 179; 186; 189; 206, bottom; 228; 230; 233; 261); David Lea Associates Limited (112); ECON Group Limited (94, bottom); The Electricity Council (180); Forest Films (8, middle); Ford (51, top); Foster Associates (22); Nick Fox (110, top); General Electric Company (217); Bob Gibbons (Natural Image) (42, top right; 88, top; 120); J H Golden (34, bottom); Hawaii Visitors Bureau (41); Holt Studios LTD (246); Honda (101); Kit Houghton (68); Hoverspeed (17); Independent Broadcasting Authority (27; 132, top); Joint European Torus (274, both); Liveaid (249); Vincent Martinelli (28, bottom; 45; 76, bottom; 116; 196); Meteorological Office (206, top; 207); Mercury Communications (146, bottom); Ministry of Defence (Navy) (73; 223; 259); National Aeronautics and Space Administration (34, middle; 57; 59, middle left; 62; 171); Motorola LTD (250); National Trust (92, bottom; 131; 265); Natural History Picture Agency (15; 51, middle right; 148; 153); New Zealand Tourist Board (93); Mark Newcombe Photography (52, top); Pacific and Orient (19; 52, bottom); Phillips (193); P & H S Architects (172); Pilkington (32); Rolls-Royce (63); Royal Greenwich Observatory (8, top; 170); St. Bartholomews Hospital (276, bottom left and right); Barrie Schwartz (269, top); Science and Engineering Research Council (262, both); Science Photo Library (1; 2, both; 3, middle; 6; 47; 62; 71; 87; 89, bottom; 111, both; 141, both; 147; 154, both; 161; 175; 204, both; 211; 237; 266); Shell (36); Sodel Photothetique (109); Statue University of New York (34, top); Syndication International (134); Tefal (97); Topham Picture Library (3, bottom); Transport and Road Research Laboratory (56, bottom); United Kingdom Atomic Energy Authority (269, bottom; 275; 276, top); United Press International (182, top); Virgin Atlantic (37); Volvo (21; 56, top); P D Waghorn (124); Wessex Water (94, top).

We are grateful to the following examining bodies for permission to reproduce questions from specimen GCSE papers and from recent joint 16+ examinations in Physics.

The London and East Anglian Group (LEAG), The Midland Examining Group (MEG), The Southern Examining Group (SEG), The Welsh Joint Education Committee (WJEC), The Northern Examining Association (Associated Lancashire Schools Examining Board, Joint Matriculation Board, North Regional Examinations Board, North West Regional Examinations Board, Yorkshire and Humberside Regional Examinations Board) (NEA), and the Northern Ireland Schools Education Committee (NISEC).

Artwork by Oxford Illustrators.

Typeset in Ehrhardt by Tradespools Ltd, Frome
Printed and bound in Italy for Hodder and Stoughton Limited, Mill Road, Dunton Green, Sevenoaks, Kent by New Interlitho S.p.A. - Milan